Particle Physics and Astrophysics

H. Mitter F. Widder (Eds.)

Particle Physics and Astrophysics

Current Viewpoints

Proceedings of the
XXVII Int. Universitätswochen für Kernphysik
Schladming, Austria, February 1988

With 100 Figures

Springer-Verlag Berlin Heidelberg New York
London Paris Tokyo

Professor Dr. Heinrich Mitter
Professor Dr. Fridebert Widder

Institut für Theoretische Physik, Karl-Franzens-Universität,
A-8010 Graz, Austria

ISBN-13: 978-3-642-74393-1 e-ISBN-13: 978-3-642-74391-7
DOI: 10.1007/978-3-642-74391-7

Softcover reprint of the hardcover 1st edition 1989

2155/3150-543210 – Printed on acid-free paper

Preface

This volume contains the written versions of some invited lectures presented at the 27th "Internationale Universitätswochen für Kernphysik" in Schladming, Austria, in February 1988.

Again the generous support of our sponsors, the Austrian Ministry of Science and Research, the Styrian government and others, made it possible to invite expert lecturers. In choosing the topics we have tried to combine both the particle and the astrophysics aspects of recent developments, which show a remarkable convergence. Thanks to the efforts of the speakers this "interdisciplinary" spirit was maintained in all lectures.

After the School the lecture notes were revised and partially rewritten in TeX by the authors, whom we thank for their efforts. Unfortunately two of the lectures could not be included in this volume (K. McDonald: "Techniques of Experimental Particle Physics as Applied to Astrophysics", R. Ruffini: "Dark Matter and Fractal Large-Scale Structure of the Universe").

Thanks are also due to Mrs. Neuhold for the careful typing of those notes, which we did not receive in TeX, and to the publishers for their patience.

Graz, Austria
October 1988

H. Mitter
F. Widder

Contents

Phase Transitions as the Origin of Large Scale Structure in the Universe

Introduction and Survey I: Concepts from Particle Physics

Peter Breitenlohner

Max-Planck-Institut für Physik und Astrophysik
Werner-Heisenberg-Institut für Physik, P.O.Box 40 12 12
D–8000 München, Fed. Rep. of Germany

1. Introduction

In this series of three lectures I shall try to present some of the concepts from elementary particle physics which play a rôle in astrophysics. These lectures shall, together with those given by G. Börner who will present concepts from astrophysics and cosmology, give an introduction to this school.

Elementary particles and their interaction are conventionally described in terms of a relativistic quantum field theory (QFT). In order to quantize such theories I shall always use a flat space-time background. The interesting but complicated aspects of QFT in a general (non flat) space-time geometry, e.g., the Hawking radiation or the stability of certain false vacua in Anti-de Sitter spaces, are thus ignored.

2. The Standard Model of Electro-Weak and Strong Interaction

In the standard model the electro-weak and strong interaction of elementary particles is described by an $SU(3) \times SU(2) \times U(1)$ gauge theory. In addition there is, of course, the gravitational interaction described (at least classically) by Einstein's General Relativity. The gauge group $SU(2) \times U(1)$ of the electromagnetic and weak interaction is spontaneously broken; only a $U(1)$ subgroup (the electromagnetic $U(1)$) remains unbroken. Strong interactions are described by Quantum Chromo Dynamics (QCD) where the forces are those of an unbroken (color) $SU(3)$. The fundamental fermion fields are the quarks (q) and leptons (ℓ) which seem to fall into the three families

$$(u, d; e, \nu_e)\ , \qquad (c, s; \mu, \nu_\mu)\ , \qquad (t, b; \tau, \nu_\tau)$$

fields	$U(1)$ charge	$SU(2)$ quantum number	$SU(3)$ representation
$\begin{pmatrix} u_L \\ d_L \end{pmatrix}$	$+\frac{1}{6}$	$\frac{1}{2}$	3
u_R	$+\frac{2}{3}$	0	3
d_R	$-\frac{1}{3}$	0	3
$\begin{pmatrix} \nu_e \\ e \end{pmatrix}$	$-\frac{1}{2}$	$\frac{1}{2}$	1
e_R	-1	0	1

differing only in their masses. The quantum numbers assigned to the fields of each family, say the first one, are such that the electric charge is the sum of the 'hyper charge' ($U(1)$) and the third component of the 'weak isospin' ($SU(2)$). In addition there are the gauge bosons: eight $SU(3)$ gauge fields (gluons), three $SU(2)$ gauge fields and one $U(1)$ gauge field. The latter four correspond to (mixtures of) the massive weak bosons $W^{\pm}$, Z^0 and the (massless) photon γ.

Finally one needs at least an $SU(2)$ doublet of scalar Higgs fields ϕ in order to achieve the spontaneous symmetry breaking of the electro-weak $SU(2)\times U(1)$ gauge group. These Higgs fields need, however, not be elementary; they could well be composite fields bilinear in the fermions. It is, unfortunately, beyond the scope of perturbation theory to determine whether such composite fields do indeed acquire a non vanishing vacuum expectation value as is necessary for the mechanism of spontaneous symmetry breaking.

Without spontaneous symmetry breaking all fermions would be massless. With spontaneous symmetry breaking each fermion acquires a mass proportional to the strength of its coupling to the Higgs field; only the three neutrinos remain massless.

The quantum numbers of the (unbroken) color $SU(3)$ are supposed to be confined, i.e., free quarks are not observable. The observable strongly interacting particles are the quark anti-quark ($q\bar{q}$) and three quark (qqq) bound states — the mesons and the baryons.

3. Spontaneous Symmetry Breaking

We have already encountered one of the many topics in elementary particle physics where spontaneous symmetry breaking plays a rôle, the standard model. In this chapter I want to give an outline of the concept of spontaneous symmetry breaking (SSB) and of some of the phenomena related to it. Generally speaking SSB is a situation where the dynamics of physical system (e.g., the Lagrangean or the field equations) admit a certain symmetry group G which is not a symmetry of the ground state of the system (vacuum). There may or may not be a non trivial subgroup $H \subset G$ which remains unbroken, i.e., is an invariance of the ground state. In the following I shall discuss a few models with SSB, which are simple enough to be treated in detail and yet exhibit the characteristic phenomena.

3.1 The Z_2 Goldstone Model

The Goldstone model, in its simplest form, describes a self interacting real scalar field $\varphi(x)$ with the Lagrangean (I am using space-time coordinates x^μ ($\mu = 0, \ldots, 3$) and a Minkowski metric with signature (+,−,−,−))

$$\mathcal{L} = \frac{1}{2}\left(\partial_\mu\varphi\partial^\mu\varphi - \mu^2\varphi^2 - \frac{1}{2}h\varphi^4\right) = \frac{1}{2}(\partial_\mu\varphi\partial^\mu\varphi - V(\varphi))$$

and the field equation

$$\partial^\mu\partial_\mu\varphi + \mu^2\varphi + h\varphi^3 = 0\,.$$

The Lagrangean and the field equations are invariant under the symmetry group $G = Z_2$ generated by $\varphi(x) \to -\varphi(x)$ and under the usual space-time symmetries (Poincaré group, parity P and time reversal T). The ground state is determined by a classical solution which minimizes the energy

$$E = \int \mathcal{H}\,, \qquad \mathcal{H} = \frac{1}{2}((\partial_0\varphi)^2 + (\nabla\varphi)^2 + V(\varphi))\,.$$

Note that the parameter h in $\mathcal{L}$ must be ≥ 0 in order that the energy density be bounded below, but for $h > 0$ the parameter μ^2 may have either sign. We are particularly interested in solutions where $\varphi(x)$ does not depend on x (translation invariance). The number of such solutions and their properties depend on the shape of the potential $V(\varphi)$.

For $\mu^2 \geq 0$ the only such solution, $\varphi = 0$, is the Z_2 invariant absolute minimum of the energy. For $\mu^2 < 0$ there are, however, the three solutions $\varphi = 0$ and $\varphi = \pm f$ with $f = \sqrt{-\mu^2/h}$. The Z_2 invariant solution $\varphi = 0$ corresponds to an unstable maximum of $V(\varphi)$, the two Z_2 non invariant solutions $\varphi = \pm f$ correspond to two degenerate minima of $V(\varphi)$. If we would consider the motion of a particle with coordinate φ in the potential $V(\varphi)$ the ground state would nevertheless be a Z_2 invariant superposition of states concentrated at the two

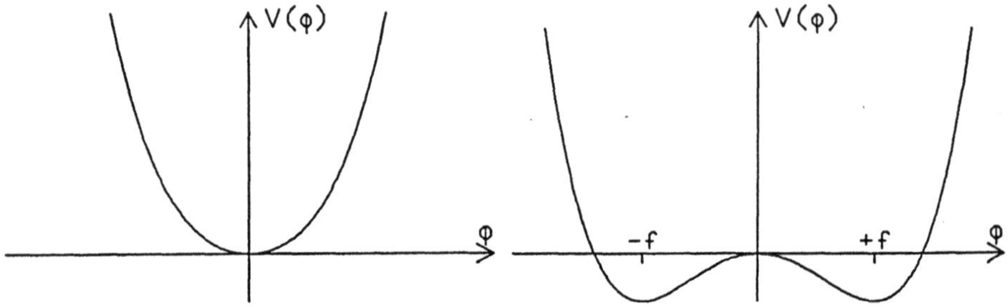

Fig. 1. Shape of the potential $V(\varphi)$ for $\mu^2 > 0$ (symmetric phase) and for $\mu^2 < 0$ (spontaneously broken phase).

minima. In a QFT with its ∞ly many degrees of freedom there are, however, two distinct solutions with lowest energy, $\varphi = f$ and $\varphi = -f$, and neither one is Z_2 invariant. In order to quantize the theory we have to select *one* ground state solution and quantize the small fluctuations around that solution. We thus arive at two different Hilbert spaces based on the two different vacua with $< 0|\varphi(x)|0 >= \pm f$.

3.2 The $U(1)$ Goldstone Model

Let us now consider a self interacting complex scalar field $\varphi(x)$ (or equivalently two real scalar fields $\varphi_1 = (\varphi + \bar{\varphi})/\sqrt{2}$ and $\varphi_2 = (\varphi - \bar{\varphi})/\sqrt{2}i)$ with the Lagrangean

$$\mathcal{L} = \partial_\mu \bar{\varphi} \partial^\mu \varphi - \mu^2 \bar{\varphi}\varphi - \frac{1}{2} h (\bar{\varphi}\varphi)^2 = \partial_\mu \bar{\varphi} \partial^\mu \varphi - V(\varphi)$$

and the field equation

$$\partial^\mu \partial_\mu \varphi + \mu^2 \varphi + h\varphi(\bar{\varphi}\varphi) = 0 \; .$$

This time Lagrangean and field equations are invariant under the group $G = U(1)$ of transformations $\varphi(x) \to \mathrm{e}^{-i\lambda}\varphi(x)$. For $\mu^2 \geq 0$ there is again only the solution $\varphi = 0$, which is $U(1)$ invariant. For $\mu^2 < 0$ all the solutions $\varphi(x) = \mathrm{e}^{-i\lambda_0} f$ are degenerate (i.e., each of them minimizes the energy) and are transformed into each other by $U(1)$. Choosing one such value for the ground state and introducing new fields ψ_1 and ψ_2 with vanishing vacuum expectation value

$$\varphi = \mathrm{e}^{-i\lambda_0} \left(f + \frac{\psi_1 + i\psi_2}{\sqrt{2}} \right) , \qquad \bar{\varphi} = \mathrm{e}^{+i\lambda_0} \left(f + \frac{\psi_1 - i\psi_2}{\sqrt{2}} \right)$$

the Lagrangean takes (up to an irrelevant constant) the form

$$\mathcal{L} = \frac{1}{2}(\partial_\mu \psi_1 \partial^\mu \psi_1 + \partial_\mu \psi_2 \partial^\mu \psi_2 - 2hf^2\psi_1^2) - \frac{hf}{\sqrt{2}}\psi_1(\psi_1^2 + \psi_2^2) - \frac{h}{8}(\psi_1^2 + \psi_2^2)^2 \; .$$

Since we have expanded around a solution of the field equations $\mathcal{L}$ contains no terms linear in the new fields. The field ψ_1 describes quanta of mass $m = \sqrt{2hf^2} = \sqrt{-2\mu^2}$ but the ψ_2-quanta are massless since the second derivative of the potential in the ψ_2-direction vanishes. This is an example for the Goldstone theorem, stating that spontaneous breakdown of an (ungauged) continous symmetry leads to the existence of massless bosons – the Goldstone bosons.

If we add a small term to $\mathcal{L}$ which explicitly breaks the $U(1)$-symmetry, e.g., a term $\epsilon(\varphi + \bar{\varphi})$, the ψ_2-quanta acquire a small mass $\sim \sqrt{\epsilon}$ (pseudo Goldstone bosons).

It should now be clear how the $Z_2 \sim O(1)$ and $U(1) \sim SO(2)$ Goldstone models can be extended to models with $G = SO(n)$ describing the self interaction of $n > 2$ real scalar fields. For $\mu^2 < 0$ there will again be a spontaneous breakdown of the $SO(n)$ symmetry by the choice of a particular ground state. In these cases a subgroup $H = SO(n-1)$ of G will, however, remain unbroken. In any case the values $|\varphi| = f$ at which the potential $V(\varphi)$ attains its minimum form the coset space G/H.

3.3 The $U(1)$ Higgs Model

This model is obtained from the $U(1)$ Goldstone model if we add an (abelian) $U(1)$ gauge field A_μ with field strength $F_{\mu\nu} = \partial_\mu A_\nu - \partial_\nu A_\mu$ and replace ordinary derivatives ∂_μ by covariant ones:

$$D_\mu\varphi = (\partial_\mu + igA_\mu)\varphi\,, \qquad D_\mu\bar{\varphi} = (\partial_\mu - igA_\mu)\bar{\varphi}\,.$$

The Lagrangean

$$\mathcal{L} = D_\mu\bar{\varphi}D^\mu\varphi - V(\varphi) - \frac{1}{4}F_{\mu\nu}F^{\mu\nu}$$

and the fields equations are now invariant under the $U(1)$ gauge transformations

$$\varphi(x) \to \mathrm{e}^{-ig\lambda(x)}\varphi(x)\,, \qquad A_\mu(x) \to A_\mu(x) + \partial_\mu\lambda(x)\,.$$

In the case $\mu^2 < 0$ the potential $V(\varphi)$ takes its minimum again for $|\varphi| = f$. A convenient way to reparametrize φ is given by

$$\varphi(x) = \mathrm{e}^{i(\lambda_0 + \psi_2/(\sqrt{2}f))}(f + \psi_1/\sqrt{2})$$

with

$$D_\mu\varphi(x) = \frac{1}{\sqrt{2}}\mathrm{e}^{i(\lambda_0+\psi_2/(\sqrt{2}f))}\Big(\partial_\mu\psi_1 + i\sqrt{2}gB_\mu(f + \psi_1/\sqrt{2})\Big)$$

where B_μ is the gauge invariant combination $A_\mu + \partial_\mu\psi_2/(\sqrt{2}fg)$. Since $F_{\mu\nu} = \partial_\mu B_\nu - \partial_\nu B_\mu$ the Lagrangean can be rewritten in terms of ψ_1 and B_μ alone in the form

$$\mathcal{L} = \frac{1}{2}(\partial_\mu\psi_1\partial^\mu\psi_1 - 2hf^2\psi_1^2) - \frac{hf}{\sqrt{2}}\psi_1^3 - \frac{h}{8}\psi_1^4 - \frac{1}{4}F_{\mu\nu}F^{\mu\nu} + \Big(g(f + \frac{\psi_1}{\sqrt{2}})\Big)^2 B_\mu B^\mu\,.$$

The last term indicates that the gauge field has acquired a mass $M = \sqrt{2}gf$ proportional to the gauge coupling constant g and to the vacuum expectation value f of the Higgs field φ. This Higgs-Kibble mechanism replaces the Goldstone mechanism in the case of gauge symmetries. In a sense the massless Goldstone boson which is no longer present has been absorbed by the formerly massless gauge field A_μ in order to provide the additional degree of freedom required for the massive vector field B_μ.

All this can easily be extended to non abelian gauge groups. In general the gauge fields for the generators of an unbroken subgroup H remain massless, all the other ones are massive (compare the standard model).

4. Topologically Stable Configurations (Defects)

Here I want to discuss certain field configurations with finite energy which are stable due to their topological structure. Typical such structures are solitons, vortices and monopoles in one, two and three space dimensions respectively and in four space dimensions (i.e., in Euclidean space-time) there are instantons. Since we live in a world with three space dimensions solitons and vortices are to be understood as solutions which are (at lest approximately) independent of two or one space coordinate, i.e., as domain walls or membranes with finite energy per unit surface or as vortex lines or strings with finite energy per unit length. As we will see such topologically non trivial configurations can occur in theories with spontaneously broken (gauged or ungauged) symmetries.

4.1 Solitons in the Z_2 Goldstone Model

In the Z_2 Goldstone model the potential $V(\varphi)$ has two minima at $\varphi = \pm f$ and a field configuration with finite energy must be close to one of these two values almost everywhere. In particular φ must go to one of these two values in the asymptotic regions $|\boldsymbol{x}| \to \infty$. In one space dimension the topologically non trivial configurations (solitons) are those where the two limits for $\boldsymbol{x} \to \pm\infty$ are different, e.g., $\lim_{\boldsymbol{x}\to\pm\infty} \varphi(x) = \pm f$. In other words the sign of $\boldsymbol{x}$ and the sign of φ are correlated for large $|\boldsymbol{x}|$. Within the manifold of continuous field configurations with finite energy such a configuration can not be deformed into the trivial one with $\varphi = f$. Among these configurations those which are stationary solutions of the field equations have the lowest energy. (There are always many such solutions centered at different points $\boldsymbol{x}$.)

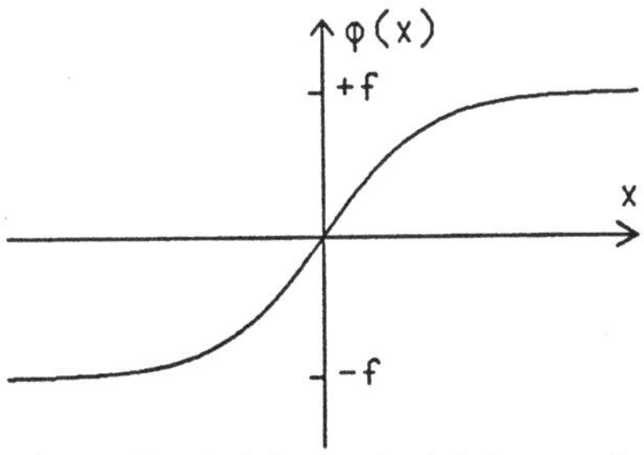

Fig. 2. Typical form of $\varphi(\boldsymbol{x})$ for a soliton solution.

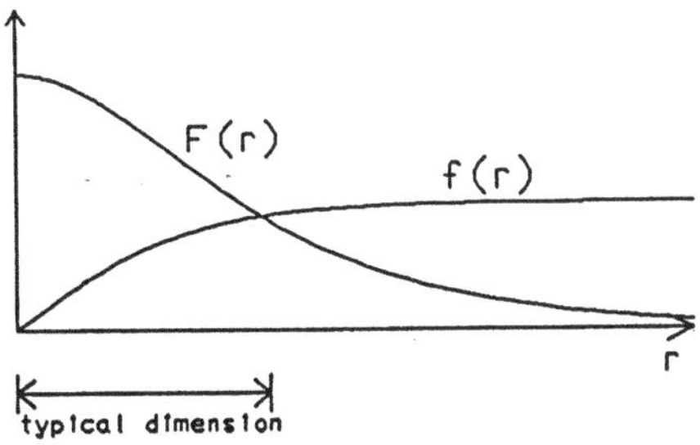

Fig. 3. Typical form of $f(r)$ and $F_{\mu\nu}(r)$ for a vortex solution.

4.2 Vortices in the $U(1)$ Higgs Model

Let us next consider the $U(1)$ Higgs model in two space dimensions. In order to have finite energy a configuration must certainly approach a vacuum configuration if we go to spatial infinity in any direction

$$\lim_{|\boldsymbol{x}|\to\infty} |\varphi(x)| = f \, , \qquad \lim_{|\boldsymbol{x}|\to\infty} (\partial_\mu + igA_\mu(x))\varphi(x) = 0 \, .$$

There may, however, be a non vanishing change in the phase of φ along a large circle around the origin. This phase change, $2\pi n$, is gauge invariant and since φ is single valued n must be an integer which is called the winding number of the configuration. Again configurations with different winding number cannot be deformed into each other. In order to find a stationary solution with winding number n we can make the ansatz $\varphi(\boldsymbol{x}) = f(r)\mathrm{e}^{in\vartheta}$ (in polar coordinates (r, ϑ)) with $\lim_{r\to\infty} f(r) = f$ and solve the resulting ordinary differential equations. The energy density of such a solution will be concentrated in a region around the origin and the size of this region is determined by the parameters of the potential $V(\varphi)$. These vortex solutions are topologically non trivial because the polar angle ϑ and the phase of φ are correlated for large r.

The solutions with winding number $n = \pm 1$ are certainly stable. Solution with, say, $n = 2$ could either be stable or could decay into two solutions with $n = 1$ again depending on the details of the potential V. An example for vortex lines (with winding number n) is given by the flux lines (with n flux quanta) in a type II super conductor.

4.3 Magnetic Monopoles ('t Hooft Polyakov)

Let us now consider an $SU(2)$ Higgs model with a triplet of real Higgs fields $\boldsymbol{\varphi} = (\varphi_1, \varphi_2, \varphi_3)$. We can choose the vacuum solution $\boldsymbol{\varphi} = (0, 0, f)$ and $\boldsymbol{A}_\mu = 0$. Here the gauge group $G = SU(2)$ is spontaneously broken but the subgroup $H = U(1)$ of rotations around the 3-axis remains unbroken and consequently the component $A_{\mu 3}$ is a massless 'photon' field. The other two components $A_{\mu 1}$ and $A_{\mu 2}$ of the non abelian gauge field describe vector particles with mass $M = gf$ where g is again the gauge coupling constant. In a topologically non trivial solution the isospin direction of $\boldsymbol{\varphi}$ is, for large r, correlated with the spatial direction of $\boldsymbol{x}$. A particular such configuration with winding number $n = 1$ is the 't Hooft Polyakov magnetic monopole obtained from the ansatz

$$\boldsymbol{\varphi} = f(r)\frac{\boldsymbol{x}}{r}, \qquad \boldsymbol{A}_i = h(r)\frac{\boldsymbol{x}\times\boldsymbol{e}_i}{gr^2}, \qquad \boldsymbol{A}_0 = 0$$

with $\lim_{r\to\infty} f(r) = f$ and $\lim_{r\to\infty} h(r) = 1$. Following 't Hooft we define

$$A_\mu^{\text{em}} = \frac{\boldsymbol{\varphi}\cdot\boldsymbol{A}_\mu}{\sqrt{\boldsymbol{\varphi}\cdot\boldsymbol{\varphi}}}$$

and the gauge invariant 'electromagnetic' field

$$\begin{aligned} F_{\mu\nu}^{\text{em}} &= \frac{\boldsymbol{\varphi}\cdot\boldsymbol{F}_{\mu\nu}}{\sqrt{\boldsymbol{\varphi}\cdot\boldsymbol{\varphi}}} + \frac{\boldsymbol{\varphi}\cdot(D_\mu\boldsymbol{\varphi}\times D_\nu\boldsymbol{\varphi})}{g(\boldsymbol{\varphi}\cdot\boldsymbol{\varphi})^{3/2}} \\ &= \partial_\mu A_\nu^{\text{em}} - \partial_\nu A_\mu^{\text{em}} + \frac{\boldsymbol{\varphi}\cdot(\partial_\mu\boldsymbol{\varphi}\times\partial_\nu\boldsymbol{\varphi})}{g(\boldsymbol{\varphi}\cdot\boldsymbol{\varphi})^{3/2}} \end{aligned}$$

and find that $F_{\mu\nu}^{\text{em}}$ asymptotically describes the field of a magnetic monopole. The mass of such a monopole solution can be estimated to be $\sim 4\pi M/g^2$.

4.4 Solitons

Finally we consider soultions of the $SU(2)$ Yang-Mills field equations in 4 dimensional Euclidean space with the property that the Euclidean action

$$I = \frac{1}{4g^2}\int d^4x\, \boldsymbol{F}_{\mu\nu}\cdot\boldsymbol{F}_{\mu\nu}$$

is finite and thus $\lim_{|x|\to\infty} \boldsymbol{F}_{\mu\nu} = 0$. Consequently the gauge field must, at far distances, approach a pure gauge: $\lim_{|x|\to\infty} A_\mu(x) = [\partial_\mu h(x)]h^{-1}(x)$. The group element $h(x)$ maps the 3-sphere at infinity into the $SU(2)$ group manifold and the winding number n of this map gives rise to the topological invariant $({}^*\boldsymbol{F}_{\mu\nu} = 1/2\epsilon_{\mu\nu\rho\sigma}\boldsymbol{F}_{\rho\sigma})$

$$n = \frac{1}{32\pi^2}\int d^4x\, \boldsymbol{F}_{\mu\nu}\cdot{}^*\boldsymbol{F}_{\mu\nu} = \frac{1}{32\pi^2}\int d^4x\, \partial_\mu k_\mu$$

where

$$K_\mu = 2\epsilon_{\mu\nu\rho\sigma}\boldsymbol{A}_\nu\cdot(\partial_\rho\boldsymbol{A}_\sigma - \frac{1}{3}\boldsymbol{A}_\rho\times\boldsymbol{A}_\sigma)$$

is the dual of the Chern-Simons 3-form. Among all field configurations with a given winding number $n > 0$ $(n < 0)$ those which are (anti) self-dual, i.e. ${}^*\boldsymbol{F}_{\mu\nu} = \pm\boldsymbol{F}_{\mu\nu}$, minimize the Euclidean action. These (anti) soliton configurations automatically satisfy the Yang-Mills field equations.

4.5 Some General Remarks on Topological Defects

In each of the cases discussed the topological stability is determined by the behaviour of the field configuration at (spatial) infinity. In one space dimension spatial infinity is topologically equivalent to the set $\{-1,+1\}$, in $d > 1$ space dimension to the $d-1$ dimensional unit sphere S^{d-1}. Toplogically non trivial configurations exist if these sets can be mapped non trivially into the manifold G/H of "vacuum configurations".

For each value of the winding number n the solution(s) which minimize the energy are the ground state(s) for a different Hilbert space. The Hilbert space for winding number $n = 0$ (vacuum sector) contains all the topologically trivial solutions. Among them are not only small deviations from the vacuum but also soliton anti-soliton or vortex anti-vortex or monopole anti-monopole pairs in one or two or three space dimensions respectively. The objects corresponding to soliton anti-soliton and vortex anti-vortex pairs in three dimensions are closed membranes and closed strings.

5. Transitions Between Different Ground States (Vacua)

We have seen that the minima of the Higgs potential $V(\varphi)$ correspond to the different possible ground states (vacua) of a theory with a spontaneously broken symmetry. Each ground state is the state of lowest energy in a different Hilbert space. One should, however, note that a Higgs potential with several minima need not be related with SSB.

As a simple example consider a model like the Z_2 Goldstone model but with a potential which is not invariant under the transformation $\varphi \to -\varphi$. If such a potential has several, say two, minima then they will in general have different depth. The example in Fig. 4 displays such a situation: The lower minimum at $\varphi = -f_1$ corresponds to the ground state (true vacuum) of one Hilbert space; the other minimum at $\varphi = +f_2$ corrsponds to the 'ground state' (false vacuum) of another Hilbert space. Naïvely one might think that this false vacuum immediately decays into the true vacuum. This is, however, not possible for two reasons. First the two vacua are in different Hilbert spaces and it is, therefore, not possible to compare their energies. Second the false vacuum is certainly meta-stable since it is a local minimum of the total energy among all configurations whith the same asymptotic value of φ.

On the other hand the false vacuum can not be absolutely stable; it can be at most meta-stable since it is not the configuration with the lowest energy. Consider a configuration (initial data with vanishing time derivative) which is close to the false vacuum outside a certain volume (e.g., a sphere) and close to the true vacuum inside, with a smooth transition between the two regions. The total energy of such a configuration (relative to the false vacuum) consists of a negative contribution $\approx (V(-f_1) - V(f_2)) \times$ volume and of a positive contribution $\eta \times$ surface, where η is the energy per unit area of a membrane for a potential similar to $V(\varphi)$ shown in Fig. 4 but symmetric under $\varphi \to -\varphi$.

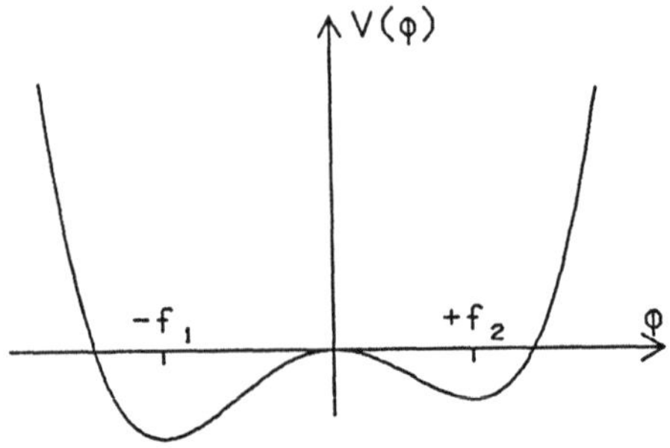

Fig. 4. Example of a Higgs potential $V(\varphi)$ with two minima of different depth. The absolute minimum at $\varphi = -f_1$ corresponds to a stable ground state (true vacuum); the local minimum at $\varphi = +f_2$ corresponds to a meta-stable 'ground state' (false vacuum).

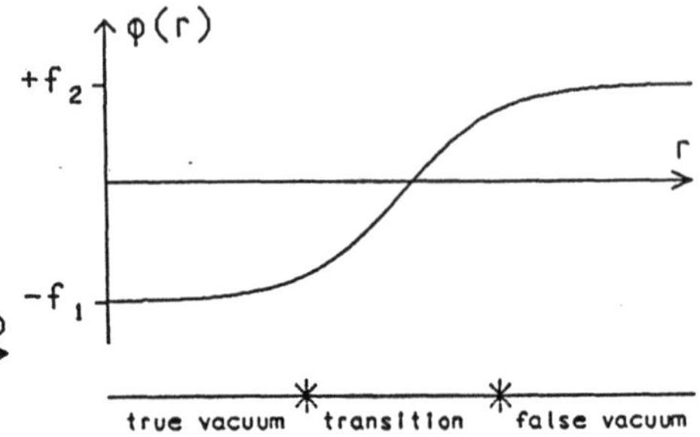

Fig. 5. Typical form of $\varphi(r)$ for a bubble of true vacuum inside a background of false vacuum.

Since the volume grows faster than the surface if we increase the volume, such bubble configurations can have arbitrarily large negative energies.

Let us consider in particular a spherically symmetric bubble configuration with vanishing energy which might have the form shown in Fig. 5. Given such initial data the transition region will rapidly start moving outwards. The region with the true vacuum will thus grow and a certain amount of energy per unit volume will become available for all kinds of excitations. If this energy is thermalized we will soon have a large region of heated true vacuum (inflationary universe).

We still have to discuss how such a bubble with vanishing energy can be created from the false vacuum [1]. This is certainly not possible classically (i.e., through a deformation of the false vacuum) since this would require a certain amount of energy. Thus some quantum mechanical tunneling has to take place. The rate for the creation of such bubbles per unit volume and unit time turns out to be

$$\Gamma/V = A\mathrm{e}^{-B/\hbar}[1 + O(\hbar)] \,.$$

In the limit when $V(-f_1) - V(+f_2)$ is small (thin wall approximation) one can even derive a closed expression for the quantity B in this expression. A spherically symmetric bubble configuration of vanishing energy (as determined by a function $\varphi(r)$ as shown in Fig. 5) can be reinterpreted as the section (at $\tau = 0$) through a spherically (i.e., $SO(4)$) symmetric solution of the field equations in Euclidean space-time (with the analytic continuation $t \to \tau = it$). It turns out that, in the thin wall approximation, B equals the Euclidean action of this solution.

6. Restoration of Spontaneously Broken Symmetries at Finite Temperatures

So far we have discussed QFT's with the vacuum as a ground state and have studied in particular the possibility of SSB. Let us now turn to some aspects of QFT at finite temperatures which are important for the dynamics of the very early universe.

Moreover QFT at finite temperature has become important for elementary particle physics as well. In recent years high energy nucleon nucleon collision experiments at accelerators have been performed. At such experiments it is expected that (in central collisions) the available energy is used to create a hot 'quark gluon plasma'.

The phenomenon of spontaneous symmetry breaking in QFT is very similar to the appearence of spontaneous symmetry breaking (e.g., spontaneous magnetization) in statistical mechanics. There SSB can occur at low temperatures; at higher temperatures a phase transition takes place and at high temperatures the equilibrium state (Gibbs state) has the full symmetry of the dynamics. One clearly expects a similar bahaviour for QFT's with spontaneous symmetry breaking: starting from a spontaneously broken ground state (vacuum) heating should lead to a phase transition and at high temperatures (i.e., high energy densities) the effects of SSB should disappear.

The quantum mechanical analogue of Gibbs states are the KMS (Kubo Martin Schwinger) states which should replace the vacuum if one wants to study QFT at finite temperatures. A Hilbert space based on a KMS state as 'ground state' is equivalent to an Euclidean quantum field theory which has periodic boundary conditions in the Euclidean time $\tau = it$ with a period $1/T$.

Without going into any details, let me try to give an intuitive picture what happens in a QFT at finite temperature. In the simplest example, the Z_2 Goldstone model, the vacuum $T = 0$ is given by $\varphi = f$. If we increase the temperature all kinds of quanta will be excited. At sufficiently high temperatures there will appear in particular closed membranes of various sizes with the vacuum $\varphi = f$ outside but with the 'other vacuum' $\varphi = -f$ inside. At first there will be few of them (per unit volume) and only small ones, but with increasing temperature their number and size will increase. Moreover they can interact, i.e., one such bubbles can split into two of them or two bubbles can merge into one. At a critical temperature we can no more distinguish whether we have bubbles of one vacuum inside a background of the other one or vice versa. In other words the spontaneous symmetry breaking has disappeared, we have a phase transition from the spontaneusly broken phase at low temperatures to a symmetric phase at high temperatures.

We can summarize the effect of a finite temperature in a spontaneously broken QFT if we replace the Higgs potential $V(\varphi)$ by an effective potential $V_{\text{eff}}(\varphi)$ which depends on the temperature. Here I just want to discuss some qualitative aspects. At $T = 0$ the effective potential is equal to the Higgs potential and is supposed to have at least two minima, the lowest one being

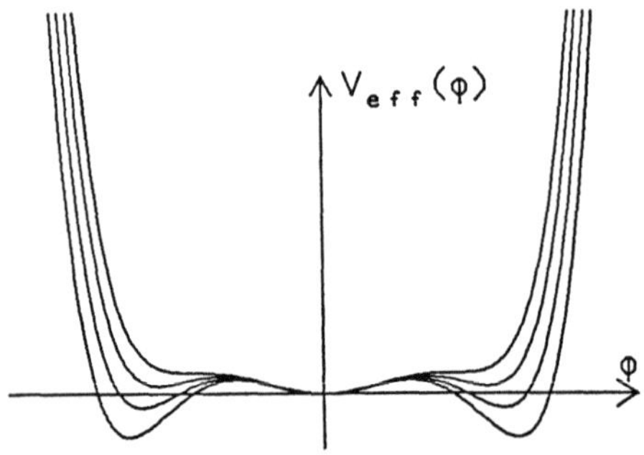

Fig. 6. Shape of the effective potential $V_{\text{eff}}(\varphi)$ for different values of the temperature T. For very low temperatures ($T \approx 0$) the effective potential might be given by the lowest curve with a false vacuum (at $\varphi = 0$) and two degenerate true vacua. With increasing temperature first the true and false vacua exchange there rôle. Finally, at still higher temperatures, there remains only one vacuum with $\varphi = 0$ (highest curve).

at $\varphi \neq 0$. At very high temperatures V_{eff} has only one minimum at $\varphi = 0$. In Fig. 6 we see how such an effective potential might look like for various values of T.

Let us apply the effective potential of Fig. 6 to the evolution of the very early universe. Starting at very high temperatures we are certainly in a symmetric phase with $\langle\varphi\rangle = 0$ corresponding to the only minimum of V_{eff} (highest curve). As the universe expands and the temperature decreases the effective potential will develop additional minima. At first they correspond to false vacua but at still lower temperatures they become the true vacua. We now have the situation that the state of the universe ($\langle\varphi\rangle = 0$) is the false vacuum. From this false vacuum there exists the possibility to create bubbles with one of the true vacua ($\varphi > 0$ or $\varphi < 0$) in their interior. As discussed earlier once such bubbles are created they will rapidly grow and will lead to a reheating of their interior regions (inflationary universe). In such a scenario the final state has large regions with one or the other of the two possible vacua and domain walls or membranes between them. Clearly similar effects can occur in a $U(1)$ Goldstone or Higgs model or in a model which has magnetic monopoles. In each case a (first order) phase transition from a symmetric phase at high temperatures to a spontaneously broken phase at low temperatures leads to the prediction that a certain number of topological defects (cosmic membranes, cosmic strings or monopoles) should be present.

7. Grand Unified Theories

We have seen that the standard model describes the electro-weak interaction with a spontaneously broken $SU(2) \times U(1)$ gauge theory (Glashow Salam Weinberg theory) whereas the strong interaction is described by an unbroken $SU(3)$

gauge theory. In the same way as the electomagnetic and weak interactions as observed at low energies (i.e., a $U(1)$ gauge theory and the Fermi theory) are unified by the standard model, grand unified theories (GUT) try to unify the electro-weak and strong interactions by a still larger gauge group G. At extremely high energies (temperatures) the gauge group G should be unbroken. At lower energies (of the order of TeV and higher!) G is spontaneously broken, such that only the subgroup $H = SU(3) \times SU(2) \times U(1)$ of G remains unbroken. Finally at energies of the order of 100 GeV and below the $SU(2) \times U(1)$ is broken as well such that only the electromagnetic $U(1)$ remains unbroken.

At first it might seem difficult to unify the $SU(2) \times U(1)$ gauge group with the $SU(3)$ gauge group because their coupling constants are so different ($\alpha_{\mathrm{em}} \approx \frac{1}{137}$ as compared to $\alpha_S \sim \frac{1}{10}$). These are, however, the values of the coupling constants measured at low energies. When they are measured at higher energies their values are different, their energy dependence being determined by the respective Gell-Mann Low β-function through the renormalization group equations. It turns out that the β-functions for $SU(3)$ and $SU(2)$ are negative (asymptotic freedom) whereas the β-function for $U(1)$ is positive. Consequently the values of the $SU(3)$ and $SU(2)$ coupling constants decrease and the $U(1)$ coupling constant increases with increasing energy. At energies of $\sim 10^{15}$ GeV all three coupling constants approximately reach the same value $\alpha(m_{\mathrm{GUT}}) \approx \frac{1}{40}$. Above this unification mass (or scale) one expects an unbroken larger symmetry group G, below this scale only the gauge group of the standard model should remain unbroken.

The simplest choice for the unification group G (which must contain $SU(3)\times SU(2) \times U(1)$ as a subgroup) is $SU(5)$, others are $SO(10)$, E_6, ... with rather similar results except that they require the existence of additional fermions which $SU(5)$ does not. In the $SU(5)$ theory the fermions of each family are grouped into two irreducible representations. For the left handed fermions (and anti-fermions) of e.g, the first family we have a representation $\overline{5}$ with ($\overline{d}_L \equiv \overline{d_R}$; e_L^-, ν_e) and a representation 10 with ($\bar{u}_L$; u_L, d_L; e_L^+). Since the group $SU(5)$ has 24 generators there are 24 gauge fields; in addition to the $8+3+1$ gauge bosons from the standard model there are 12 heavy ones with masses of the order of the unification scale. They have rather unusual couplings as they

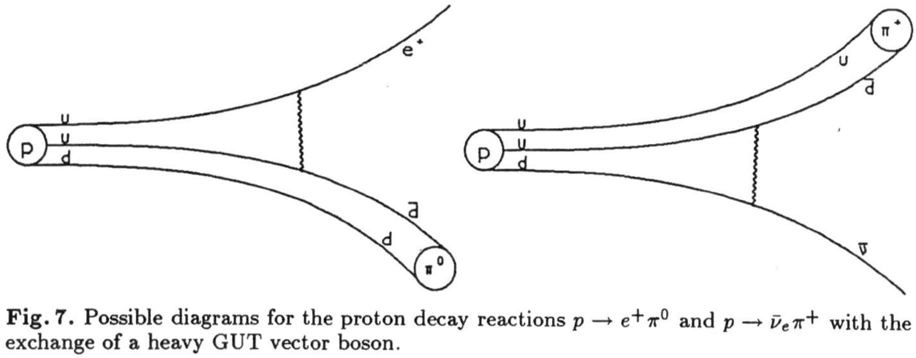

Fig. 7. Possible diagrams for the proton decay reactions $p \to e^+\pi^0$ and $p \to \bar{\nu}_e\pi^+$ with the exchange of a heavy GUT vector boson.

can induce the transitions

$$\begin{pmatrix} u \\ d \end{pmatrix}_L \leftrightarrow e^+_L , \qquad d_R \leftrightarrow \begin{pmatrix} \bar{\nu}_e \\ e^+ \end{pmatrix}_R , \qquad \begin{pmatrix} \bar{d} \\ \bar{u} \end{pmatrix}_R \leftrightarrow u_R$$

which violate the separate conservation of baryons and leptons (the difference $B - L$ between baryon and lepton number is, however, still conserved). This allows the proton to decay into $e^+\pi^0$ or $\bar{\nu}_e\pi^+$ via the exchange of one of the new bosons (see Fig. 7). Due to the very large mass of the new bosons this decay is rather slow, simple estimates lead to a proton life time of $\sim 10^{33}$ years.

The GUT prediction of the proton decay has triggered several experiments to detect this decay but until now none of them has found a clear signal for proton decay.

The lack of baryon number conservation may have important implications for the early universe. It has always been an open problem to explain why our universe contains a large number of baryons but seems to contain no anti-baryons. Without baryon number conservation this might be a consequence of the evolution. Elementary particle experiments have revealed a very small violation of CP invariance. According to the famous CPT theorem (which states that the combined operation of charge conjugation, parity and time reversal is a symmetry for every relativistic local quantum field theory) there must be a small violation of T invariance as well.

Let us consider the combined effect of baryon number non conservation and T non invariance on the very early universe. First of all in a very early state of the universe there was a huge number of baryon anti-baryon (i.e., quark anti-quark) pairs in thermal equilibrium. Any difference between the number of baryons and anti-baryons which might initially have existed must have disappeared in this thermal equilibrium due to the baryon number non conservation. In a somewhat later state of the universe the quark anti-quark pairs have started to decouple from the thermal equilibrium and thus the T non invariance became effective. Together with baryon number non conservation this must have led to a certain baryon anti-baryon asymmetry. Depending on details of the cosmological evolution it may thus be possible to explain the observed number of baryons.

8. Photon Splitting in a Strong Magnetic Field

Next I want to discuss an effect from rather old fashioned QED which might have astrophysical implications. Consider the propagation of photons in a strong external magnetic field. Adler [2] has computed the one loop contributions to this process exactly but in order to get estimates it suffices to compute the lowest order contribution from box diagrams (Delbrück scattering). Moreover for slowly varying external fields and low photon frequencies one can use the effective Lagrangean derived more than half a century ago [3]

$$\mathcal{L}_{\text{eff}} = -\frac{1}{4}F \cdot F + \frac{\alpha}{192B_{\text{cr}}^2}\left((F \cdot F)^2 + (F \cdot {}^*F)^2\right)$$

with $F \cdot F = F_{\mu\nu}F^{\mu\nu} = 2(\boldsymbol{B}\cdot\boldsymbol{B} - \boldsymbol{E}\cdot\boldsymbol{E})$ and $F \cdot {}^*F = 1/2\epsilon^{\mu\nu\rho\sigma}F_{\mu\nu}F_{\rho\sigma} = 4\boldsymbol{E}\cdot\boldsymbol{B}$. The quartic term is due to the creation and anihilation of virtual electron positron pairs and is negligible for magnetic fields much smaller than the critical field $B_{\text{cr}} = m^2/e = 4.41 \times 10^{13}$ G.

According to conventional electrodynamics one computes the response to the variation of the electric and magnetic field

$$\boldsymbol{D} = \frac{\partial \mathcal{L}}{\boldsymbol{E}} = \boldsymbol{E} + \ldots, \quad \boldsymbol{H} = -\frac{\partial \mathcal{L}}{\boldsymbol{E}} = \boldsymbol{B} + \ldots$$

and the tensors of dielectric and magnetic susceptibility as well as a tensor ρ_{ij} which can give rise to optical activity

$$\epsilon_{ij} = \frac{\partial D_i}{\partial E_j} = \delta_{ij} + \ldots, \quad \mu_{ij}^{-1} = \frac{\partial H_i}{\partial B_j} = \delta_{ij} + \ldots, \quad \rho_{ij} = \frac{\partial D_i}{\partial B_j} = -\frac{\partial H_j}{\partial E_i} = 0 + \ldots .$$

From these tensors one can compute the propagation of photons and finds that a strong magnetic field leads to double refraction. Photons with polarization parallel and perpendicular to the external magnetic field have slightly different indices of refraction.

From the same effective Lagrangean one can compute transition amplitudes for the splitting of one photon into two photons. The opening angle for such processes tends to zero for small external fields but the phase space remains finite provided the splitting process is kinematically allowed. Adler has found that there is only one process allowed by this kinematic constraint as well as by the CP selection rule: the splitting of one photon with polarization parallel to the magnetic field into two photons both with perpendicular polarization. Thus a very strong magnetic field which is homogeneous over a sufficiently large region will convert unpolarized photons into polarized ones. It has been speculated that such magnetic fields might be present in the vicinity of neutron stars and that they might be responsible for the observed polarization of photons from pulsars.

9. Candidates for Dark Matter from Particle Physics

Elementary particle physics theory has is a long tradition to predict new particles on purely theoretical ground (mostly consistency or simplicity of the theoretical description). Some of these postulated new particles have subsequently been confirmed by experimental observation (e.g., the antiparticles e^+ and $\bar{p}$, the neutral particles n and ν and, more recently, the massive vector mesons Z^0 and $W^\pm$) or are theoretically well established (e.g., quarks and gluons). Others (e.g., t-quarks, gravitinos, axions, ...) have not (yet?) been observed.

Likewise there is a long tradition and a lot of imagination has been invested to explain why this or that particle has not yet been observed.

Here I want to discuss some of the particles proposed by various theories which have not yet been observed in experiments. Some of these somewhat unconventional particles could be the constituents of dark matter.

9.1 Neutrinos with a Non Vanishing Mass

Let us start rather conventionally with the possibility that neutrinos could have a non-vanishing rest mass. The present experimental limits from particle physics experiments

$$m_{\nu_e} \lesssim 10 - 15 \text{ eV} , \qquad m_{\nu_\mu} \lesssim 150 \text{ keV} , \qquad m_{\nu_\tau} \lesssim \text{a few keV}$$

leave room for plenty of speculations and will be discussed extensively in other lectures of this school.

9.2 New Particles from Supersymmetry or Supergravity

Let me start with a short answer to the questions: What is super supersymmetry (SUSY) and what is supergravity (SUGRA)? SUSY is a (global) symmetry which puts bosons and fermions (i.e., particles with different spin) into the same multiplet. SUGRA is gauged SUSY, i.e., SUSY with space-time dependent parameters. Equivalently one might say that SUGRA is SUSY in the presence of gravitons or SUSY on a curved space-time manifold because each one of these three definitions implies the other ones.

9.2.1 Supersymmetry

In the past there have been many attempts to find symmetries which unify particles with different spins in the sense that they occur in the same multiplet. Although this can be achieved in a non relativistic theory all such attempts turned out to be incompatible with relativistic QFT. Later on O'Raifeartaigh, Coleman and Mandula [4] have proved 'no-go' theorems which state that the only possible generators of symmetries in a relativistic local QFT are the generators of the Poincaré group plus scalar charges, and thus all particles in a multiplet must have the same mass *and spin*. These theorems always (implicitely) assume that the generators (charges) of the symmetry are bosonic and therefore transform bosons into bosons and fermions into fermions. They do not apply to symmetries such as SUSY with fermionic generators transforming bosons into fermions and vice versa. Nevertheless these theorems impose restrictions on the structure of SUSY algebras because they do apply to the bosonic part of such algebras [5].

The simplest ($N = 1$) SUSY algebra is generated by a fermionic two component spinor charge Q_α and its adjoint $\bar{Q}_{\dot{\alpha}}$ satisfying the (anti) commutation relations

$$\{Q_\alpha, \bar{Q}_{\dot{\alpha}}\} = \sigma^\mu_{\alpha\dot{\alpha}} P_\mu \,, \quad \{Q, Q\} = \{\bar{Q}, \bar{q}\} = [Q, P] = [\bar{Q}, P] = 0 \,,$$
$$[J_{\mu\nu}, Q_\alpha] = \frac{i}{2}(\sigma_{\mu\nu})_\alpha{}^\beta Q_\beta \,, \quad [J_{\mu\nu}, \bar{Q}_{\dot{\alpha}}] = \frac{i}{2}(\bar{\sigma}_{\mu\nu})_{\dot{\alpha}}{}^{\dot{\beta}} \bar{Q}_{\dot{\beta}} \,.$$

Due to these commutation relations each SUSY multiplet must contain the same number of bosonic and fermionic states. The detailed structure is, however, different for $m \neq 0$ and for $m = 0$.

For non vanishing rest mass ($m \neq 0$) a multiplet can be characterized by an integer or half integer spin $j = 0, 1/2, 1, \ldots$ and contains four states with spins $j - 1/2$, j, j and $j + 1/2$, where the two states with spin j have opposite parity. As usual the state with spin $j - 1/2$ is absent for $j = 0$. The simplest such multiplet with $j = 0$ contains three states with spin parity 0^+, 0^- and $1/2$ corresponding to the fields $A(x)$, $B(x)$ and $\psi(x)$ of the Wess-Zumino model. Here A is a real scalar, B a real pseudo scalar and ψ a Majorana spinor, i.e., ψ satisfies a reality condition and describes particles which are their own antiparticles. The next multiplet with $j = 1/2$ is a massive vector multiplet with states 0^+, $1/2$, $1/2$ and 1^-.

For vanishing rest mass ($m = 0$) the multiplets are characterized by an integer or half integer heliciy $\lambda = \ldots, -1, -1/2, 0, 1/2, 1, \ldots$ and contain two states with helicities λ and $\lambda + 1/2$. Invariance under P or C requires that the multiplet with helicities $-\lambda - 1/2$ and $-\lambda$ exists as well. Consider, e.g., the super Yang-Mills (SYM) multiplet which contains, for each generator of the gauge group, states with helicities $\pm 1/2$ and ± 1 described by a Majorana spinor field $\boldsymbol{\chi}(x)$ and the gauge field $\boldsymbol{A}_\mu(x)$.

Extended SUSY algebras are generated by $N > 1$ different super charges Q^I_α ($I = 1, \ldots, N$) satisfying

$$\{Q^I_\alpha, \bar{Q}^J_{\dot{\alpha}}\} = \delta^{IJ} \sigma^\mu_{\alpha\dot{\alpha}} P_\mu \,,$$

and give rise to larger multiplets. For $m = 0$ a multiplet contains states with helicities in the range $\lambda_{\rm min} \leq \lambda \leq \lambda_{\rm max}$ where $\lambda_{\rm max} - \lambda_{\rm min} = N/2$ and where the state with helicity λ has multiplicity $\binom{N}{2(\lambda - \lambda_{\rm min})}$. Again we may have to add the P (or C) conjugate multiplet unless the multiplet is self conjugate ($\lambda_{\rm max} = N/4$). If we require $|\lambda| \leq 1/2$ (absence of gauge fields) $N \leq 2$ must be satisfied. From $|\lambda| \leq 1$ (gauge fields without gravity) follows $N \leq 4$ and $|\lambda| \leq 2$ (theories with gravity) implies $N \leq 8$. For $N > 8$ (in $d = 4$ space-time dimensions), i.e., with $|\lambda| > 2$, no consistent interacting theory has been constructed and there are indications that no such theory can exist.

Unlike than for other symmetries (e.g., $SU(2)$, $SU(3)$, ...), the trivial representation of SUSY is of no interest in particle physics. If a (boson or fermion) field $\varphi(x)$ transforms trivially under SUSY

$$[Q, \varphi(x)]_\pm = [\bar{Q}, \varphi(x)]_\pm = 0 ,$$

then

$$[\{Q, \bar{Q}\}, \varphi(x)] = [\sigma \cdot P, \varphi(x)] = -i\sigma^\mu \partial_\mu \varphi(x) = 0 ,$$

i.e., $\varphi(x)$ is constant and describes no particles. If nature is supersymmetric then each particle must have its SUSY partner. The SUSY partners of bosons (e.g., the photino and gluino for the photon and gluon) are fermions, the partners of fermions (e.g., the squark and selectron for the quark and electron) are bosons; all the internal quantum numbers are the same for both partners (at least in the case of $N = 1$ SUSY).

Nature is certainly not supersymmetric. One might assume that SUSY is explicitely broken, an assumption with very little predictive power. If SUSY is spontaneously broken there must be a massless Goldstone fermion with helicity $\lambda = \pm 1/2$ (goldstino). It was once speculated that this could be the neutrino, but it turned out that such an assignment leads to phenomenologically inacceptable predictions. The simplest mechanisms for spontaneous SUSY breaking lead to predictions for the mass splitting within SUSY multipletts.

So far none of the SUSY partners of known particles has been found. It might be that SUSY breaking occurs at a very high mass scale and thus the mass splittings are large. In this situation very little, if anything, of SUSY would be left at low energies.

There are models for, e.g., supersymmetric QED, which are, however, (in my opinion) not terribly convincing.

9.2.2 Supergravity

If we consider SUSY with space-time dependent parameters (gauged SUSY) we are led to SUGRA. Gravity becomes involved because space-time dependent SUSY transformations imply, via the SUSY algebra, space-time dependent translations, i.e., infinitesimal diffeomorphisms.

This argument can also be reversed: the combination of SUSY and gravity yields SUGRA. First the graviton must have a SUSY partner with helicity $\lambda = \pm 3/2$ (the other possibility $\lambda = \pm 5/2$ does not lead to a consistent theory), the gravitino. This gravitiono is described by a Rarita-Schwinger field ψ_μ which is the gauge potential for SUSY. Moreover, the concept of a constant spinor is meaningless in a generic curved space-time background and thus the parameters of SUSY transformations necessarily must be space-time dependent in the presence of gravity.

The minimal $N = 1$ SUGRA theory describes a graviton and a gravitino. Various matter multiplets can be coupled to these, essentially by substituting super covariant derivatives for ordinary ones.

The maximal $N = 8$ SUGRA theory in $d = 4$ space-time dimensions can conveniently be derived via dimensional reduction from a much simpler $N = 2$, $D = 10$ or $N = 1$, $D = 11$ theory. The $N = 8$ multiplet consists of 128 bosonic and 128 fermionic states; in four dimensions these are 1 graviton with $\lambda = \pm 2$, 8 gravitinos with $\lambda = \pm 3/2$, 28 vector (gauge) fields with $\lambda = \pm 1$ 56 Majorana spinors with $\lambda = \pm 1/2$, 35 real scalars and 35 real pseudoscalars with $\lambda^P = 0^\pm$.

The 56 vector fields might be the gauge fields of (a subgroup of) $SO(8)$. Note, however, that the gauge group $SU(3) \times SU(2) \times U(1)$ of the standard model is not contained in $SO(8)$.

If SUGRA is spontaneously broken a supersymmetric version of the Higgs-Kibble mechanism leads to a massive gravitino instead of a massless goldstino.

Originally it was hoped that SUGRA would solve the problem that gavity (considered as QFT) is non-renormalizable. SUGRA, indeed, yields finite one- and two-loop scattering amplitudes but three-loop divergences can not be excluded. By power counting SUGRA is as non-renormalizable as gravity.

9.3 New Particles from String Theories

The fundamental objects of local QFT are point particles. In string theory they are replaced by elementary (not cosmic) strings which are extended in one space dimension. The motion of such a string in time yields a two-dimensional world sheet instead of the familiar world line. In order to describe this world sheet embedded into space-time one uses coordinates ξ^α and a metric $\gamma_{\alpha\beta}(\xi)$ for the two-dimensional world sheet, and coordinates X^μ and a metric $g_{\mu\nu}(X)$ for D-dimensional space-time (possibly curved). The dynamics of a bosonic string is described by an action equal to the area of its world sheet:

$$S = \int d^2\xi \sqrt{\gamma(\xi)}\gamma^{\alpha\beta}(\xi)\frac{\partial X^\mu(\xi)}{\partial \xi^\alpha}\frac{\partial X^\nu(\xi)}{\partial \xi^\beta} g_{\mu\nu}(X(\xi)) \, .$$

This action has the same form as for a two-dimensional non-linear σ-model.

If we add SUSY, either in the two-dimensional ξ-space or equivalently in the D-dimensional X-space, we obtain superstrings.

The classical (super) string action is invariant under reparametrizations of the world sheet and under conformal transformations $\gamma_{\alpha\beta}(\xi) \to \rho^2(\xi)\gamma_{\alpha\beta}(\xi)$. The quantized theory has, in general, a conformal anomaly (trace anomaly). The precence of such an anomaly leads to inconsistencies in the physical interpretation, e.g., ghost states. The conformal anomaly is, however, absent if we choose (in the case of flat X-space) $D = 26$ for the bosonic string or $D = 10$ for the superstring. One usually assumes that the 10-dimensional X-space of super string theories can be reduced to a 6-dimensional internal space and a 4-dimensional space-time through a compactification of the Kałuza-Klein type.

The string action describes both the motion of the center of mass of a string and string excitations (internal motion). It is the sum over all these excitations which might lead to a consistent quantum theory in the presence of gravity. In the 'zero slope' approximation the string excitations are extremely heavy. The remaining 'massless' excitations are those of a (super) Yang-Mills theory for open (super) strings and those of (super) gravity for closed (super) strings.

There are, in fact, several versions of superstring theory.

- Type I superstrings have $N = 1$ SUSY in $D = 10$, corresponding to $N = 4$ in $d = 4$.
- Type IIa and type IIb superstrings have two different versions of $N = 2$ SUSY

in $D = 10$, both corresponding to $N = 8$ in $d = 4$.
– Finally heterotic strings have different assignments for left and right moving excitations.

For reasons of consistency the heterotic strings require a gauge group $E_8 \times E_8$. One E_8 may be useful as grand unification group with a breaking scheme $E_8 \supset E_7 \supset E_6 \supset SO(10) \supset SU(5)$ but the other E_8 is phenomenologically completely superfluous. All known particles or fields transform under representations $\mathit{any} \times 1$ of $E_8 \times E_8$. Why is this so? What is the rôle of the other fields transforming with representations $1 \times \mathit{any}$ or $\mathit{any} \times \mathit{any}$ of $E_8 \times E_8$? One possibility would be that these excitations do exist, but do not interact with the known particles, except via gravitation. Does such a 'shadow world' with 'shadow states' or 'shadow particles' exist? And if so, could they be candidates for dark matter?

9.4 Axions

Let me conclude this discussion of possible dark matter candidates with the axion, a light pseudo scalar particle postulated on purely theoretical grounds.

First consider the ground state (vacuum) in a theory with non-abelian gauge fields. It turns out that there are, in fact, infinitely many 'θ-vacua' which differ by the value of a parameter θ which is not present in the classical Lagrangean [6]. If we describe the ground state in terms of a path integral over Euclidean field configurations this parameter is due to the existence of (anti) soliton configurations: we can attribute a phase factor $e^{in\theta}$ to all configurations with soliton or winding number n.

Next consider in addition scalars, pseudo scalars and spinors coupled such that the classical action is invariant under a chiral $U(1)$ group. In the quantized theory there is a chiral anomaly proportional to $\boldsymbol{F} \cdot {}^*\boldsymbol{F}$, i.e., the integrated chiral anomaly is proportional to the winding number. A chiral rotation is therfore equivalent to a shift in θ. Thus if all fermions are massless then all θ-vacua are physically equivalent. If the fermions are, however, massive such a chiral rotation transforms a scalar mass term ($\sim \mathbf{1}$) into a pseudo scalar one ($\sim \gamma_5$).

It is experimentally obvious that we live in a world where P and CP are good symmetries at the level of strong interactions. This means that in our world we have simultaneously scalar mass terms and $\theta = 0$. In a general theory this can only be understood as the result of some fine tuning, an idea which theorists find very unattractive. One can, however, obtain the same result in a 'natural' way if an additional pseudo scalar particle, the axion, exists [7]. This axion should have a mass in the range 100 KeV – 1 MeV or could even be nearly massless; it could be a candidate for dark matter and would have other experimentally observable consequences [8]. Such an axion has, however, not (yet?) been observed experimentally.

References

1. S. Coleman and F. de Luccia, Phys. Rev. D 21 (1980) 3305.
2. S.L. Adler, J.N. Bahcall, C.G. Callan and M.E. Rosenbluth, Phys. Rev. Lett. 25 (1970) 1061; S.L. Adler, Ann. Phys. (N.Y.) 67 (1971) 599.
3. H. Euler and B. Kockel, Naturwissenschaften 23 (1935) 246; W. Heisenberg and H. Euler, Z. Phys. 98 (1936) 714; V. Weißkopf, Kgl. Danske Vidensk. Selsk. Mat.-Fys. Medd. 14 (1936) No. 6.
4. S. Coleman and J. Mandula, Phys. Rev. 159 (1967) 1251.
5. R. Haag, J.T. Lopuszanski and M. Sohnius, Nucl. Phys. B 88 (1975) 257.
6. G. 't Hooft, Phys. Rev. Lett. 37 (1976) 8 and Phys. Rev. D 14 (1976) 3432.
7. R.D. Peccei and H.R. Quinn, Phys. Rev. Lett. 38 (1977) 1440.
8. S. Weinberg, Phys. Rev. Lett. 40 (1978) 223; F. Wilczek, Phys. Rev. Lett. 40 (1978) 279. J.E. Kim, Phys. Rev. Lett. 43 (1979) 103.

This article was processed by the author using the TEX Macropackage from Springer-Verlag.

Introduction and Survey II: Concepts from Astrophysics and Cosmology

G. Börner

Max-Planck-Institut für Physik und Astrophysik,
Institut für Astrophysik, Karl-Schwarzschild-Str. 1,
D–8046 Garching / München, Fed. Rep. of Germany

Introduction

Astrophysics and astronomy try to understand the objects found in the sky: the stars, their assembly into large systems, the galaxies and their clustering, the various radiation fields, the cosmic ray particles, and the global underlying structure of the universe. Information about these objects is collected through patient observation of the stream of radiation and particles that reaches telescopes and detectors on or near earth. This soft way of looking at things which does not influence or change their nature has quite a different quality from the hard way in which nature is forced by lab experiments to reveal her secrets. An obvious drawback is the unique nature of events, such as e.g. the supernova explosion in the Large Magellanic Cloud in 1987 – this was an experiment that cannot be repeated. Astronomers, and neutrino astronomers as well, will have to wait for a long time before they'll have another neutrino burst to analyse. Another interesting aspect is the way in which new instruments have changed our view of the world. The picture of an eternally unchanging universe filled uniformly with stars prevailed until the early 1920's (even Einstein had this view originally, as is testified by his attempts to find a static solution of his field equations of general relativity through the introduction of a cosmological constant), when Hubble made use of the new large telescopes at Mt. Palomar and discovered the existence of other galaxies and the general expansion of galaxies. Radio observations by Penzias and Wilson in 1964 established the existence of a 3K background radiation field which is interpreted as a relic of a hot and dense early phase of the universe. The concept of an unchanging sky broke down completely, when the first x-ray satellites went to work in the early 1970's, and demonstrated that rapid changes, flare activity, explosions were the common way of life in this spectral range. Since then with each new instrument new and sometimes seemingly exotic phenomena have been found.

Astrophysical concepts are simply extrapolations of the laws of physics known from laboratory experiments to stellar and cosmological phenomena. Every now and then the suggestion is made that some new physics is needed to explain the behaviour of certain objects (remember e.g. the "redshift-controversy" of the quasars), but as yet there has not been an absolutely compelling case for new physical laws. This does not mean that we understand all the objects of astronomy. But in many cases, especially the mysterious ones, the conservative way of relying on known physical concepts has by no means been exhausted.

In principle the systems are very simple. They consist of a large number of baryons (10^{57} for stars, 10^{69} for galaxies, $\sim 10^{80}$ for the universe) with various ingredients such as magnetic fields, radiation etc.. The big complication is introduced by the all-pervading force of gravitation. Negligible on laboratory scales, it becomes the dominating force on astronomical dimensions, where it determines the behaviour of these large assemblies of baryons. Einstein's theory of gravitation leads to the concept of a maximum stable mass for baryonic matter, as long as the average properties of matter are determined by known laws of physics. In the case of neutron stars the density can be ten times the density in a nucleus, and thus some extrapolation from the regime known from lab experiments is involved.

Large mass configurations will probably end in a black hole, a singularity of space time. Since general relativity predicts the existence of black holes from the normal properties of matter, it also points to its own limitations. Singularities are unphysical, and they must be avoided in a more fundamental theory – perhaps in a unified description of general relativity and quantum field theory. At present such a unified theory does not exist, and we must discuss black holes and other singularities, such as the initial space-time event of a cosmological model, in terms of classical space-time geometry. You will hear more about black-holes in the lectures by HAJICEK and STRAUMANN.[1]

Black holes look quite decent and reasonable from the outside, because their singular interior is veiled to external observers by a horizon, a surface which encloses the singularity. Signals cannot penetrate outwards through the horizon. Thus astrophysically black holes are dark objects, but they influence matter by their gravitational field. Infalling matter will heat up and radiate (mainly in x-rays for a black-hole of stellar mass), and so the black hole may appear as a compact source of radiation to an outside observer. Its observational appearance inside is an experience that cannot be shared, it is unique to the observer falling into the black hole.

We believe that we understand quite well the properties and time evolution of agglomerates of $\sim 10^{56}$ to 10^{59} baryons. These form stars in the mass range of $0.1 M_\odot$ to $100 M_\odot$ (, $1 M_\odot \equiv 2 \times 10^{33}$g), and some aspects of their life history will be discussed in Section 1.

On the other hand we must admit that the formation and subsequent evolution of larger systems (10^{68} baryons, corresponding to a typical galaxy) is not understood too well. Especially the formation of large-scale structures in the universe is still a mystery. Some aspects of this problem will be discussed in the talks by RUFFINI, TUROK, and in Section 3. Section 2 is reserved to a short description of the Friedmann-Lemaître models of the universe. These highly symmetric solutions of Eistein's theory of general relativity can accomo-

[1] Names in capital letters refer to other lectures at this school.

date cosmological observations quite well, but it is clear that they cannot be the true picture of the system of 10^{80} baryons which is the universe. Successes and shortcomings of the classical hot big bang model will be pointed out. Some attempts to include particle physics concepts in a description of the early universe will also be discussed, although these concepts are very speculative and only loosely related to observations.

1. Stellar Evolution

1.1 The Fusion Reactor

The stars can be arranged in a diagram where colour (or effective surface temperature) is plotted versus luminosities. Such a Hertzsprung-Russel diagram (Fig. 1) shows several distinct features: Most stars lie on a line from the upper left to the lower right in such a diagram, the 'main sequence' which ranges from red stars of low luminosity to blue stars of high luminosity. In addition there is a branch of stars in the upper right section, stars of red colour and high luminosity, the red giants. The lower left section is populated by white stars (of high surface temperature) with low luminosity, the white dwarfs.

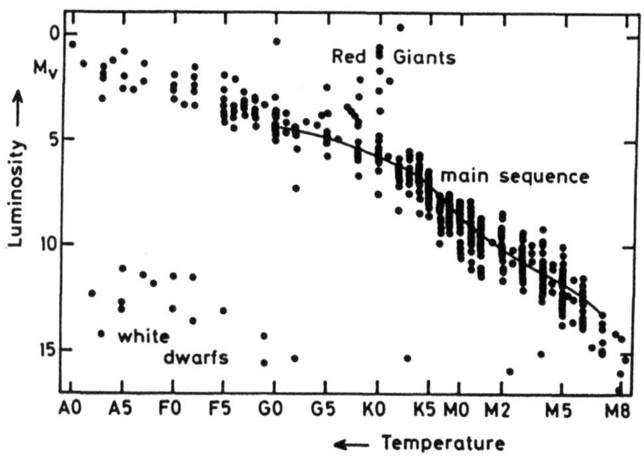

Fig. 1. Hertzsprung-Russel diagram for stars in the solar neighbourhood (within 10 parsec). The labels A0 to M8 on the horizontal axis refer to spectral types. One can read this as a temperature scale with increasing temperature from right to left. (After Ref. [1])

This distribution of stars can be understood from the physical processes and the evolution of the inner regions (an up-to-date introduction to this subject is [2]). Stars are selfgravitating gas spheres of large mass. The contraction of a cloud of interstellar gas may lead to temperatures high enough for the fusion of hydrogen into helium in the core (by the CNO cycle). These fusion reactions supply the energy for the star's luminosity:

Thermal energy ($\sim c_v TM$) and gravitational potential energy $\left(-GM^2/R\right)$ are too small:

For the sun $E_T \sim 5 \times 10^{48}$ erg and
$E_G \sim 4 \times 10^{48}$ erg
are used up in a typical time (the Kelvin-Helmholtz time)

$$\tau_{KH} = \frac{E_T}{L} = -\frac{E_G}{L} = \frac{GM^2}{RL} \sim 10^{15}\text{sec.} \tag{1.1}$$

This is much smaller than the age of the sun of 4.6×10^9 years. The total mass-defect of the sun due to fusion over its lifetime is

$$\Delta M = \frac{L\,\Delta t}{c^2}, \text{ i.e. } \left(\frac{\Delta M}{M_\odot}\right) = \frac{L_\odot}{M_\odot}\,\frac{\Delta t}{c^2} = 3 \times 10^{-4}. \tag{1.2}$$

Fusion reactions do not appreciably change the mass. The stars on the main sequence burn hydrogen in the core. They have a maximum of temperature, energy production, and consumption of fuel at the center. Stars with a radiative central region, i.e. with mass $M \leq 1.2M_\odot$ burn out inhomogeneously. Stars with a convective central region ($M > 1.2M_\odot$) burn out the central core homogeneously, a helium core (a mixture of hydrogen and helium) appears. Once the fuel is exhausted in the center, the fusion processes move into regions where there still is fuel – a shell source around the core develops which burns hydrogen and eats itself into the outer regions. The outer layers of the star expand enormously when the star tries to find a new equilibrium configuration. The star changes into a red giant. The time of hydrogen core burning can be estimated: The shell source develops when about 12 % of the hydrogen have been burnt. The energy produced is 0.7 % of the rest mass per reaction, i.e.

$$\Delta E = \Delta M c^2\, 0.007.$$

The luminosity L is the energy loss per unit time. Thus the time on the main sequence (hydrogen burning) is

$$t_{m.s.} = 0.1Mc^2\, 0.007/L = 10^{10}\,(M/M_\odot)\,(L_\odot/L)\,\text{years}$$

($M_\odot = 2 \times 10^{33}$g, $L_\odot = 4 \times 10^{33}$erg sec^{-1} are mass and luminosity of the sun.)

Mass-luminosity relations for main sequence stars are

$$L/L_\odot = (M/M_\odot)^4 \qquad \text{for } L < L_\odot$$
$$L/L_\odot = (M/M_\odot)^{2.8} \qquad \text{for } L > L_\odot$$

Therefore

$$t_{m.s.} = 10^{10}\,(M/M_\odot)^{1.8} \qquad \text{for} L < L_\odot$$
$$t_{m.s.} = 10^{10}\,(M/M_\odot)^{-3} \qquad \text{for} L > L_\odot$$

$$\text{Sun: } t_{m.s.} = 10^{10}\text{years; } t_{m.s.} = 10^7\text{years for } M = 10M_\odot,$$
$$t_{m.s.} = 3 \times 10^{10}\text{years for } M = 0.5M_\odot.$$

The burnt out core undergoes an almost homologous contraction. A sphere of radius r, pressure $P \simeq P_c$, and density $\rho \simeq \rho_c$ with a general equation of state $\rho \sim P^\alpha T^{-\delta}$ follows evolution equations

$$\left.\begin{aligned} \frac{\dot\rho}{\rho} &= \alpha\frac{\dot P}{P} - \delta\frac{\dot T}{T} \\ \frac{\dot\rho}{\rho} &= -3\frac{\dot r}{r} \\ \frac{\dot P}{P} &= -4\frac{\dot r}{r} \end{aligned}\right\} \Rightarrow \frac{\dot T}{T} = \frac{4\alpha - 3}{3\delta}\,\frac{\dot\rho}{\rho}\,. \tag{1.3}$$

Two different cases:

a) Ideal gas:

$$\alpha = 1,\ \delta = 1 \quad \text{and}$$
$$\frac{\dot T}{T} = \frac{1}{3}\,\frac{\dot\rho}{\rho} = -\frac{\dot r}{r}. \tag{1.4}$$

Contraction leads to an increase of the central temperature and to the ignition of the next nuclear burning stage, unless the central region becomes degenerate. Schematically we have

start of a new nuclear burning phase	$\rightarrow$	exhaustion of fuel
$\uparrow$		$\downarrow$
heating up of central core	$\leftarrow$	contraction of center

The virial theorem

$$E_G \text{ (gravitational energy)} = -2E_T \text{ (thermal)}$$

Contraction:

$$\delta E_T = -\frac{1}{2}\delta\, E_G > 0 \text{ (heating)}$$

$$\nearrow$$

$$-\delta\, E_G$$

$$\searrow$$

$$-\frac{1}{2}\delta\, E_G \text{ (radiated away)}$$

This scheme works, in principle, until the core consists of ^{56}Fe, i.e. as long as nuclear reactions produce energy. Not all stars go all the way to a ^{56}Fe core. Eventually matter in the core becomes degenerate and the degeneracy is increased by the homologous contraction. Small mass stars have a core closer to degeneracy; and therefore small mass stars leave the nuclear burning cycles earlier than massive ones.

b) Degeneracy: complete nonrelativistic degeneracy for $a = \frac{3}{5}$; $1 \gg \delta > 0$. Since $\dot{T}/T < 0$ when $\alpha < \frac{3}{4}$, $\delta > 0$, the core cools when it contracts. There is a maximal central temperature (depending on the mass), and ignition of the following nuclear burning is prevented or postponed until a thermal runaway reaction in degenerate matter can occur ('He-flash; C-flash').

Examples. (Cf. Tab. 1.1.)

Table 1.1. The path of a 15 $M_\odot$ star is documented through various phases of nuclear burning. Such stars are rare (born only every 50 to 100 years in our galaxy) and they end as supernovae.

	General				15 $M_\odot$		
Fusion process	temperat. (K)	density (gcm^{-3})	mass ($M_\odot$)	main products	duration (years)	temperat. (K)	radius (10^6km)
Hydrogen fusion (H→He)							
p-p chains	10^7	10	<2-3	^4He			
CNO cycle	$(2\text{-}4)\times10^7$	10	>2-3	$^4He, ^{14}N, ^{17}O, ^{13}C$	1.2×10^7	3.4×10^7	3.2
Helium burning							
(He→C,O)	2×10^8	10^3	<2	^{12}C, ^{16}O	1.4×10^6	$1{,}6\times10^8$	22
s-process	3×10^8	10^3	≈ 4	A>56			
Carbonburning (C→Mg,Ne)	7×10^8	10^5	8	Ne, O, Mg, C, Si Na,Al	8.8×10^4	6.2×10^8	370
Neonburning $^{20}Ne(\nu,\alpha)^{16}O$ $^{20}Ne(\alpha,\gamma)^{24}Mg$	1.3×10^9	10^6	>8	Mg, Cl, S, O	7	1.3×10^9	390
Oxygenburning (O→Si,S,...)	2×10^9	10^7	> 8	Si, S, Ar, K, Ca	1.7	1.9×10^9	390
Siliciumburning (Si→Fe,Ni)	3×10^9	10^8	>10	Fe, Ni, Mn, Cr, Ti	4 days	3.1×10^9	390

No hydrogen burning, if $M \leq 0.08M_\odot$. Stars become degenerate when they contract towards the main sequence.

No helium burning, if $M_{He} \leq 0.35M_\odot$; stars are degenerate after hydrogen burning.

No Carbon burning, if $M_c \leq 0.9M_\odot$; stars are degenerate after He burning. (Stars with a degenerate He core can reach normal He fusion by the so-called He-flash.)

Stars with $M > 10M_\odot$ run through all cycles without degeneracy complications and develop a ^{56}Fe core (cf. Fig. 2).

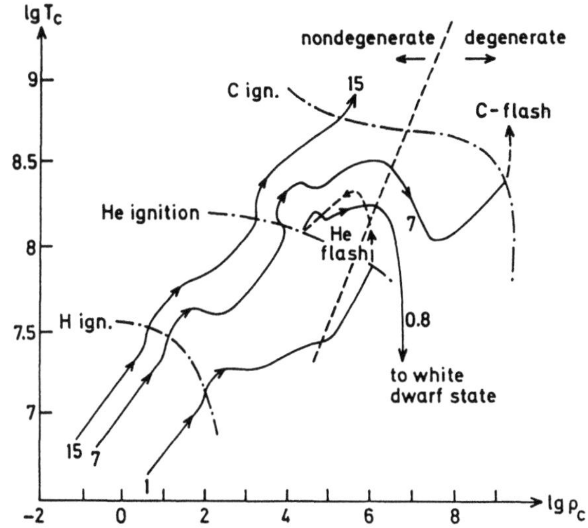

Fig. 2. A plot of the core density and temperature of stars with different inital masses. The numbers at the top and bottom of each curve give the mass of the star. Stars of $1M_\odot$ and $2M_\odot$ can ignite helium only via a helium flash. A star of $15M_\odot$ reaches carbon burning in a nondegenerate phase. The degeneracy parameter for the electrons is indicated by the dotted line.

In models where only spherical configurations in hydrostatic equilibrium are considered the star develops an onion structure with the products of different fusion processes arranged in spherical shells (Fig. 3).

The endpoint of thermonuclear evolution for stars with $M \geq 8$ to $10\ M_\odot$ is a Fe-Ni core with a mass slightly above the Chandrasekhar limit (see white dwarfs). Contraction of the Fe-Ni core is followed by efficient neutrino cooling, and by a rapid decrease in electron pressure by electron capture reactions. The core collapses eventually on a free fall time scale $\tau_{ff} \sim (G\rho)^{-1/2} \sim 1$ sec.

The timescales after the central carbon burning are very short, typically ≤ 100 years, shorter than the thermal timescale of the envelope. During the last stages stars do not change their position in the Hertzsprung-Russel-diagram, they appear as red giants until the final core collapse and the supernova explosion. But the progenitor of SN1987A was a blue star!?

This standard picture of stellar evolution will change somewhat, when mass loss, rotation, effects of a binary system, and different abundances of heavy elements are taken into account. Rotation can lead to a very deep mixing of the shells containing different nuclear burning products. Mass loss occurs quite frequently, and it will certainly change the outward appearance, and also the inner structure, if a substantial fraction of the initial main sequence mass is removed by a stellar wind, or by Roche lobe overflow in a binary system, on a time scale less than the nuclear burning time. Fig. 4 [3] gives an impression of the effect of mass loss on the final mass after hydrogen and helium burning phases.

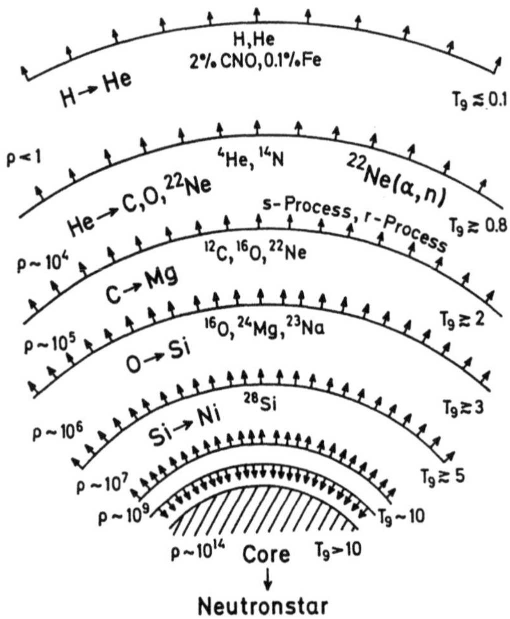

Fig. 3. The onion structure of an evolved star of $25M_\odot$.

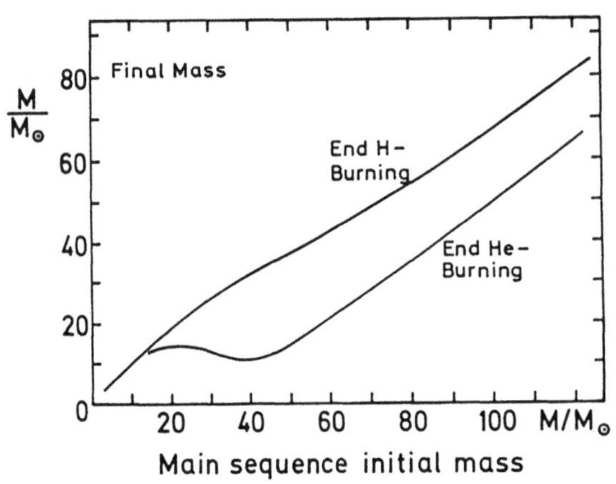

Fig. 4. Final versus inital mass for stellar evolution with mass loss. Stars of up to $40M_\odot$ have only 10 to $12M_\odot$ after He burning (after [3]).

Research in stellar evolution concentrates at present especially on the effects of mass loss. The old problem of convection (i.e. large scale internal motions) is, of course, still there. Rotation and other complications are generally not yet included in stellar evolution calculations.

There are two models to explain the blue progenitor of SN 1987A: Metal abundances in the Large Magellanic Cloud are only 25 % of solar ones, and evolutionary calculations show that stars with such an abundance initially can reach the helium core burning stage as blue stars [5]. The second possibility is a change from a red giant to a blue star due to rapid mass loss prior to the supernova explosion [6]. Observations of LMC indicate the existence of many blue stars in the transition region. This speaks in favour of the star never leaving the blue phase, because in the case of mass loss the transition must be short compared to the nuclear burning epochs from helium onward. The problem has not been solved.

The evolution is also modified in binary systems, where mass-loss to the companion star can be very large (see observations of neutron stars, Section (1.5)).

1.2 Explosive Nucleosynthesis

In supernova events the heavy elements above ^{56}Fe are formed by rapid processes of neutron capture. The general understanding is as follows:

Collapse of the iron-nickel core to densities $\sim 3 \times 10^{14}$g cm^{-3} and a radius of 100 km occurs within seconds, then the nuclei dissolve to a gas of protons and neutrons. The pressure is sufficient to stop the collapse (for $M_c \leq 3M_\odot$), a hot "proto-neutron star" appears, which cools by neutrino emission. Outer layers fall onto the star (with 0.2 c), a shock front develops, moves outwards, and expels the envelope (cf. STRAUMANN).

Inner zones with $T \geq 6 \times 10^9$ K and $\rho \geq 10^8$ g cm^{-3} reach nuclear statistical equilibrium, especially Fe-type elements are built ($\tau \sim 0.04$ sec).
Explosive Si burning: $T = 4 \cdot 10^9$ K, $\rho = 10^8$ g cm^{-3};
timescale 0.1 sec, products ^{32}S, ^{36}Ar, ^{40}Ca, ^{54} Fe, ^{52}Cr, ^{54}Mn.
Explosive O and C burning $T \sim (2-4)\times 10^9$ K; $(1-2) \times 10^5$ g cm^{-3}
timescale: 0.5 – 1 sec; products: ^{28}Si, S, Cl, Ar, Cn, Ti, Ne, Mg, Al.

Solar abundances of elements and isotopes in $20 \leq A \leq 65$ can be explained (^{46}Ca and ^{48}Ca don't come out right – ratio ^{46}Ca/^{48}Ca too high).

Rapid neutron capture: The shock wave of the SN explosion runs through the He burning shell – $T \sim 10^9$ K, high neutron densities $\geq 10^{19}$ cm^{-3} from ^{22}Ne $(\alpha, n)^{25}$Mg for 0.5 to 1 sec. Build-up of heavy elements (r-process synthesis of elements).

1.3 Final States of Stellar Evolution

(A) White Dwarfs. Equilibrium configurations where gravitation is balanced by the pressure of a degenerate electron gas ($P \sim \rho^{5/3}$ for nonrelativistic, $P \sim \rho^{4/3}$ for relativistic degenerate electron gas). There is a maximum stable mass, the Chandrasekhar mass

$$M_{CH} = \frac{5.76}{\mu_e^2} M_\odot \,.$$

For He, C, O, Ne, Mg: $\mu_e = 2 \quad M_{CH} = 1.44 M_\odot$.

For Fe: $\mu_e \neq 2$, because of neutronization; negative contribution by Coulomb forces of the ions. $(M_{CH})_{Fe} = 1.24 M_\odot$.

Why only white dwarfs with $M < M_{CH}$?
Hydrostatic equilibrium:

$$\frac{dP}{dr} = -\frac{G\,M_r}{r^2}\rho\;;$$

$$1 = \text{const.}\frac{M^2}{R^4\bar{P}}\;.$$

$P = \text{const.}\ \rho^{5/3}$, then $\bar{P} = \text{const.}\,M^{5/3}/R^5$ and $1 = \text{const.}\,M^{1/3}R$.

For each mass there is a radius such that hydrostatic equilibrium is achieved.

$P = \text{const.}\ \rho^{4/3}$ (complete relativistic degeneracy)

$1 = \text{const.}\ M^{2/3}$, independent of R.

There is only one mass (M_{CH}) for which hydrostatic equilibrium is possible (1931: controversy Eddington-Chandrasekhar about this fact. Chandrasekhar did not get public support from physicists with respect to the equation of state of degenerate relativistic electrons!).

The formation of white dwarfs is not well understood. After central H-burning stars are in the upper right of the HR diagram. White dwarfs are in the lower left. Since no stars are in between, there must be a rapid evolution leading to white dwarfs.

Typical parameters

$$\begin{aligned}\langle M\rangle &= 0.58M_\odot,\\ \langle R\rangle &= 0.012R_\odot,\\ \langle\bar{\rho}\rangle &= 4.7\times10^5\text{g cm}^{-3}.\end{aligned}$$

The progenitors must have been more massive, rapidly evolving stars which have lost part of their mass (stellar wind, planetary nebula ejection). Determination of the maximum initial mass from observations of open star clusters: The initial mass M_i of white dwarf progenitors must be greater than the "turn-off mass" (i.e. the mass of stars which are just leaving the main sequence).

Observationally

$$\text{Max}\,(M_i) = (8\pm2)M_\odot\;.$$

(B) Neutron Stars. Stars of $M \geq 10M_\odot$ develop a (non-degenerate) Fe-Ni core with $M_c \geq M_{CH}$. When $\rho_c \geq 10^{10}\text{g cm}^{-3}$ β-capture reactions

$$p + e^- \rightarrow n + \nu.$$

Collapse, because of removal of electrons. Contraction stops, when repulsive nuclear forces and degeneracy pressure of neutrons becomes important. Neutron star – remnant of type II Supernova explosion; typically: $M = 1M_\odot$, $R = 10\,\text{km}$, $\bar{\rho} \sim 10^{15}\,\text{g}\,\text{cm}^{-3}$.

Theoretically (Fig. 5) neutron stars exist with central densities between 3×10^{14} and $5\times10^{15}\,\text{g}\,\text{cm}^{-3}$; for any equation of state there is a maximum stable mass:

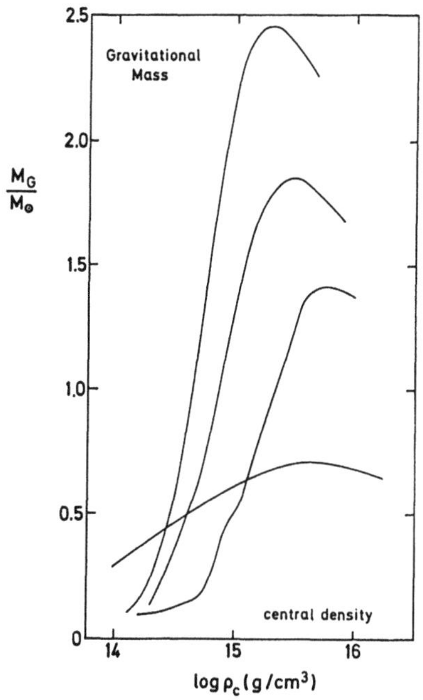

Fig. 5. Gravitational mass vs. central density of neutron stars for various equations of state. Stable configurations are on the graphs rising from left to right. The maximum mass is determined by the repulsive part of the 2-particle interaction used. The lowest graph considers only statistical effects of many different particle types with a Fermi distribution, but no repulsive interaction. Such an equation of state can be excluded, because observations show that neutron stars with masses $\geq 1 M_\odot$ exist (after Ref. [7]).

$M \sim 1.5$ to $2.5 M_\odot$ (strong dependence on the equation of state). Stars of $8 M_\odot$ to $25 M_\odot$ can be the progenitors of neutron stars.

(C) Black Holes. For core masses greater than the maximum stable mass of a neutron star the collapse cannot be stopped by repulsive nuclear forces. Core collapse proceeds to a black hole – a singularity of spacetime (HAJICEK). Practically: The core disappears behind its gravitational radius

$$R_s = \frac{2GM}{c^2} = 3\text{km}\,(M/M_\odot)$$

(Schwarzschild Radius).

Observations of black holes can only be made indirectly: accreted material (from a companion in a binary system) radiates (mainly in X-rays), and in a binary system masses can be determined. Whenever the mass of the compact component is very large there is a black hole candidate. It seems very likely that many black holes exist, since already for a neutron star of $2 M_\odot$ the radius of ~ 10km is between $1 R_s$ and $2 R_s$.

(D) Strange Stars. Bulk quark matter phase may have a lower energy per baryon than ordinary nuclei [8]. It may be absolutely stable, has a density comparable to that of atomic nuclei; can exist in lumps ranging in size from a few fermis up to "strange stars" of radius $\sim 10\,\text{km}$ [9].

$$\begin{aligned} M \sim R^3 \quad \text{for} \quad M \leq 1 M_\odot \\ M \text{ between 1 to } 2M_\odot : R \approx 10 \text{ km.} \end{aligned} \tag{1.7}$$

Conversion of neutron stars to strange stars?

One must create 10^{57} strange quarks, each reaction involving a weak interaction (many simultaneous weak int. processes).

1.4 Bounds on Axions from Stellar Evolution

The hypothetical axion [10] may couple to photons, $(a \to 2\gamma)$

$$L_{a\gamma\gamma} = -\frac{G}{4} F_{\mu\nu} \tilde{F}_{\mu\nu} a \tag{1.8}$$

$F_{\mu\nu}$: electromagnetic field strength tensor; G $[\text{energy}]^{-1}$: coupling constant.

Axion emission is an additional cooling mechanism and thus influences the evolution of stars. For instance, the axion luminosity of the sun must be less than the surface photon luminosity, because otherwise the sun would have turned off the main-sequence already. Therefore [11]

$$G \leq 2.5 \times 10^{-9} (\text{GeV})^{-1}. \tag{1.9}$$

The axion mass

$$m_a = 6.9\text{eV} \left(G / 10^{-9} \text{GeV}^{-1} \right), \tag{1.10}$$

i.e. the limit corresponds to $m_a \leq 17$ eV.

Axion cooling also reduces the helium burning phase of stars. For a $1.3\, M_\odot$ star the helium burning lasts $t_{He} = 1.2 \times 10^8$ years; with $G_9 = 0.3$ it is reduced by an order of magnitude

$$t_{He} = 1.6 \times 10^7 \text{ years.}$$

1.5 Observations of Neutron Stars

The existence of neutron stars has been established by various observations since the discovery of the first radio pulsars in 1968. For single pulsars it is the rapid rotation which confirms their nature as dense and compact stars. Many other interesting objects have been discovered especially in the X-ray and gamma ray spectral range: X-ray bursters, gamma ray burst sources, X-ray flare stars, X-ray pulsars, and so on (for reviews see e.g. [12]). In the following I want to point out a few general properties of the X-ray binaries, as one example of this variety of phenomena.

(A) The Binary X-Ray Sources. Since X-rays are absorbed by the earth's atmosphere the window to the X-ray sky has really been opened up for us when the first satellites began collecting data in this frequency range. Earlier rocket and balloon flights had already given glimpses of the existence of X-ray point sources and an unresolved X-ray background, but the systematic exploration by the X-ray satellites has revealed (and continues to do so) many new, unexpected – unpredicted by theorists too! – and exciting features. One big surprise was the discovery of many point sources which could not be identified with known stars or supernova remnants already by the first X-ray satellite, named Uhuru, which was launched by NASA in 1970, and which could detect X-rays in the range of $2-20$ keV. The lack of identifications was partly due to a low precision of the X-ray position measurement, but also – as it turned out later – due to the fact that inconspicuous optical stars apparently were associated with X-ray sources whose X-ray luminosity exceeded the sun's total luminosity by factors of 10^3-10^5. Clearly these X-rays could not come from the hot chromosphere of a star. Subsequent optical identifications and the analysis of the optical and/or X-ray lightcurves have established for many X-ray stars that they are members of a binary system. The data lend themselves to a fairly obvious, but not necessarily compelling general interpretation:

The rapid variability of the X-rays implies a small emission area, and this suggests that a compact object (white dwarf, neutron star, black hole) is involved. The compact X-ray source orbits the optical, normal star and collects the gas flowing over. The accretion of this matter in the deep gravitational well of the compact object provides the energy source for the X-rays. This accretion picture for the X-ray sources is by now widely accepted.

Most of the infall energy is liberated deep in the gravitational potential well, i.e. on the stellar surface if the compact object is a neutron star, or within a few Schwarzschild radii if the object is a black hole of a few solar masses (in both cases distances are $\simeq 10^6$ cm). If 10^{36} erg/sec are thermally radiated away from areas with such typical dimensions, one finds temperatures $T \simeq 10^7 - 3\times 10^7$ K, i.e. X-rays in the keV range.

The efficiency of producing X-rays from black holes and neutron stars can be very high: about 10% of the rest mass energy of the infalling gas are converted into radiation. Therefore the accretion rates do not have to be very large: $10^{-8} M_\odot/\mathrm{yr}$ ($\simeq 10^{18}$ g/sec) to $10^{-10} M_\odot/\mathrm{yr}$ ($\simeq 10^{16}$ g/sec) are sufficient to give the powers observed.

White dwarfs have radii around 10^9 cm, so black-body temperatures and accretion efficiencies are lower. It is much harder to make keV X-rays, and their luminosities are limited to a few times 10^{36} erg/sec. There are many objects which clearly are not white dwarfs: The source Her X-1 (first X-ray source discovered in the constellation Hercules), as well as the source SMC X-1 in the Small Magellanic Cloud, show very regular short pulsations (1.24 sec period for Her X-1, 0.71 sec for SMC X-1) which indicate that the rotation of a neutron star provides the time keeping mechanism. We see pulses, because the strong magnetic field of the neutron star makes the geometry of the accretion process sufficiently asymmetric to produce one or more hot spots on the surface which move through our field of vision. This picture is strongly supported by the measurement of a spectral line of Her X-1 which can be interpreted as cyclotron emission, implying a magnetic field strength of $B = 5 \times 10^{12}$ gauss near the surface. Such strong fields can exist only on the surface of neutron stars, but not on white dwarfs.

Other possible neutron star candidates are the regularly pulsating sources Cen X-3 (4.84 sec) and Vela X-1 (282 sec). The period of Vela X-1 is, of course, also accessible to a white dwarf, but here, as in the case of Cen X-3 one has to favour a neutron star picture because of the mass determinations and the time variation of the pulse periods; the period of Cen X-3 decreases significantly with time ($\Delta P/P \simeq -3 \times 10^{-4}$ (year)$^{-1}$). If one attributes this speed-up to the angular momentum transferred by the accreting material, then the mass-flow rates estimated from the luminosity ($\simeq 10^{17}$ g/sec) lead to transfer rates of angular momentum that can explain the observed speed-up for a neutron star, but not for a white dwarf. Furthermore, the lever arm of the momentum transfer has to be large, about 100 times the stellar radius. Therefore it must be the neutron star's magnetic field which transmits angular momentum from the accreting matter to the star. For Vela X-1 similar arguments can be given.

The mass-flow rates required to power a neutron star or a black hole as an X-ray source can easily occur in close binary systems. Three different mechanisms may be important:

(i) A massive companion overflowing its Roche Lobe (i.e. the critical equipotential surface between the 2 stars) would produce rates of $10^{-6} M_\odot$ (year)$^{-1}$, and could give an accretion rate of $10^{-9} M_\odot$ (year)$^{-1}$. That large a mass-flow may suffocate the X-ray source, because column densities above 1 g cm^{-2} are opaque for keV X-rays. Therefore this mechanism may not be operative in the observed X-ray sources.

(ii) A stellar wind that is the outflow of a hot corona, as expected from O or B stars, can have a magnitude of $10^{-6} M_\odot$(year)$^{-1}$, and could give an accretion rate of $10^{-9} M_\odot$(year)$^{-1}$ onto the neutron star. The outflow may show deviations from complete spherical symmetry due to the presence of the other star, and resemble somewhat case (i). Then even a lower outflow rate could give the same accretion rate.

(iii) The mass transfer may also be induced by X-rays, through heating of the outer layers of the optical star. The outflow produced has to be large enough to sustain the X-rays. Such a self-sustaining mechanism may be applicable to Her X-1.

In a close binary system the gas leaving the optical star possesses – just from the orbital motion – a certain amount of angular momentum with respect to the compact object. This will in many cases lead to the formation of an extended flat disk around the compact oject. The disk must in principle extend out to a fair fraction of the critical Roche radius until angular momentum can be efficiently removed from its outer edge by tidal interaction. The physics of the accretion disk will then determine many features of the X-ray source.

For a high velocity, spherical stellar wind very little angular momentum may be present, and close to the neutron star the infall will be almost spherical, too.

In Fig. 6 two canonical pictures are schematically represented of a regularly pulsating X-ray source, as they emerge from this general discussion of the

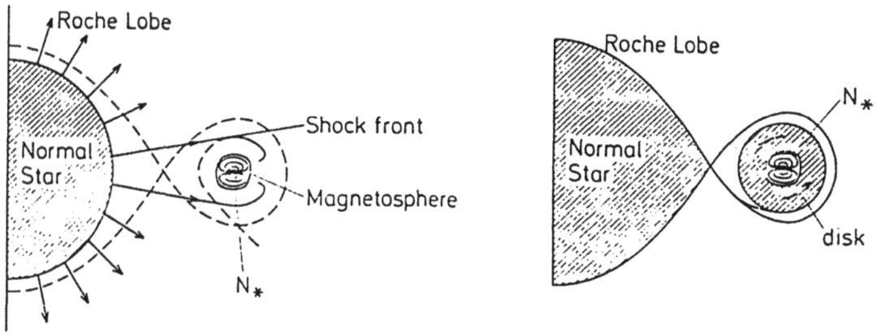

Fig. 6. Schematic diagram of wind and disk accretion.

observations: A disk of hot gas in Keplerian motion or a stream of plasma in almost radial motion surrounds the X-ray source, and at a certain radius r_M begins to interact strongly with the rotating magnetic field of the neutron star. Inside this radius the X-rays are produced and modified by the interaction with the inflowing plasma.

It was a big surprise to find neutron stars in binary systems, because this means that the binary system must have survived a supernova explosion. The evolution of systems such as HZ Her-Her X-1 (which contains only 3 $M_\odot$) is ill-understood. For the massive systems with a massive blue supergiant as the companion an evolution for an initial configuration of $20M_\odot$ and $8M_\odot$ may have occurred (Fig. 7) as follows:

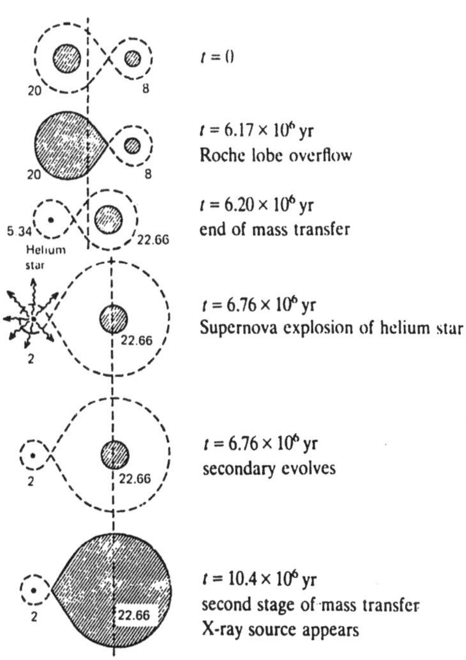

Fig. 7. A possible evolutionary path of a binary system containing a neutron star.

After the $20M_\odot$ star, the primary, has burnt its hydrogen core, a stage of core contraction and hydrogen shell burning sets in, during which the hydrogen envelope of the primary is almost completely transferred to the secondary. The primary is then a helium star with helium core burning, and a mass of $\simeq 5M_\odot$. Its lifetime after the onset of oxygen burning is much less than the thermal time scale of the envelope. Thus the core collapses before the star reaches its critical Roche lobe, and the primary explodes as a supernova. At that time the secondary is still unevolved with a much larger mass than the primary, i.e. the system will probably not be disrupted.

The remnant is a black hole or a neutron star which may for a short time appear as a pulsar. The secondary then evolves away from the main sequence to a blue supergiant stage, produces a strong stellar wind, and accretion onto the compact primary brings the X-ray source into being.

This evolutionary path provides a nice answer to the question as to how a binary system can survive a supernova explosion, but it has difficulties too: The binary period of $\simeq 2$ days for a typical binary system cannot be obtained in this model. The orbital period after the supernova explosion is always too large. Probably one has to appeal to additional mass loss to the system to achieve lower orbital periods.

As the secondary evolves further it eventually overflows its own Roche lobe and covers the X-ray source. The radiation will then emerge only after many scatterings, at a lower frequency, and probably sometimes in bursts of harder radiation when Rayleigh-Taylor-like instabilities permit the photons to escape more directly.

The final outcome is unclear. It may be that the secondary eventually also undergoes a supernova explosion, in which a second neutron star is formed. The mass loss normally will disrupt the system and set free 2 neutron stars. But it may be that the binary system survives even the second supernova explosion, if the compact primary has radiatively driven away the envelope of the secondary. Then a system such as the binary pulsar PSR 1913 + 16 might come into being.

(B) Masses of binary X-ray sources. The analysis of the motion in a binary system allows to determine the masses of the components. The problem consists in finding simultaneously the 2 star masses and the inclination angle i of the line of sight to the normal of the orbital plane. In principle there are enough observationally determined quantitative relations to solve for those three quantities: The measurement of the X-ray star's radial velocity – possible for the regularly pulsating sources – gives an X-ray mass function

$$F(M_{opt}, M_x, \sin i) = M_{opt}^2 \sin^3 i / (M_{opt} + M_x)^2. \tag{1.11}$$

From spectral Doppler shifts a similar optical mass function – with M_{opt} and M_x interchanged – can be obtained. A third relation comes from the duration of the X-ray eclipse, or from light variations of the optical star due to its deformation by the companion. There are, however, uncertainties involved in

the determination of these relations. The optical mass function, e.g., has to be found from the motion of the center of mass of the optical star (Newtonian two-body problem), and the observations of stellar absorption lines contain, of course, all kinds of additional effects such as X-ray heating of the photosphere, tidal distortions of the surface, streaming of gas in the system. These intrinsic effects as well as uncertainties in the measurements produce uncertainties in the masses obtained.
In Table 2 we have listed the current estimates for the masses of several binary X-ray sources. The limits given for Cen X-3, SMC X-1, Her X-1 can be narrowed down if strict corotation of the optical star is assumed. Then the masses for the various sources are

$$M(\mathrm{CenX}-3) \leq 3.1 M_\odot; \quad M(\mathrm{SMCX}-1) \leq 1.8 M_\odot; \quad M(\mathrm{HerX}-1) \leq 1.7 M_\odot. \tag{1.12}$$

From observations of optical 1.24 sec pulsations in Her X-1 which are interpreted as X-ray reprocessed pulses in the atmosphere of HZ Her or in surrounding matter, a mass of 1.33 $M_\odot$ for Her X-1 has been deduced. This value, however, depends sensitively on the physics of the reprocession of the X-rays, and on the assumed strict corotation. A less stringent limit of 1.1 $M_\odot \leq M$ (Her X-1) $\leq 1.6 M_\odot$ may be less model dependent.

Source	$M_x/M_\odot$	$M_{opt}/M_\odot$
Her X-1	0.4–2.2	1.4–2.8
Cen X-3	0.7–4.4	14–20
SMC X-1	0.5–1.8	13–22
VELA X-1	1.0–3.4	19–32
1700–37	>0.6	>10
CYG X-1	8.5–14	23–45
1538–522	1.5–2.5	16–24

The best known mass of a neutron star is that of Vela X-1 for which both an optical and an X-ray mass function have been measured. $M_x = 1.85 M_\odot$, if strict corotation is assumed. This mass is rather high. But in any case the neutron star masses obtained from the lightcurve analysis fit nicely with the theoretically computed masses using Einstein's theory of gravitation, and a specific equation of state derived from 2-particle potentials. They also bracket the maximum mass for a stable white dwarf ($1.2 M_\odot$) – the Chandrasekhar mass – which is at present a fashionable value for a collapsed remnant among supernova theorists. (The preference for remnant masses of this magnitude does not seem to have too much substance, since the collapse calculations invariably assume spherical symmetry and neglect rotation and magnetic fields.)

These mass determinations through the analysis of lightcurves have received special attention in the case of the irregularly varying source Cyg X-1. The mass of the compact source was found to lie between 9 and 15 $M_\odot$. The estimate of the optical companion's mass strongly affects these values. In the lectures by

STRAUMANN this case and two other possible candidates will be discussed in detail; especially the dependence on theoretical assumptions will be quantified. The large amount of theory involved in the mass determination leaves room for doubts, and a direct confirmation of the black hole character by some peculiar manifestation of black hole accretion would be highly desirable.

(C) Pulsars. The Binary Pulsar PSR 1913+16 has allowed an extremely precise mass determination from relativistic effects

$$M = 1.41 \pm 0.01 M_{\odot}. \tag{1.13}$$

The Millisecond Pulsar 1937 + 21 has kept a precise pulsation period for the 4 years of observation so far [14]. This allows to place a limit on any stochastic gravitational wave background. The density parameter for such a gravitational wave noise is limited:

$$\Omega_g < 4 \times 10^{-7} \quad (\text{at}\,\nu \sim 0.23\,(\text{year})^{-1}).$$

1.6 Very High Energy Gamma Ray Sources

This is another exotic phenomenon involving neutron stars. Detection of very high energy (VHE, 10^{11} to 10^{14}eV) and ultrahigh energy photons (UHE, above 10^{14}eV) from some X-ray binary systems, especially from binary X-ray pulsars Her X-1, Vela X-1, 4UO115+63, and also from Cyg X-3 [15], seems to have received some confirmation recently (see McDONALD for a discussion of the experimental set-up)

VHE photons are detected from the Cerenkov light emitted by the relativistic electrons and positrons as they move through the atmosphere. Ground dimension of the Cerenkov light: a few 100 m diameter.

UHE photons are detected by the secondary particles that penetrate to the ground and are registered by particle detector arrays. Angular resolution $\sim 1^0$. The signal is weak, only $\sim 1\%$ of the cosmic ray background. Signals are identified as narrow angular enhancements (Cyg X-3) or by a characteristic periodic behaviour (X-ray pulsars, Cyg X-3). In Table 1.3 the characteristics of 4 sources are listed.

Her X-1: Identification by period of spin, but deviations. Episodes of emission 30 to 100 min., typically 1/10 of observing time. 10 to 500 TeV.

$F_\gamma(> \text{TeV}) \sim 3 \times 10^{10}\ \text{cm}^{-2}\ \text{sec}^{-1}$.

4UO115+63: Discovered during outburst in 1984.

$F_\gamma(> 1\text{TeV}) = 7 \times 10^{-11}\ \text{cm}^{-2}\ \text{sec}^{-1}$.

Vela X-1: Several detections (usually weak) one outburst ($\sim$ 90 seconds duration) reached the background level.

Cyg X-3: Identification of signals by angle and 4.8 hr orbital period modulation.

$F_\gamma(> 2\text{TeV}) = 1.4 \times 10^{-11}\ \text{cm}^{-2}\ \text{sec}^{-1}$

$F_\gamma(> 2\text{PeV}) = 7.4 \times 10^{-14}\ \text{cm}^{-2}\ \text{sec}^{-1}$.

Table 1.3.

Source	Period		Cyclotron line	Luminosity (erg sec^{-1})	
	Spin	Orbit	(keV)	X-ray	VHE Gamma
Her X-1	1.24s	1.7d	30 to 50	10^{35} to 10^{37}	10^{35} to 10^{37}
4UO115+63	3.61s	24d	11.5,23	10^{33} to 10^{37}	10^{35} to 10^{36}
Vela X-1	283s	8.97d		10^{34} to 10^{36}	10^{34}
Cyg X-3	12.59ms?	4.8hrs		10^{37} to 10^{38}	10^{35} to 10^{37}

Models. A beam of accelerated particles (electrons or protons) strikes a target somewhere in the binary system (plasma cloud, surface of companion) and gamma rays are produced by bremsstrahlung, synchrotron radiation or π^0 production and subsequent decay.

No consensus about a correct model exits at present. The requirement on the cosmic accelerator are high, about to the point of improbability.

Cosmic Ray Production. The source Cyg X-3 has a luminosity in gamma rays (10^{15} to 10^{16} eV) of $\sim L = 4 \times 10^{36}$ erg sec^{-1} (d = 10 Kpc, isotropic emission). Assuming an efficiency of the conversion of particle energies to gamma rays of 10%, the cosmic ray luminosity is $L_{CR} = 4 \times 10^{38}$ erg sec^{-1}. The energies of the particles: 10^{16} to 10^{17} eV. The total power of cosmic rays in the galaxy in this energy band is $\sim 3 \times 10^{38}$ erg sec^{-1}. Only one source like Cyg X-3 would be sufficient to maintain the observed cosmic ray power! Cosmic rays observed at earth are, however, not due to binary sources: A 10^{15}eV proton has a Larmor radius of 1 light year in the magnetic field (10^{-6} Gauss) of the galaxy.

2. The Standard Big Bang Model of the Universe

2.1 Basic features

The standard model of cosmology rests on a few fundamental observations: the redshifted spectral lines of the distant galaxies [16] indicate that there is an overall expansion. The rate of this expansion (the "Hubble" constant) gives the scale of our universe. Age determinations of the elements and of the oldest stars are in agreement with this scale. Finally a cosmic radiation field is present [17] with a black-body spectrum of a temperature of 3K. This background radiation is highly isotropic, i.e. within observational accuracy it arrives with equal intensity from all directions.

These basic features can be approximated quite well in a simple picture, where the matter (and radiation) content is described by an ideal fluid characterized by a spatially homogeneous and isotropic density ρ and pressure p. It was a gratifying result of early cosmological research that with such a matter configuration very simple solutions of Einstein's field equations of General Relativity could be found by Friedmann and Lemaître [18] which exhibit an isotropic and homogeneous overall expansion.

In these models a time coordinate can be introduced such that the surfaces of constant time are 3-spaces of constant curvature. The line element can be written

$$ds^2 = dt^2 - R^2(t)d\sigma^2 \qquad (c = 1). \tag{2.1}$$

The 3-space line element $d\sigma^2$ can be written

$$d\sigma^2 = d\chi^2 + f^2(\chi)(d\theta^2 + \sin^2\theta d\phi^2) \tag{2.2}$$

where $f(\chi)$ depends on the sign of the curvature. By a suitable rescaling of the factor $R(t)$ the curvature can be normalized to $K = +1, \ -1, \ 0$. The function

$$f(\chi) = \begin{cases} \sin\chi & \text{for} \quad K = +1 \\ \chi & \text{for} \quad K = 0 \\ \sinh\chi & \text{for} \quad K = -1 \end{cases} \tag{2.3}$$

determines how the area of a sphere $\chi =$ const. changes with radius χ. The coordinate χ runs from 0 to ∞ if $K = 0$ or $K = -1$. If $K = +1$ it varies from 0 to π. When $K = -1$ the 3-spaces are said to have a hyperbolic or Lobatschewskian geometry, and when $K = 0$ a planar or Euclidean geometry. The curvature determines the ultimate fate of the universe model: For $\Lambda = 0$, models with $K = 0$, $K = -1$ will expand forever, while models with $K = +1$ will eventually recollapse.

It should be pointed out here that the curvature does not determine whether the universe is infinite or finite in extent. The statement that if the universe is finite its geometry must be locally spherical and that if the geometry is locally hyperbolic the universe must be infinite is wrong. There are many counterexamples to this belief – the global topology of locally homogeneous spaces can be quite involved.

The metric (2.2) requires that the energy-momentum tensor is of the perfect fluid type

$$T_{\mu\nu} = \rho V_\mu V_\nu - p(g_{\mu\nu} - V_\mu V_\nu).$$

The density and pressure are functions of t only, and the flow lines with tangent V_μ are the curves (χ, θ, ϕ) – constant. The coordinates chosen in (2.2) are "comoving". The cosmic time t corresponds to the proper time of observers following the mean motion V_μ.

The expansion factor $R(t)$ changes with time like the spatial distance of any two particles of the perfect fluid, i.e. of 2 "galaxies" without peculiar motions.

Such comoving galaxies are considered as the representative particles following the perfect fluid motion. It must be tested observationally, of course, whether a gas of galaxy particles can be approximated well by a perfect fluid. For two points with comoving coordinates (χ, θ, ϕ) and (χ_0, θ, ϕ) the spatial distance at a time t is:

$$d = R(t)(\chi - \chi_0)$$

the corresponding kinematical relative speed is thus $V_{kin} = (\dot{R}/R)d$. Light propagates along the null-geodesics of space-times. This can most easily be studied by changing to a new time coordinate η:

$$dt = R(t)d\eta$$

Then

$$ds^2 = R^2(\eta)\left(d\eta^2 - d\chi^2 - f^2(\chi)d\Omega^2\right) .$$

("conformal coordinates": $d\Omega^2 \equiv d\theta^2 + \sin^2\theta d\phi^2$)

"Radial" propagation (no loss of generality) of signals, i.e. $\theta = \varphi =$const. gives $d\eta = \pm d\chi$ along the light cone: for rays towards an observer at $\chi = 0$

$$\chi = \eta_0 - \eta$$

where η_0 denotes the present epoch. Two signals emitted at times η_s and $\eta_s + d\eta_s$ by a source following the mean motion at χ_s will be received at times $\eta_0 + d\eta_0$ at $\chi = 0$, and $d\eta_s = d\eta_0$ (Fig. 8).

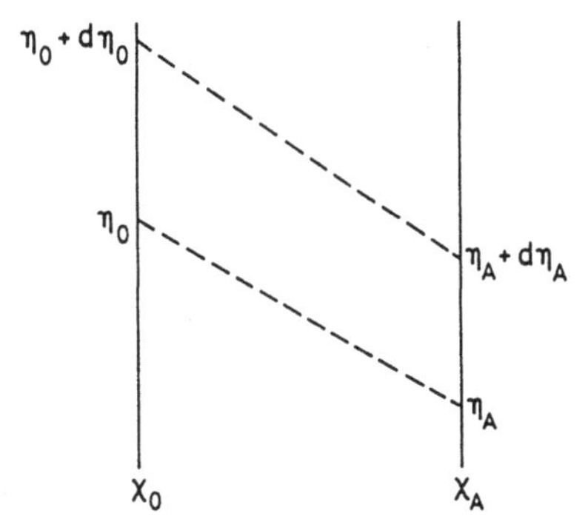

Fig. 8. Comoving observers at two different positions measure the same intervals of conformal time $d\eta$. This corresponds to a cosmic effect on proper time intervals dt – the cosmological redshift.

Expressed in the proper time t this condition reads

$$\frac{dt_0}{R(t_0)} = \frac{dt_s}{R(t_s)};$$

or expressed in terms of a frequency $\omega \propto (dt)^{-1}$

$$\omega_0 R(t_0) = \omega_s R(t_s).$$

The emitted frequency ω_s differs from the observed one ω_0:

$$\frac{\omega_s}{\omega_0} = \frac{R_0}{R(t_s)}.$$

This corresponds to a shift of spectral lines

$$z = \frac{\lambda_0 - \lambda_s}{\lambda_s} = \frac{\omega_s}{\omega_0} - 1 = \frac{R_0}{R(t_s)} - 1.$$

We know from Slipher's and Hubble's observations that this shift is a redshift:

$$1 + z = \frac{R_0}{R(t_s)} > 1. \tag{2.4}$$

This redshift of the spectral lines arises from the relative motion of source and observer, and may therefore be called a Doppler shift. In general space-times the frequency shift cannot be separated into a part due to gravitational effects and a part due to motion.

For nearby galaxies we can see that (2.4) corresponds to the elementary nonrelativistic Doppler formula:

Expand $R(t) = R_0 + (t - t_0)\dot{R}_0$ and

$$z = \frac{R_0}{R_0 - \Delta t \dot{R}_0} - 1 = \Delta t \dot{R}_0 = d\frac{\dot{R}}{R_0} \equiv v. \tag{2.5}$$

Hence $v = z = d\, H_0$ where

$$H_0 = \frac{\dot{R}_0}{R_0}, \text{ the "Hubble constant".}$$

Eq. (2.5) is "Hubble's Law", the linear relation between redshift and distance. It leads to the conclusion that the universe expands. Einstein's field equations can now be written as:

$$\ddot{R} = -\frac{4\pi}{3}(\rho + 3p)GR + \frac{1}{3}\Lambda R; \tag{2.6}$$

$$\dot{R}^2 = \frac{8\pi G}{3}\rho R^2 + \frac{1}{3}\Lambda R^2 - K; \tag{2.7}$$

$$\frac{d}{dt}(\rho R^3) + p\frac{d}{dt}(R^3) = 0. \tag{2.8}$$

Whenever $\dot{R} \neq 0$ any two of eqs. (2.6), (2.7), (2.8) imply the third. These are the Friedman equations [18], and they determine the dynamics of the cosmological model if an equation of state, e.g. $p = f(\rho, t)$ or $p = f(\rho)$ is given.

At the present epoch $t = t_0$ the pressure is very small $p/\rho c^2 \leq 10^{-6}$. Let us take $p = 0$, then from (2.8)

$$\rho R^3 = const. \equiv \frac{3M}{4\pi}$$

where M is a constant mass. (2.7) can then be written as

$$\dot{R}^2 - \frac{2MG}{R} - \frac{1}{3}\Lambda R^2 = -K \equiv E/M$$

with $(E \equiv -KM)$.

This is an energy conservation equation for a comoving volume of matter; the constant E is the sum of the kinetic and potential energy, the term ΛR^2 is a kind of oscillator energy (for $\Lambda > 0$).

At the present epoch $t = t_0 (p_0 = 0)$ the equations (2.6) and (2.7)

$$\ddot{R}_0 = -\frac{4\pi}{3}\rho_0 G R_0 + \frac{1}{3}\Lambda\, R_0,$$
$$\dot{R}_0^2 = \frac{8\pi G}{3}\rho_0 R_0^2 + \frac{1}{3}\Lambda\, R_0^2 - K,$$

can be rewritten in terms of the Hubble constant $H_0 = \dot{R}_0/R_0$ and the "deceleration parameter" $q_0 \equiv -(\ddot{R}_0 R_0)/\dot{R}_0^2$ as

$$\frac{1}{3}\Lambda = \frac{4\pi}{3}G\rho_0 - q_0 H_0^2,$$
$$\frac{K}{R_0^2} = H_0^2(2q_0 - 1) + \Lambda.$$

Substituting Λ:

$$\frac{K}{R_0^2} = \frac{H_0^2}{2}(3\Omega_0 - 2q_0 + 2)$$

where

$$\Omega_0 \equiv \frac{\rho_0}{\rho_c} \equiv \frac{8\pi G}{3}H_0^{-2}\rho_0. \tag{2.9}$$

$\rho_c = (3H_0^2)/(8\pi G)$ plays the role of a critical density. The sign of K is determined by the term $3\Omega_0 - 3q_0 - 2$. For $\Lambda = 0$, the relation is especially simple: $\Omega_0 = 2q_0$, and then

$$\begin{aligned} K &= +1 \quad \text{for} \quad \rho_0 > \rho_c \\ K &= \;\;0 \quad \text{for} \quad \rho_0 = \rho_c \\ K &= -1 \quad \text{for} \quad \rho_0 < \rho_c \end{aligned} \tag{2.10}$$

Since the present mean density ρ_0 is in principle a measureable quantity, it is possible to determine whether our universe is a hyperbolic or spherical FL spacetime. Indeed for $\Lambda = 0$, H_0, and q_0 determine the spacetime and the present age completely: One finds

$$\left(\frac{\dot{R}}{R_0}\right)^2 = H_0^2\left\{1 - 2q_0 + 2q_0\frac{R_0}{R}\right\};$$

substituting $x = R/R_0$

$$t - t_M = H_0^{-1}\int_0^x \left(1 - 2q_0 + \frac{2q_0}{y}\right)^{-\frac{1}{2}} dy \tag{2.11}$$

where t_M is a time in the past for which the approximation $p = 0$ still holds. $t_M \ll t_0$ and $R_M \ll R_0$ lead to the approximate relation (for times close to the present epoch t_0):

$$t \approx H_0^{-1} \int_0^x \left(1 - 2q_0 + \frac{2q_0}{y}\right)^{-\frac{1}{2}} dy.$$

The approximation $p \ll \rho$ is reasonable now and for most of the past. But the universe contains the 3 K background radiation field, and because of its equation of state, $p = 1/3\rho$, the continuity equation leads to an increase of the energy density as $\rho_\gamma R^4 = \text{const.}$. Thus at a sufficiently early epoch radiation will be the dominant form of energy. This leads inevitably to a hot origin of the universe. The crucial assumption is the interpretation of the 3 K microwave signals as a cosmic background radiation.

Without solving the Friedmann equations it is seen from (2.6) that the present expansion $\dot{R}_0 > 0$, together with the requirements $\rho + 3p - \frac{\Lambda}{4\pi G} \geq 0$, i.e. $\ddot{R}_0 \leq 0$, leads to a concave graph of $R(t)$ vs. t. If the condition $\rho + 3p \geq 0$ holds for all times, then $\ddot{R} \leq 0$ always, and $R(t)$ must necessarily be zero at some finite time in the past. Moreover the time since then is bounded by H_0^{-1} the "Hubble time" ($t_0 \leq H_0^{-1}$).

Let us call this instant $t = 0$. Thus $R(0) = 0$, and our present epoch t_0 is the cosmic time elapsed since the initial singularity $t = 0$. This "point" is a true singularity of the space-time: The energy density $\rho \to \infty$ for $R \to 0$; one can construct an invariant quantity from the components of the space time curvature tensor (even if $K = 0$) which will also diverge as $R \to 0$ [19].

The "point" $R = 0$ does not belong to the space-time, and it is not clear at present how close to $R = 0$ we should accept the validity of the classical theory of GR. Most physicists would agree that the state of matter in a classical model of the universe can still be described reasonably accurately up to temperatures and densities close to the limits of terrestrial lab experiments. This would correspond to $t \approx 10^{-4}$ sec.

The condition $\rho + 3p - \frac{\Lambda}{4\pi G} \geq 0$ is sufficient, but not necessary for the existence of a singularity. Observations do not unambigously give $\rho + 3p - \frac{\Lambda}{4\pi G} \geq 0$ now, but if this inequality was satisfied at some past epoch, this implies a singularity, provided $\rho + 3p$ was never smaller before that epoch.

The inequalities $\rho > 0$ and $p \geq 0$ are certainly satisfied by the kinds of matter known to us. The inequality $\rho + 3p > 0$ is certainly satisfied up to nuclear matter density. The singularity can be avoided only by a special choice of the cosmological constant Λ, and then $\ddot{R}$ is always positive. This would make it difficult to explain the background radiation and would make it virtually impossible to explain the cosmic abundance of helium as a pregalactic phenomenon.

The inescapable conclusion of an initial singularity is that the description of the universe by a FL model leads to the existence of an early epoch of high density, when all the structure we see now – galaxies, stars, etc. – did not yet exist.

What was there before the initial singularity? This question cannot be answered within the context of classical GR, because the concepts of this theory cease to be applicable at the singularity.

2.2 Cosmological Observations

(A) The Hubble Constant. The measurements of the Hubble constant are in principle very simple: Only the distance and the redshift of some objects must be determined precisely. In practice the situation is a bit more involved. Reliable distance measurements are only possible for local regions ($z < 0.001$), but the galaxies follow the smooth Hubble expansion probably only at distances beyond the Virgo cluster ($z > 0.01$). In the local regions large peculiar velocities must be expected. Sandage and Tammann [20] find values of

$$H_0 = (55 \pm 5)\ \text{km s}^{-1}\text{Mpc}^{-1} \quad \text{or}$$
$$H_0 = (50 \pm 7)\ \text{km s}^{-1}\text{Mpc}^{-1}$$

de Vaucouleurs [22] gives

$$H_0 = (100 \pm 10)\ \text{km s}^{-1}\text{Mpc}^{-1}.$$

The discrepancy of a factor 2 can be traced back to the distance estimate to the Virgo cluster, where the two groups differ by a factor 2 (i.e. $\Delta M = 1.5$) Fig. 9 shows a Hubble diagram for type I Supernovae [21].

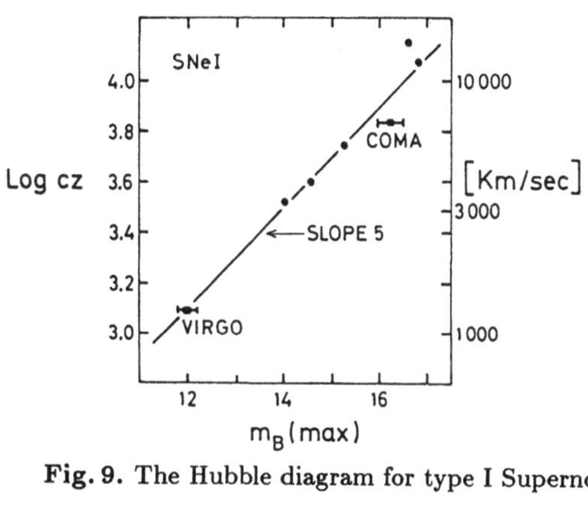

Fig. 9. The Hubble diagram for type I Supernovae [21]; $H_0 = 50 \pm 7$ km s^{-1}Mpc^{-1}.

For $K = 0$ cosmologies the age of the universe $t_0 = \frac{2}{3}H_0^{-1}$, and taking $H_0 = h \times 100$

$$t_0 = \frac{2}{3h} \times 10^{10}\text{years} \quad , \text{i.e.}$$
$$7{\times}10^9 y < t_0 < 1.3 \times 10^{10}\text{years} \tag{2.12}$$

According to eq. (2.11) t_0 becomes larger for $\Omega_0 < 1$. The measurements of the age of the elements and of the oldest stars give values of t_0 betwen 10 and

20 billion years. It is remarkable that these different characteristic time scales agree so well.

(B) The Mean Density. The distribution of matter in our neighbourhood is very inhomogeneous: Galaxies are made up of stars, they are arranged in groups, in clusters, and in superclusters. The large-scale structures e.g. [23], [24], (see Fig. 15) have dimensions of 10 Mpc up to 500 Mpc, usually a flattened shape. The spatial structure is obtained by assuming an undisturbed Hubble expansion for the members of the supercluster. Typical masses of the luminous matter in superclusters are $10^{15} - 10^{16} M_\odot$ ($10^{18} M_\odot$ in the structure found by Tully). String-like structures are observed. The neighbouring superclusters to our local supercluster are at a distance of $\simeq 100$ Mpc. In between the clusters and chains of clusters there are giant volumes almost empty of visible objects [25]. The picture of the large-scale structure of the universe that emerges from these observations is a distribution of the luminous matter in filaments and on the surface of spherical bubbles. The interior of the bubbles (of 30 Mpc typical dimension) does not contain luminous matter, the walls consist of a thin layer of galaxies, sometimes arranged in strings or clusters.

All these complex structures do not really look like a uniform density distribution. Can these observations be fitted into a homogeneous model of the universe? In my opinion there is only evidence for structure, not for homogeneity, from the observations of luminous matter. The data do not contradict, however, the assumption that the mean density, found by averaging over scales of ≥ 500 Mpc, can be used as the homogeneous density of a FL universe. This belief is somewhat supported by the apparently uniform distribution of bright radio sources, and the isotropy of the radiation background.

The observations involved in determining a mean mass-density ρ_0 are discussed in [26]. The galaxies in a certain angular cross-section of the sky are measured down to a limiting magnitude, and a spatial distribution is constructed by using the redshifts as distance indicators via the Hubble law. The luminosity density L of the sample is then derived from a well-tested luminosity function $\phi(L)$. In addition a mass-to-lumiosity ratio M/L for galaxies is used to obtain $\rho_0(G) = L(M/L)$ g/cm^3, the mean density corresponding to luminous matter. A typical M/L value for spiral galaxies is ≈ 4 h, and a typical value obtained from this analysis is

$$\Omega_0 \equiv \frac{\rho_0}{\rho_c} = 0.01 \qquad \left(\simeq 5 \times 10^{-32} \left(\frac{H_0}{50} \right)^2 \text{g cm}^{-3} \right) . \tag{2.13}$$

Now in rich clusters of galaxies the application of the virial theorem – i.e. viewing the cluster as a gravitationally bound system – shows that the usual M/L values do not give enough mass. This "missing mass" – rather missing luminosity (i.e. nonluminous matter) – changes the M/L ratio. This indicates that the luminous matter does not give a true picture of the gravitationally active mass distribution. Theoretical considerations of the stability of rotating flat disks (unstable if $E_{rot}/E_{Grav} \geq 0.27$) suggest a much larger M/L, i.e. a large fraction of unseen mass. If $v^2 = (GM)/r =$ const., then $M(r)$ (the mass within radius r) is proportional to r. Finally in rich clusters a higher background density is obtained, if the mass in galaxies is extrapolated according to $M(r) \propto r$ – as suggested by the rotation curves – to the edge of the neighbouring galaxies. If this is done for all galaxies then [26]

$$\Omega_0 = 0.1 \quad \text{to} \quad 0.3 \,. \tag{2.14}$$

This is rather close to $\Omega_0 = 1$, the critical value for a ($\Lambda = 0$) FL model. We must conclude that the observations point to $\Omega_0 < 1$, but $\Omega_0 = 1$ or even $\Omega \geq 1$ cannot be excluded, since any background density distributed more homogeneously than the large clusters is not accounted for in these measurements. Therefore a decision whether we live in a closed or open universe cannot be made yet.

There may be an enormous amount of non-luminous matter in galaxies. What kind of matter? Black holes, neutrinos of nonzero rest-mass, dark stars of small mass, rocks have all been mentioned. We see that the value of the mean density is extremely uncertain, and that part of this uncertainty is due to a lack of knowledge of local physics.

(C) The 3 K Cosmic Black-Body Radiation. In 1964 Penzias and Wilson discovered an excess radio background at $\lambda = 7.25$ cm corresponding to a black-body temperature of 2.5 to 4.5 K. Since its publication in 1965 this discovery has been interpreted as a cosmological background signal indicating the existence of a hot, dense early stage of the universe, when radiation and ionized matter had been in equilibrium. The history of this discovery is told in Weinberg's book "The first 3 minutes".

In the last few years accurate measurements have verified the high frequency tail of the Planck distribution of the radiation [27] (Fig. 10). The Nobel Prize in physics has been awarded to Penzias and Wilson in 1978 – a further acknowledgement of the importance of their measurement. The recent observations have established the high isotropy, the black-body spectrum, and the absence of polarization – a correspondence as close as one could hope to the simple hypothesis that the radiation is the remnant of a homogeneous and structureless primeval explosion.

One must keep in mind that although this fits beautifully to the concept of a FL universe, the measurements do not really prove that the 3 K background signal is the relic radiation from a hot big origin. To date, however, no serious challenge to the canonical bang interpretation has been given. Attempts to explain the radiation as the result of the superposition of the microwave emission of many individual sources have all crumbled beneath the weight of evidence produced in recent years. The measurements satisfy all the test for a cosmic background radiation that have been proposed until the present time.

There are many excellent reviews on the experimental and theoretical situation [28], [29], [30]. One important fact about a Planck radiation field in a homogeneous FL universe is that its spectrum keeps the Planck shape during the expansion: The adiabatic expansion law $T_\gamma(R^3)^{\kappa-1} = \text{const.}$ gives for $\kappa = 4/3$ (photons) $T_\gamma R = \text{const.}$ Thus the Planck spectrum

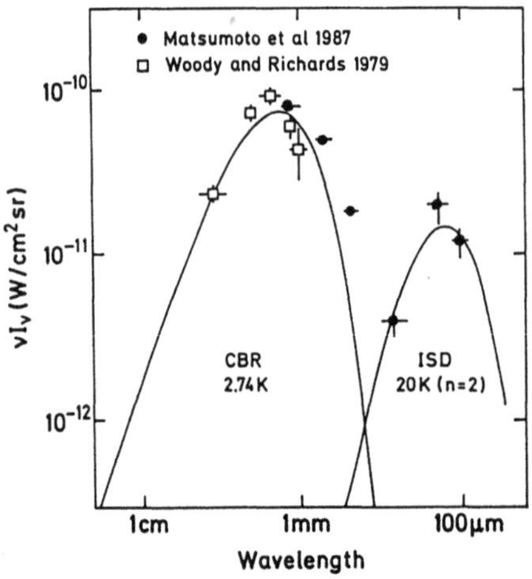

Fig. 10. Recent measurements of the 3 K background (graph marked CBR establishes nicely the blackbody shape of the spectrum, except for a mysterious deviation at wavelengths shorter than 1 mm. The measurements of the emission of interstellar dust (ISD curve) peaking at 100 μm are shown for comparison.

$$\rho_\gamma d\nu = 8\pi h \left(\frac{\nu}{c}\right)^3 d\nu \left[\exp\left(\frac{h\nu}{k_B T}\right) - 1\right]^{-1}$$

stays a Planck spectrum, since $T_\gamma = T_{\gamma 0}(1+z)$ and $h\nu = h\nu_0(1+z)$. The number density of photons is

$$n_\gamma = \int \frac{\rho_\gamma}{h\nu} d\nu = \frac{0.37a}{k_B} T_\gamma^3 .$$

Therefore the ratio of photon to baryon number density is a constant in FL cosmologies:

$$\frac{n_\gamma}{n_B} = \frac{0.37aT_\gamma^3}{k_B\, n_B} = \frac{0.37a\; T_{\gamma 0}^3}{k_B\; n_{B0}} \frac{(1+z)^3}{(1+z)^3} = \frac{n_{\gamma 0}}{n_{B0}} . \tag{2.15}$$

In Fig. 10 the spectrum of the radiation is shown. The maximum intensity is at $\lambda = 2\,\mathrm{mm}$, and the temperature of the Planck spectrum is $T_{\gamma 0} = 2.7\,\mathrm{K}$. Therefore the energy density is $0.25\,\mathrm{eV/cm^3}$ i.e. $n_{\gamma 0} \approx 400/\mathrm{cm^3}$. It is the dominating radiation energy density $\rho_\gamma = 3 \times 10^{-34}\,\mathrm{g/cm^3}$. Thus the number

$$\frac{n_{\gamma 0}}{n_{B0}} = 10^{9\pm 1} \tag{2.16}$$

(the large uncertainty in n_{B0} is reflected here) is a constant, characteristic for our univere, if it is a FL space-time. n_γ/n_B has to be set down as an initial condition in any classical FL cosmology. Modern particle theories applied to the early universe may find a way to remedy this undesirable situation of having to introduce a net baryon asymmetry at the beginning (cf. Section 2.5).

The measurements show deviations from a black-body spectrum: At higher frequencies beyond the maximum the intensity is above a 2.7 K blackbody curve, it is not clear yet whether this is an experimental or a theoretical problem.

Such a background radiation field has important consequences: Since the energy density of matter $\rho_M = \rho_{M0}(1+z)^3$ and of radiation $\rho_\gamma = \rho_{\gamma 0}(1+z)^4$, we have that before a redshift z_R between $z_R \approx 10^3$ and $z_R = 10^4$ (uncertainty of ρ_{M0}!) the universe was dominated by hot radiation.

Before the time t_{dec} when the recombination of hydrogen was finished ($T \approx 4500\,\mathrm{K}$), i.e. $z_{dec} \approx 1500$, matter and radiation have been in thermodynamic equilibrium. It is a remarkable coincidence that these two times may concide $z_R \approx z_{dec}$.

The universe must have been completely different for $t < t_R$ from its present appearance: This radiation dominated phase of the early universe was a hot and dense gas of radiation and matter. Structures, such as galaxies and stars did not yet exist.

Although at z_{dec} the mean free path of photons is drastically increased, there is still a small interaction with matter via Thomson scattering. Therefore the photons that we observe come from a surface of their last scattering, which has a smaller redshift than $z_R \approx 1500$.

This surface of last scattering is not necessarily sharply defined: Photons seen along one line of sight may have been last scattered in various uncorrelated regions. One expects nevertheless to see inhomogeneities in the matter distribution reflected in a variation of the temperature [19].

For adiabatic fluctuations, $S/n_B = $ const., i.e.

$$\frac{\delta n}{n} = -3\frac{\delta T_\gamma}{T_\gamma}.$$

The angular scale associated with a typical galactic mass of $M = 10^{11}\,M_\odot$ at z_R is $\theta = 40''$, and $\theta \propto (M)^{1/3}$. The high isotropy of the 3 K background proves that fluctuations in the homogeneous background around t_R must have been quite small. In fact, these limits seem to destroy the possibility to let galaxies grow from small adiabatic fluctuations (cf. Section 3).

On the other hand the 3 K background can be viewed as a perfectly isotropic medium, defining a perferred frame of reference. Any peculiar motion with respect to this frame can be measured. The expected anisotropy has a $\cos\theta$ behaviour:

$$T(\theta) = T_0\left(1-\frac{v^2}{c^2}\right)^{\frac{1}{2}} \Big/ \left(1-\frac{v}{c}\cos\theta\right) = T_0\left(1+\frac{v}{c}\cos\theta\right). \tag{2.17}$$

Observations at 33 MHz with a U-2 plane showed indeed such a cosine behaviour of $T(\theta)$. This is a measurement of the peculiar motion of the earth in the cosmic ether of the microwave background. The amplitude $\delta T/T \approx 10^{-3}$ corresponds to

$$v_\oplus = 361\left(\frac{3K}{T}\right)\,\mathrm{km/sec}$$

and adding the galactic rotation we find the proper velocity of our galaxy

$$v_G = (500 \pm 75)\ \text{km/sec}$$

in the direction r.a. $= 11.3 \pm 0.1$ (hr.), $\delta = 4 \pm 2$ deg. [39]. The large magnitude of this velocity is disturbing, since the peculiar velocities of nearby galaxies – from optical observations – are small ~ 80 km/sec to 200 km/sec. The galactic motion implied by the dipole anisotropy has a component toward Virgo (~ 400 km/sec) and components of 300 km/sec tangent and normal to the plane of the local supercluster.

2.3 Thermodynamics of the Radiation Dominated Phase

The existence of a thermal radiation field with an energy density increasing proportional to R^{-4} has the immediate consequence that there is an early phase where the universe was hotter and denser. Matter and radiation must have been in thermal equilibrium in this hot initial stage.

For an adiabatic expansion (this should be a rather good approximation except for some very special epochs) $T_\gamma R =$ const., and the temperature increases beyond any bound as the big bang is approached, $R \to 0$. All the particles present acquire thermal energies as high as one likes, if the big bang is approached close enough. In that sense the early universe is the ultimate accelerator – effective and inexpensive. Unfortunately the observational possibilities are limited. Our present universe is only a cold relic of the hot early phase. Connections can only be found by a patient and imaginative scrutiny of theoretical ideas. Thus the early universe is a playground of theorists.

This interplay of cosmology and particle physics has brought about various interesting consequences. It has turned out, for instance, that it is not quite straightforward to implement a grand unified theory (GUT) of elementary particles in the early phase of a FL universe: Due to the horizon structure of the FL model there will be too many heavy monopoles, and the universe will recollapse. In that respect the horizon structure is a "deficiency" of the standard model.

In the following some of these concepts of particle physics and the early universe will be discussed. But at first I want to introduce some of the concepts of the standard classical big bang.

As long as the constituents of the radiation background can be treated as noninteracting the temperature increases beyond any limit as the "hot" singularity is approached. The energy density increases as the fourth power of the temperature $\propto T^4$, or equivalently proportional to R^{-4}. The curvature term and the cosmological constant become negligible compared to the potential energy term $\propto \rho$ in the Friedmann equation. The expansion factor evolves according to

$$\dot{R}^2 = \frac{8\pi G}{3} \rho\, R^2. \tag{2.18}$$

Since $\rho R^4 =$ const. it follows

$$\frac{32\pi}{3} G\, \rho\, t^2 = 1 \tag{2.19}$$

(in the following we use a choice of units $\hbar = c = k_B = 1$).

A crucial assumption of all attempts to describe the early universe is a phase of thermal equilibrium at some epoch. One expects the weak interactions to be in equilibrium for temperatures above 10^{11} K. (The nature of the interactions determines the equilibrium. The reaction rates τ_{coll}^{-1} computed from the theory are compared to the expansion rate H(t). Whenever $\tau_{coll}^{-1} > H(t)$ thermal equilibrium is possible. For GUT theories, e.g., one finds that a phase of thermal equilibrium is possible for temperatures corresponding to the characteristic GUT energy scale. Typically $k_BT \sim m_{GUT} \sim 10^{15}$ GeV. Interestingly for higher and lower temperatures GUT reactions are *not* in thermal equilibrium).

In thermal equilibrium the number density n_i of particles of type i – and their energy density ρ_i – at the temperature T can easily be derived [33], [34], [35]:

$$
\begin{aligned}
n_i &= \frac{g_i}{2\pi^2} T^3 \int_0^\infty \left\{ \exp\left(\frac{E_i - \mu_i}{T}\right) \pm 1 \right\}^{-1} z^2 \, dz \\
\rho_i &= \frac{g_i}{2\pi^2} T^4 \int_0^\infty \left\{ \exp\left(\frac{E_i - \mu_i}{T}\right) \pm 1 \right\}^{-1} z^2 \frac{E_i - \mu_i}{T} \, dz;
\end{aligned}
\tag{2.20}
$$

where $z \equiv |\boldsymbol{p}|/T$; $E_i \equiv \{\boldsymbol{p}^2 + m_i^2\}^{1/2}$.

g_i is the number of spin degrees of freedom, i.e. the number of helicity states. The $+(-)$ sign in the integral applies to fermions (bosons). Antiparticles are treated as independent particle species.

The chemical potentials μ_i are unknown parameters in this state of thermodynamic equilibrium. There are as many independent chemical potentials as there are conserved additive quantum numbers. In the standard model there are four such quantities (total charge, baryon number, e-lepton number, μ-lepton number) which are usually set equal to zero (the reason is that charge density and baryon or lepton number density are all very small compared to the photon number density). In most GUT theories there are no conserved quantum numbers, and therefore all chemical potentials are zero. Photons are bosons with $g_i = 2$ and $m_i = 0$.

Hence from (2.20) one obtains

$$
\begin{aligned}
n_\gamma &= \frac{2\xi(3)}{\pi^2} T^3 = \frac{2.404}{\pi^2} T^3 \\
\rho_\gamma &= \frac{6\xi(4)}{\pi^2} T^3 \cdot T = 2.7T \, n_\gamma \quad .
\end{aligned}
\tag{2.21}
$$

With all chemical potentials equal to zero only those particles can be present in appreciable numbers for which $mc^2 \ll k_BT$. Those particles can all be treated in good approximation as ultra-relativistic, $c|p| \gg mc^2$. This leads to the following relations:

For bosons

$$
n_B = (g_B/2) \, n_\gamma; \qquad \rho_B = (g_B/2) \, \rho_\gamma; \tag{2.22}
$$

and for fermions

$$n_F = \frac{3}{8}\, g_F\, n_\gamma; \qquad \rho_F = \frac{7}{16}\, g_F\, \rho_\gamma;$$

The total energy density in the radiation dominated phase is then

$$\rho = \rho_B + \rho_F \equiv \frac{1}{2} g(T) \rho_\gamma \tag{2.23}$$

where $g(T) \equiv g_B + (7/8) g_F$. The expansion time can therefore be written as

$$t_{exp} = \left(\frac{3}{32\pi}\right)^{1/2} G^{-1/2} \left(\frac{g}{2}\right)^{-1/2} \rho_\gamma^{-1/2} = \left(\frac{45}{8\pi^3}\right)^{1/2} (g\,G)^{-1/2}\, T^{-2} \tag{2.24}$$

Let us plot this relation around $T \sim 1$ MeV, i.e. for $g = \frac{43}{4}$ (Fig. 11) (3 types of light neutrini, electrons and positrons, photons).

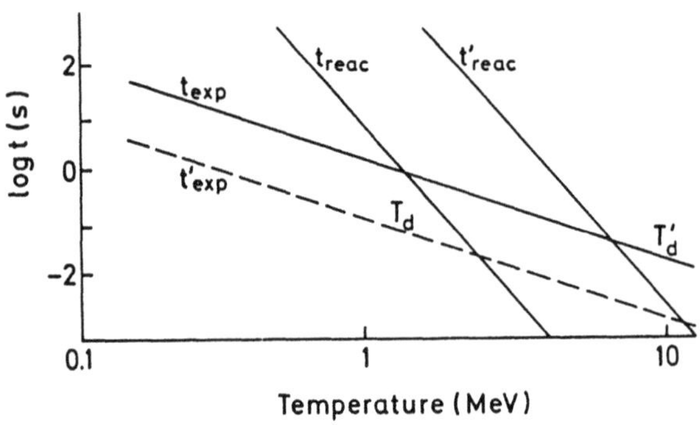

Fig. 11. The relations between the expansion time (t_{exp}) and the cosmic temperature, and between the mean reaction time of the weak interactions and the cosmic temperature are shown in this diagram. At temperatures below the crossing of the curves (defining the decoupling temperature $T_d \sim 1$ MeV) the weak interactions can no longer establish thermal equilibrium. Thus also the neutron-proton equilibrium abundance is lost at this point. An increase in the mass of the W-boson would decrease the value of the Fermi constant ($G_F \propto m_W^{-2}$) and shift the t_{reac} curve to the right (t'_{reac}) leading to a higher decoupling temperature. Changes of $g(T)$ or G would also change the position of T_d. Thus the big bang nucleosynthesis is a sensitive test of the values of these different constants.

Additional families (such as new neutrino types or other types of light, weakly interacting particles) increase the value of $g(T)$, consequently also the total density, and the rate of expansion $t_{exp}^{-1} \propto g^{1/2} T^2$. A new curve of t'_{exp} vs. temperature results (plotted in Fig. 11 for $g' > g$), which is below the standard graph. Weak interaction rates (for e.g. the capture of a neutrino by a neutron and the decay into a proton and an electron) are of the form

$$t_{reac}^{-1} \propto G_F^2 T^5 \tag{2.25}$$

(G_F is the universal Fermi coupling constant, $G_F = 10^{-5} m_p^{-2}$ (GeV)$^{-2}$). The two curves t_{exp} and t_{reac} cross at the decoupling temperature T_d. Below T_d the reactions are no longer fast enough to establish thermal equilibrium. From then on the neutrons decay freely (with a lifetime of $\sim 10^3$ seconds). Around

$T \simeq 0.1$ MeV (about 100 sec later) deuterium nuclei manage to resist photodisintegration. At that point a chain of fast reactions leads to the build up of the light elements, mainly ^{4}He. Essentially all the surviving neutrons are used up in this nucleosynthesis epoch.

For very high temperatures (larger than the neutron-proton mass difference Q = 1.29 MeV) the abundance ratio of neutrons X_n ($X_n = N_n/(N_n + N_p)$) is given by the equilibrium value

$$X_n = (1 + \exp Q/T)^{-1}. \tag{2.26}$$

For $T \to \infty$ $X_n = 1/2$. The hot big bang model determines the initial abundance ratio as $X_n = 1/2$, $X_p = 1/2$ – no initial condition has to be posed. Thus the results of primeval nucleosynthesis are honest predictions of this model.

At the decoupling time $X_n = 0.16$ in the standard model. We can see from Fig. 11 qualitatively, that additional families which raise $g(T)$ lead to an increase of T_d, hence to a higher neutron abundance, and consequently a higher abundance of helium. Let us discuss these features of the hot big bang model in a bit more detail.

2.4 Nucleosynthesis

The big bang nucleosynthesis leads to the formation of D, ^{3}He, ^{4}He, a small amount of ^{7}Be, ^{7}Li. Heavier elements are not produced in significant amounts, because at A = 5, A = 8 there is no stable nucleus.

The value of the helium abundance $Y \equiv X(^4\mathrm{He})$ depends crucially on the pressure of a hot radiation field which prevents the formation of D before $T = 0.1$ MeV. Thus the fact that at present most of the matter is hydrogen, i.e. not all the matter is transformed into ^{4}He in the big bang, is another point speaking in favour of the existence of a primeval cosmic background radiation.

Detailed calculations of D and He synthesis have led to the following quantitative picture: (first calculations by Hayashi [36]; recent reviews [37], [38]). In Fig. 12 and Fig. 13 some of these results have been collected. The resulting abundance of ^{4}He – plotted and listed always as the ratio of the ^{4}He mass density to the overall mass density (Fig. 13) – varies very little with baryon density, and hence with the baryon density ρ_{B_0} of the present epoch (since $\rho_B \propto R^{-3}$). For 3×10^{-30} g/cm^3 one finds a varation of X(^{4}He) between 0.23 and 0.25. These values are much larger than the abundance that would be expected from fusion processes in stars. Models of the galactic evolution estimate the amount of helium produced in stars to be a few percent:

$$Y = (0.01 \div 0.04) \quad \text{from} \quad \text{stellar} \quad \text{fusion.} \tag{2.27}$$

This agreement between the prediction of the big bang models and observations is another impressive point in favour of such simple cosmological models.

Contrary to ^{4}He, the predicted deuterium abundance is strongly dependent on the baryon density. This is quite plausible, since deuterium can only survive, if the reactions leading to He formation are not rapid enough to use up all the

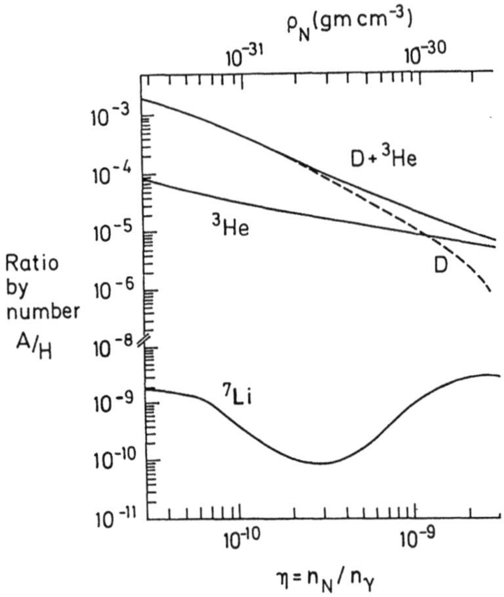

Fig. 12. The abundance of light elements as a function of the cosmic baryon density. The broad valley of the ^{7}Li abundance is remarkable.

D-nuclei. With increasing density two-body collisions are more frequent, and fewer D-nuclei can escape the fusion into helium. The graphs show that the deuterium abundance is so sensitive a function of the mean density, that a determination of the cosmic abundance of deuterium could fix the uncertain baryon density of the present epoch – at least the density of all the baryons that have participated in the primeval nucleosynthesis.

Results of various observations are listed in Table 2.1. The last column in Tab. 2.1 gives the range of the parameter η_{10} ($\Omega_B = 0.00353 h_0^{-2} \vartheta^3 \eta_{10}$) for which observation and prediction from the standard big bang model are in agreement. For $3 < \eta_{10} < 5$ all light elements can have been synthesized in the big bang.

This corresponds to limits of $0.01 < h_0^2 \Omega_B < 0.018$ ($0.04 < \Omega_B < 0.07$ for $H_0 = 50$). If we just take into account the upper limit in the D abundance, $\eta_{10} < 10$, an upper limit of

$$\Omega_B < 0.14 \tag{2.28}$$

results (for $h_0 = 1/2,\ \vartheta = 1:\ \vartheta \equiv T_{\gamma 0}/(2.7\,\mathrm{K})$).

At least all the baryons which participated in primeval nucleosynthesis are subject to this limit. $\Omega \leq 0.14$ leads to a smaller mean density than that derived from – rather extreme – assumptions about the dynamics of the clusters of infrared galaxies ($\Omega = 0.80 \pm 0.15$) [39]. But otherwise the allowed range for Ω can accomodate the "missing mass" problems derived from cluster dynamics and from the rotation curves of spiral galaxies.

If Ω_{B0} was equal to the average density, this would mean that we live in an open universe with curvature $K = -1$. However, unseen forms of matter may

Table 2.1. Limits on primeval abundance ratios from observations [38]

	Remark	range of η_{10} ($N_\nu = 3, \tau_n = 10.6$ min.)
$(D/H)_p > (1 \text{ to } 2) \times 10^{-5}$	interstellar	$\eta_{10} > 7$ to 10
$\left[(D + {}^3He/H\right]_p < (6 \text{ to } 10) \times 10^{-5}$	a)	$\eta_{10} > 3$ to 4
$({}^7Li/H)_p = (0.7\text{to } 1.8) \times 10^{-10}$	pop*II* stars	$1.5 < \eta_{10} < 5$

a) The survival of ^{3}He, the conversion of D to ^{3}He is taken into account; the small abundance of newly synthesized ^{3}He is ignored.

exist invalidating this conclusion. It should be stressed that the small value of the deuterium abundance, and the large variations in the observations, should keep us open-minded toward local production mechanisms. It seems a bit like erecting a colussus on clay feet, if one draws the most far-reaching conclusions on the cosmic density from the small value of the D and Li abundance.

We have mentioned already that an interesting conclusion can be drawn from the ^{4}He abundance (e.g. [38]). The decoupling temperature of the weak interactions T_d determines the n/p ratio. An increase of T_d leads to an increase of n/p at the onset of nucleosynthesis, and thus to a larger primordial helium abundance Y.

In Fig. 13 the effect of different numbers of neutrino families on the helium abundance is shown.

(The Hubble constant $H_0 = 100\ h_0$ km sec^{-1} Mpc^{-1}
and thus $\rho_{B0} = 2 \times 10^{-29}\ \Omega_{B0}\ h_0^2$ g cm^{-3}.)

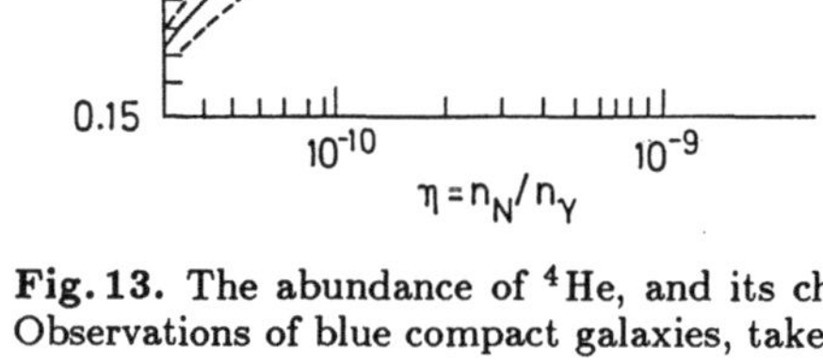

Fig. 13. The abundance of ^{4}He, and its change with different numbers of neutrino types. Observations of blue compact galaxies, taken at face value, require $N_\nu = 3$.

The primordial abundance of helium increases roughly by 1% for each additional neutrino family.

If the primeval abundance ratio is $Y \leq 0.25$ there can be only 3 families of neutrini. Thus besides the electron-, μ-, and the τ-neutrino (if it is light,

$\leq$ 100 eV) there can be no other light neutrino, or other weakly interacting neutrino-like supersymmetric particle (photino, gravitino).

The very conservative upper limit of $Y \leq 0.28$ leads to $\Delta N \leq 5$. The number of different families of neutrinos is therefore strongly limited by the helium abundance, if we accept the framework of the nuclear synthesis in a highly symmetric Friedmann model.

We can also turn the argument around and state that the existence of three neutrino species and the results on ^{4}He abundance give us some confidence that the time evolution of the scale factor as described by the Friedmann equations, namely $R(t) \propto t^{\frac{1}{2}}$, is experimentally verified for $t \sim 1$ to 10 seconds after the inital singularity. Since the ^{4}He production is so sensitive to a change in the expansion rate, a deviation from the law $R(t) \propto \sqrt{t}$ would be clearly imprinted on the ^{4}He abundance.

2.5 Baryon Synthesis

There seems to be little chance to achieve a physical separation of baryon and antibaryon phases in an initially baryon-symmetric cosmological model. If baryon-number was exactly conserved – as it is assumed to be in the standard model – the small asymmetry necessary for our existence must be postulated initially.

Grand unified theories offer the possibility to create this small asymmetry from physical processes.

Already in 1967 A. Sakharov [40] has looked into the conditions necessary to create a baryon-excess in an initially B-symmetric universe. Clearly there must be B-violating interactions. In addition the symmetry between particles and antiparticles must be disturbed, i.e. a violation of the CP symmetry must occur. Besides these symmetry violations there must be epochs in the early universe where thermodynamic equilibrium was not established. Loosely speaking, the CPT-invariance of local, relativistic field theories and thermodynamic equilibrium would imply the invariance under CP, because in thermodynamic equilibrium there is no arrow of time.

The appearance of supermassive gauge bosons or Higgs bosons is a natural outcome of any scheme of grand unified theories. The collisions and decays of these particles violate B and L conservation. It seems likely that in simple GUT schemes the dominant mechanisms generating a baryon-antibaryon asymmetry are the decays and inverse decays of such supermassive bosons, so-called X-particles, just as in the SU(5) scheme. The generation of a net baryon number may have been quite complicated with several different baryon number violating processes contributing or competing (for a review see [41]).

Until the temperature has dropped down to the decoupling temperature T_D of the X-particles, the density n_x is proportional to T^3, and $n_x \simeq n_\gamma$. Then at $T = T_D$.

$$n_{xD} \simeq N_x \frac{1}{\pi^2} \xi(3)\, T_D^3 \tag{2.29}$$

where N_x counts the X and $\bar{\text{X}}$ spin degrees of freedom.

The total entropy density of all other relativistic particles is

$$s_D = g(T)\frac{1}{2}\,\frac{4\pi^2}{25}T_D^3 \qquad (2.30)$$

and therefore the ratio of baryons to s is

$$\frac{n_B}{s} \simeq 0.01\Delta B \qquad (2.31)$$

ΔB is the baryon number produced per decay of an $(\mathrm{X}\bar{\mathrm{X}})$ pair. The baryon number produced per decaying pair of particle and antiparticle is in principle completely determined by particle physics. For the creation of a net, nonzero baryon number the decays of particle and antiparticle must be different. Thus the CP symmetry must be violated, but the degree of violation is an unknown parameter.

The following range of values has been estimated [41]: If the decays of Higgs bosons are the dominant mechanism, then

$$\Delta B \sim 10^{-6} \text{ to } 10^{-2}$$

If gauge bosons dominate

$$\Delta B \sim 10^{-10} \text{ to } 10^{-6}.$$

We see that these estimates cover the observationally determined $n_B/s \simeq 10^{-9\pm1}$. The main impact of the result is, perhaps, that it seems possible to calculate a baryon excess within GUT theories which does not largely deviate from the observed value. One can understand why the excess is small and nonzero, but the numerical agreement can hardly be called impressive.

Any initial baryon asymmetry is strongly damped by the thermal equilibrium phase around $T \sim m_x$ [42]. If the difference of baryon number and lepton number is strictly conserved, $\Delta(B-L) = 0$, as required by certain GUT theories (e.g. the minimal SU(5)), then the damping is large enough – diluting any initial asymmetry $(n_B - n_{\bar{B}})$ by a factor $\sim 10^{-28}$ – to restore complete baryon-antibaryon symmetry. We cannot resort to appropriate initial imprints of baryon asymmetry, if we take these GUT theories seriously. They not only offer a mechanism to create a net baryon number, but they also destroy any initial asymmetry. Therefore within these theories one is forced to explain the observed baryon-assymmetry by physical processes.

The discrepancies of some of the calculated ratios (n_B/n_γ) with the observed value can perhaps serve to eliminate certain models of GUT.

Recently the possibility has been discussed that nonlinear soliton effects at the electroweak phase-transition can wash out any nonzero $(B+L)$ (baryon + lepton number), leaving $(B-L)$ unchanged [43]. If this speculative idea turns out to be true the problem of creating a baryon asymmetry after the electroweak phase transition around 200 GeV raises its ugly head. But since we have a baryon asymetry now, there must be a way to get past this difficulty. One way would be to have $(B-L) \neq 0$ from a GUT theory. The end of this fascinating story will probably not be written during the next few Schladming schools.

2.6 The Inflationary Universe

(A) The Horizon Problem. The uniform temperature of the 3 K background radiation implies in the context of the standard model a uniform temperature of the "cosmic photosphere", where radiation and matter decouple. This uniformity is a trivial consequence of the isotropy assumed for the standard cosmological model. But to explain the observed isotropy without assuming isotropy initially is a difficult problem. Two antipodal points on the cosmic photosphere are separated by a distance large enough to prevent causal contact before the decoupling epoch (each point is outside of the particle horizon of the other). But both points have the same radiation temperature!

Generally in a Friedmann-Lemaître model the number n_{AB} of causality lengths between antipodal emission points A and B is given by

$$n_{AB} = \int_{t_R}^{t_0} \frac{dt}{R(t)} \left(\int_{t_{pl}}^{t_R} \frac{dt}{R(t)} \right)^{-1} \tag{2.32}$$

The numerator of this expression depends only on the expansion factor at times later than the time of decoupling (or "recombination") t_R and is rather well known. The denominator depends on the behaviour of $R(t)$ in the very early universe. In a radiation-dominated FL model, $R(t) \propto \sqrt{t}$, and $n_{AB} \sim 60$. An inflationary phase, however, implies a dramatic exponential increase of $R(t)$ in an early epoch, and leads to $n_{AB} \ll 1$. Thus a solution of the horizon problem seems possible, since antipodal points become causally connected. The fact that $n_{AB} \simeq 60$, is related to the value of the entropy S: The radius of the observable universe L_0 shrinks to a length $L(t)$ at an earlier epoch t, and $L_0 = \frac{T}{T_0} L(t)$ (T is the temperature). During the radiation dominated era a causally connected volume has the size of the horizon

$$r_H = 2t \approx (G\rho)^{-\frac{1}{2}} \approx M_{pl} T^{-2}.$$

Therefore the ratio of the two volumes at t_R is

$$\left(\frac{L(t_R)}{r_H(t_R)} \right)^3 \simeq (L_0 T_0)^3 \left(\frac{T_R}{M_{pl}} \right)^3 .$$

Another aspect of the horizon problem in FL models is the different time behaviour of the horizon scale, $r_H = 2t$, and any length scale $L = L_0 \frac{R}{R_0} \propto \sqrt{t}$. For t approaching the singularity at $t = 0$ the horizon is shrinking much faster than any length scale. This means that any finite scale now has been through an acausal history in the early universe where its endpoints could not influence each other by physical processes. The importance of this property for galaxy formation is evident. How can a reasonable understanding of an initial fluctuation spectrum be found, if the fluctuations in early phases were spread over causally disconnected regions?

(B) The Flatness Problem. It is a remarkable fact that the present mean density in the universe is rather close to the critical value required for a 3-space of curvature zero. The density parameter Ω_0 is certainly between the limits

$$0.1 \leq \Omega_0 \leq 4$$

The deviation of Ω from the critical value $\Omega = 1$ depends on the value of the entropy S. One can derive ($\Omega(t) \equiv \rho\, 8\pi G/(3H^2)$)

$$\frac{\Omega - 1}{\Omega} = \frac{3}{8\pi G} K \frac{s^{\frac{2}{3}}}{S^{\frac{2}{3}}\rho} \tag{2.33}$$

$$\frac{\Omega - 1}{\Omega} = 0.2K\, g^{-1/3}\, S^{-2/3} \left(\frac{M_{pl}}{T}\right)^2 . \tag{2.34}$$

At temperatures above 1 MeV this inequality expresses the phantastic fine-tuning necessary in the standard model to arrive at the present value:

At $T = 1$ MeV it leads to $|(\Omega - 1)/\Omega| < 10^{-15}$.

At a typical unification energy of $T \sim 10^{14}$GeV $|\Omega - 1| \leq 10^{-49}$.

This precise fine tuning to $\Omega = 1$ expresses the fact that an "old" universe, as ours with an age of $t_0 \simeq 10^{60} t_{pl}$, can have Ω_0 close to one only if the entropy S has a large numerical value – a fact often refered to as the "flatness" problem, since $\Omega = 1$ corresponds to the flat Euclidean space ($K = 0$). For $S \approx 1$ the universe would be very empty at epochs $T \ll M_{pl}$ in the case of an hyperbolic 3-space ($K = -1$). Stars and galaxies could not form in this case. A positive curvature ($K = +1$, elliptical 3-space) would lead to a recollapse shortly after t_{pl}.

(C) The Monopole Problem. Another problem that arises in the classical standard cosmology is the large number of monopoles produced when a GUT model in the early universe is at work. Magnetic monopoles are topological defects in the Higgs field (a scalar field responsible for symmetry breaking), which appear whenever the GUT symmetry is spontaneously broken down to a smaller symmetry group containing a U(1) factor. The correlation length of the Higgs field, ξ, is limited by the causality length in the early universe

$$\xi \leq r_H(t) \leq 2t.$$

A lower limit in the density of monopoles can thus be obtained by having one monopole per horizon volume $r_H^3(t)$. Then the ratio of monopole density to entropy density is

$$\frac{n_M}{s} \simeq \left(\frac{T_c}{M_{pl}}\right)^3 = 10^{-12} \tag{2.35}$$

(To derive this formula use $n_M \propto r_H^{-3}$ and $r_H = 2t = \left(45/(\pi^3 g)\right)^{\frac{1}{2}} \left(M_{pl}/T^2\right)$).

The temperature T_c can be expected to be around the typical GUT scale $T_c \simeq 10^{14}$ GeV. Comparing n_M/s to n_B/s, we find that the mass-density of monopoles

$$\rho_M = n_M m_M = 10^{-2}\ m_p n_B \left(\frac{m_M}{m_p}\right) = 10^{14} \rho_B \tag{2.36}$$

(The monopole mass $m_M \approx 10^{16} m_p$; consult [45] for more details on monopoles).

This terrible dominance of monopoles would of course violate all observational limits on the mean density of the universe. This is the monopole problem, the original motivation for the concept of the inflationary universe.

(D) The Inflationary Universe – A Qualitative Account. The basic qualitative picture is very simple. Assume that at some epoch in the early universe the potential energy of a scalar field ϕ, $V_{eff}(\phi)$, was the dominant form of energy density for some time. This energy would appear as a huge cosmological constant $\lambda = 8\pi\ G\ V_{eff}(\phi)$. At the present epoch any cosmological constant must be very small, it cannot exceed the critical density by much, i.e.

$$\frac{\lambda_{now}}{8\pi G} \leq \rho_c = 10^{-46}\ (\mathrm{GeV})^4\ . \tag{2.37}$$

But at the unification epoch, for instance, the energy of the Higgs field could lead to an energy density

$$\rho_v \equiv \frac{\lambda}{8\pi G} = T_c^4 \simeq (10^{14}\mathrm{GeV})^4\ .$$

A small region of space-time, for simplicity assumed to be isotropic and homogeneous, filled with this huge energy density would give a Hubble parameter

$$H = H_v = \left(\frac{8\pi\rho_0}{3M_{pl}^2}\right)^{\frac{1}{2}} \simeq \frac{T_c^2}{M_{pl}} \simeq 10^{10}\mathrm{GeV}. \tag{2.38}$$

As long as this constant energy density dominates there will be an exponential expansion, the deSitter expansion

$$R(t) \propto e^{H_v t}.$$

It is simply a result of taking ρ_v on the right-hand side of the Friedmann equation. This is the "inflationary" phase of the universe. The characteristic time scale is H_v^{-1}, i.e. 10^8 to 10^{10} Planck times. The rapid expansion damps exponentially initial inhomogeneities and anisotropies in a time of $\sim H_v^{-1}$. If the expansion factor $R(t)$ increases by at least a factor 10^{29} the horizon, flatness, and monopole problems can be solved. At the end of the inflationary phase a reheating and reentry into a radiation dominated Friedmann universe must occur. This is something like a new beginning: A smooth radiation field is established at least in that part of the universe which has undergone inflation. Baryons have to be synthesized afresh, because any objects present initally have been diluted according to the conservation law nR^3 = const. Monopoles have disappeared from this region. Thus, if this reheated part embraces our observable universe the cosmological puzzles have been solved.

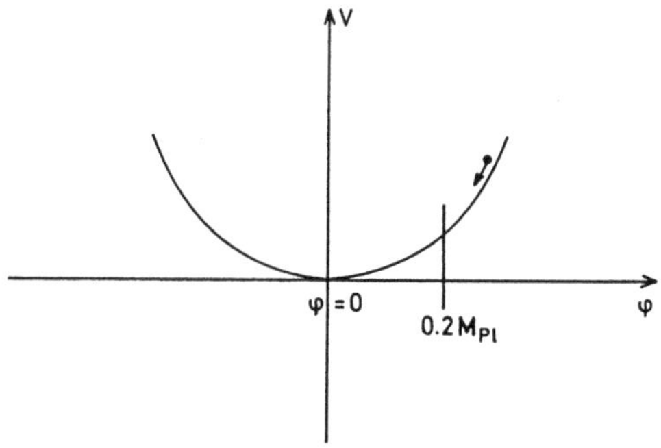

Fig. 14. The flat effective potential of chaotic inflation. The field ϕ initially has a high value $\gg M_{pl}$. As it rolls down past 0.2 M_{pl}, oscillations begin to transform the potential energy into heat.

It is a difficult problem to find out, how and under what conditions particle theories give rise to inflation, and how it works in quantitative detail.

(E) Chaotic Inflation. The model of chaotic inflation [46] does not use a fundamental theory, but treats the early universe as filled by massive scalar fields. For such a massive self-interacting field ϕ the equation of motion is

$$\ddot{\phi} + 3H\dot{\phi} = -\frac{\partial V}{\partial \phi} . \tag{2.39}$$

$V(\phi)$ is the effective potential; a simple example is

$$V(\phi) = \frac{1}{2}m^2\phi^2 ; \tag{2.40}$$

and the Friedmann equation reads

$$H^2 + \frac{K}{R^2} = \frac{8\pi}{3M_{pl}^2}\left[\frac{1}{2}\dot{\phi}^2 + V(\phi)\right] . \tag{2.41}$$

If ϕ is large initially, then ϕ(t) and R(t) rapidly approach the asymptotic regime

$$\phi(t) = \phi_0 - \frac{mM_{pl}}{2(3\pi)^{\frac{1}{2}}}t \tag{2.42}$$

$$R(t) = R_0\exp\left[2\pi\ M_{pl}^{-2}\left[\phi_0^2 - \phi^2(t)\right]\right] . \tag{2.43}$$

During a time $\tau \sim \phi(mM_{pl})^{-1}$ the value of ϕ remains almost constant, the universe expands quasiexponentially.

$$R(t + \Delta t) \sim R(t)\exp(H\ \Delta t)$$
$$H \sim \frac{m}{M_{pl}}\phi, \quad \Delta t \le \tau .$$

As ϕ is rolling down towards $\phi = 0$ (see Fig. 14) it starts to oscillate for $\phi < 0.2M_{pl}$ transforming the energy V($\phi \sim 0.2\ M_{pl}$) $\sim m^2M_{pl}^2$ into heat. The reheating temperature is $T_R \sim (m\ M_{pl})^{\frac{1}{2}}$.

Such a classical description is possible only for $|T_{\mu\nu}| < M_{pl}^4$, hence $\partial_\mu\phi\partial^\mu\phi \leq M_{pl}^4$ or $\frac{1}{2}m^2\phi^2 \leq M_{pl}^4$ have to be satisfied. A typical initial value is thus $\phi_0 \sim M_{pl}^2/m$ (e.g. $m \sim 10^{-5}M_{pl}$); a typical size of $\mathcal{O}(M_{pl}^{-1})$ is blown up to

$$L \sim M_{pl}^{-1}\exp(2\pi\ M_{pl}^{-2}\ \phi_0^2) \sim M_{pl}^{-1}\exp(2\pi\ m^{-2}\ M_{pl}^2).$$

With $m \sim 10^{-5}M_{pl}$

$$L \sim M_{pl}^{-1}\exp(10^{11}) \geq 10^{10^{10}}\text{cm!} \tag{2.44}$$

(F) Some Puzzles of Inflation Models. The simplicity of the inflationary universe concept is somewhat deceptive, since fundamental problems creep up, when its basis in particle physics and in general relativity is considered.

i) Particle Physics Input. The problem in particle physics is to find a scalar field in a fundamental theory with the required properties. Detailed models, so far, have made use of the concept of a phase transition connected with the spontaneous breaking of a symmetry. The scalar field is identified as the Higgs field ϕ with a quartic selfinteraction. In gauge field theories there are problems with such a semiclassical description. In several models, calculated on a lattice, the Higgs phase (with a nonzero expectation value of ϕ) does not occur [47].

In equilibrium $V_{eff}(\phi)$ must be convex, and thus $dV_{eff}/d\phi = 0$ in the region of interest. Therefore the equation of motion of ϕ cannot be defined as in Eq. (2.39).

Linde's chaotic model is only subject to this criticism, when an underlying quantum theory is considered. This has not been done up to now. Another undesirable feature of the chaotic model is the fact that it should work close to the Planck time, when the use of a classical spacetime geometry can no longer be justified.

ii) General Relativity Input. The claim of the inflationary models to explain the homogeneity and isotropy of the universe without resorting to specific initial conditions must be substantiated by quantitative investigations. Anisotropic and inhomogeneous models could be used as a first step, instead of the Friedmann equation commonly employed. It will be interesting to see whether chaotic initial conditions exist which prohibit inflation. A more fundamental problem is the description of solutions of Einstein's equations with a complicated, inhomogeneous phase structure.

Very little has been done in this respect (for spherical inhomogeneities [48]; for anisotropic models [49]).

(G) Some Remarks on Classical Explanations of Homogeneity and Isotropy. While most of the questions of inflationary universe models must be answered by elementary particle theories, the aspects of homogeneity and isotropy have also been investigated along other paths. The "mixmaster universe" [50] proposes to smooth out a chaotic initial state by dissipative effects (e.g. neutrino viscosity, hadron collisions, particle creation), the aim is to produce the observed entropy per baryon of $\sim 10^{9\pm1}$ and the presently observed isotropy. This program suffers from the difficulty that there are specific initial anisotropies which are not smoothed out by the dissipation mechanisms.

A closer look at the thermodynamic system which is presented by the expanding, time-asymmetric universe may lead one to conclude that all these attempts are quite unsatisfactory [51]. Penrose argues against an initial chaos and in favour of a uniform big bang singularity. The initial state of the expanding universe should have a lower entropy than the final one. Penrose points out that the entropy per baryon of $10^{9\pm1}$ is not a large number when gravitational entropy is taken into consideration.

To get an idea of the entropy of gravitation we may consider the entropy of a black hole. There is a formula which sets the intrinsic entropy S proportional to the surface area A of a black hole's absolute event horizon

$$S/k_B = A\ c^3/(4\ h\ G) \tag{2.45}$$

($A = 16\pi G^2 M^2/c^4$ for a Schwarzschild black hole).

Using this formula we can easily find out that a black hole of $1M_\odot$, formed from baryons, has an entropy per baryon of $\sim 10^{18}$. The central part of a galaxy of $10^{10}M_\odot$, therefore, collapsing to a huge central black hole, leads to a value of 10^{28}. (S is proportional to M^2, hence to N_B^2).

Finally a closed universe containing 10^{80} baryons which undergoes collapse by a mutual swallowing process of black holes reaches an entropy per baryon of $\sim 10^{40}$. This large value is mainly produced by the irregularities, the ripples and wrinkles, of spacetime. The initial entropy per baryon could well have been close to $\sim 10^{40}$ had the universe started out in a completely chaotic state. The problem would be then to explain the absurdly low value of 10^9 observed now. Why is only a fraction $\sim 10^{-31}$ of the available entropy used? This enormous discrepancy seems to indicate that the explanation for $n_B \sim 10^9$ does not lie in the properties of the gravitational interaction. It seems more reasonable to look for an explanation in terms of particle physics.

Penrose concludes that these values suggest a complete lack of chaos in the initial geometry. A low entropy constraint must be imposed on the initial state. But the matter and radiation in the early universe apparently were in thermal equilibrium, so the matter distribution cannot account for the low entropy. Rather a very special initial space-time geometry must be assumed. A precise local condition can be formulated which expresses this idea. But to explain this local condition an appeal must be made to a more fundamental theory unifying quantum mechanics and general relativity.

3. Formation of Large-Scale Structure in the Universe

This topic is discussed by RUFFINI and TUROK. Therefore I shall confine myself to few introductory remarks.

3.1 Problems of the Gravitational Instability Picture

A natural way to explain the formation of galaxies (or clusters) is to assume that initially small density perturbations grow and collapse finally due to gravitational instability. This idea meets the following difficulty:
In linear approximation small density contrasts $\delta \equiv \delta\rho/\rho$ grow after recombination as $\delta \propto t^{\frac{2}{3}}$. At the present time

$$\delta(t_0) = (1 + z_R)\delta(t_R) \ , \tag{3.1}$$

where

$$z_R \sim 10^3 \text{ for } \Omega_0 = 1$$
$$z_R \sim 10^2 \text{ for } \Omega_0 \ll 1.$$

Initial perturbations $\delta(t_R)$ can thus increase their value by factors 100 to 1000, depending on Ω_0. The initial amplitudes are limited, on the other hand, by the high degree of isotropy of the 3 K background. Observations of the temperature in different directions $\boldsymbol{n}$ give

$$T(\boldsymbol{n}) = T_0(1 + \Delta T(\boldsymbol{n})) \ .$$

Possible variations are analysed in terms of the autocovariance function

$$C(\theta) \equiv \langle \Delta T(\boldsymbol{n}) \Delta T(\boldsymbol{n}') \rangle \tag{3.2}$$

(here θ is the angle between $\boldsymbol{n}$ and $\boldsymbol{n}'$: $\cos\theta = \boldsymbol{n} \cdot \boldsymbol{n}'$). In terms of spherical harmonics on the celestial sphere we have for $C(\theta)$ and $T(\boldsymbol{n})$:

$$\begin{aligned} T(\boldsymbol{n}) &\equiv T_0 \Sigma\, a_{lm}\, Y_{lm}(\boldsymbol{n}), \\ C(\theta) &\equiv \Sigma(2\ell + 1)c_l\, P_l(\cos\theta). \end{aligned} \tag{3.3}$$

For Gaussian random fields

$$(2\ell + 1)c_l = \sum_{-l}^{+l} |a_l^m|^2.$$

The sky distribution is folded with the resolution function of the telescope; a typical Gaussian shape with a characteristic angular fall off θ_s is:

$$\frac{1}{2\pi\theta_s^2} \exp\left(-|\boldsymbol{n} - \boldsymbol{n}'|^2/2\theta_s^2\right) .$$

The resulting effective sky distribution is then (after subtraction of the monopole $\ell = 0$ and dipole $\ell = 1$ contribution) compared to stochastic data with e.g. a maximum likelihood method [52]. Best upper limits at various angular scales are derived this way.

$$\frac{\Delta T}{T} \leq 4 \times 10^{-5} \quad \text{at} \quad 3'; \qquad [53]$$
$$\frac{\Delta T}{T} \leq 1.5 \times 10^{-5} \quad \text{at} \quad 7'; \qquad [54]$$
$$\frac{\Delta T}{T} \leq 3.7 \times 10^{-5} \quad \text{at} \quad 8^0. \qquad [55]$$

A mass M at recombination time subtends an angle

$$\theta = 40''(h\ \Omega_0)\left(\frac{M}{10^{11}M_\odot}\right)^{\frac{1}{3}} \tag{3.4}$$

i.e. these limits correspond to limits on the density fluctuations on cluster scales. Indeed for adiabatic perturbations $s = (4a/3k_B)(T_\gamma^3/n_B)$ stays constant, and therefore

$$\frac{\delta n}{n} = -3\frac{\delta T_\gamma}{T_\gamma}.$$

Limits on $\delta n/n$ are obtained from the 3 K background isotropy:

$$\text{e.g.} \quad \frac{\delta n}{n} < 5 \times 10^{-5} \ \text{at } 7' \tag{3.5}$$

hence on the scale of $7'$

$$\delta(t_0) \leq 5 \times 10^{-2} \quad \text{for} \quad \Omega_0 = 1,$$
$$\delta(t_0) \leq 5 \times 10^{-3} \quad \text{for} \quad \Omega_0 \ll 1.$$

(From light element formation in the hot big bang there is the limit $\Omega_{B,0} \leq 0.2$).

There is not enough growth in the small perturbations! The universe stays homogeneous, and therefore the simple model of galaxy formation starting from adiabatic perturbations has to be abandoned. Various loopholes are discussed in the literature.

Basically there seem to be three different ways out:

i) isothermal fluctuations;
ii) a period of reionization;
iii) a background of nonbaryonic dark matter.

Let us comment on each one in turn:

i) Isothermal fluctuations require $\delta T = 0$ by some (more or less artificial) constraint on the fluctuations. No reasonable physical model for such a constraint has been found so far. Recently it has been shown that isothermal fluctuations start to grow only when adiabatic fluctuations of the same magnitude have developed [56]. Thus isothermal fluctuations can no longer be considered a reasonable loophole.

ii) Scattering processes at an epoch of reionization at redshift z_i (perhaps the epoch of the first formation of radiating objects) would smear out δT_γ on small angular scales. The horizon scale is an upper bound for this effect: The angular extent of the horizon length is $\theta_H \simeq 30^0$. On angular scales larger than θ_H the fluctuations are not affected by the reionization epoch. Limits on $\delta T_\gamma/T_\gamma$ at $\theta \geq \theta_H$ provide constraints on this idea, but there is an even more drastic constraint: It has been found that new fluctuations are caused at the reionization epoch z_i by the bulk motion of the collapsing matter [57]. The cross-terms $\delta \cdot \boldsymbol{v}$ of a perturbation analysis are formally small of second order (δ is the density contrast, $\boldsymbol{v}$ the velocity disturbance in linear approximation), but effectively they give rise to fluctuations of the radiation field on arcminute scales as large as the original ones. Therefore Vishniac's calculations – if correct – exclude the second way out.

iii) The effect of nonbaryonic dark matter is the following: Fluctuations do not couple strongly to the photon field (only by the gravitational potential is the photon propagation affected – Sachs-Wolfe effect), and can therefore have larger amplitudes at the decoupling time. The obvious constraints on the baryonic model are removed. At present there is, however, no experimental evidence for a good candidate particle or object. RUFFINI and TUROK will talk about different models, and I will point out a few basic concepts in the following. Let me end this Section with a quotation from Sherlock Holmes:

"You will not apply my precept", he said, shaking his head. "How often have I said to you that when you have eliminated the impossible, whatever remains, however improbable, must be the truth?" (The Sign of Four).

3.2 Basic Information from Observations

The new catalogues of galaxies (e.g. the CfA catalogue [58]) contain already complete samples of objects (down to some limiting magnitude) with position and redshift measured. Therefore in certain regions of the sky a 3-dimensional picture of the galaxy distribution can be constructed by using the Hubble law to obtain the radial distances from the redshift. The transition from redshift space to real space may depend on the model used for the matter distribution (see RUFFINI), peculiar velocities may also disturb the picture. Despite these cautions, there is much evidence already for interesting structures in the galaxy distribution.

In a small angular strip in the Coma cluster [59] a distribution is found with large empty regions (voids), with galaxies seemingly concentrated on the surface of spherical bubbles. (You can see this picture in the contributions by RUFFINI and TUROK.)

Another impressive structure was discovered by Tully [23] (Fig. 15). This is a superstructure connecting 3 superclusters of extent ~500 Mpc and with a mass of $10^{18} M_\odot$. The largest object seen so far. Remember that its extent is already $\sim 10\%$ of the typical size $2c/H_0$.

How can such structures be formed in a model theory? At present this question cannot be answered. Strong emphasis now is on simple statistical quantities which must be reproduced in models of galaxy formation. One of

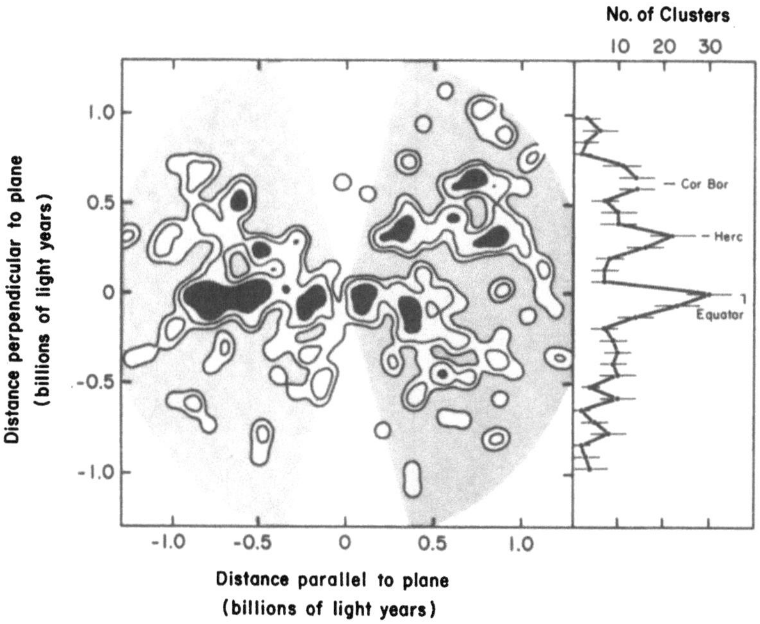

Fig. 15. The superstructure found by Tully with a nearest neighbour statistical analysis. The object connects the local supercluster, Hercules, Pisces - Cetussuperclusters. It has a flattened shape with the orientation parallel to the small local system. 48 Abell clusters are contained within this object, about $10^{18} M_{\odot}$. Typical extent is 500 Mpc.

the main tools is the analysis of 2-point correlations [26]. The number of pairs within distance r of each other in a given section of space is counted (actually one determines the number of objects in a spherical shell between radii r and $r + \Delta r$ around a given object, then averages over the reference object) as $n_s(r)$. Then this number is computed for a random distribution within the same volume and for the same number of points. The average value $\langle n_r(r) \rangle$ for the random distribution is determined. The function

$$\xi(r) \equiv (n_s(r)/\langle n_r(r) \rangle) - 1 \tag{3.6}$$

is called the "2-point correlation function". It measures the deviations of the actual distribution from a random one.

For the average galaxy distribution [26] $\xi(r)$ shows a power law behaviour

$$\begin{aligned} &\xi(r) = (r_0/r)^{\gamma} && \text{for } 0.1\text{Mpc} \leq hr \leq 9\text{Mpc} \\ &\gamma = 1.77 \pm 0.04 && r_0 = (4.3 \pm 0.3)\text{h}^{-1}\text{Mpc}. \end{aligned} \tag{3.7}$$

Recent investigations have indicated modifications on large scales (stronger correlations for clusters) as well as for different types of galaxies. In Fig. 16 one example [60] is shown, in which galaxies from the CfA catalogue have been divided into 4 different classes which correspond to 4 intervals in intrinsic luminosity. The strongest correlations are found for the high luminosity objects with a power law of $\gamma = 1.38$. Such results may favour a scheme of galaxy

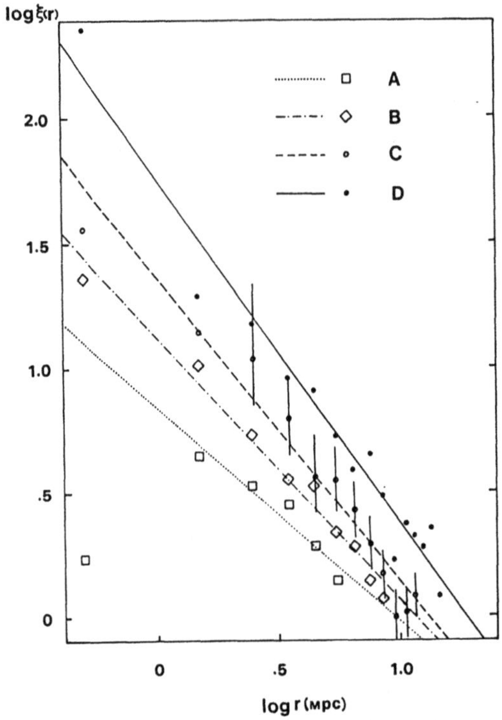

Fig. 16. The spatial two-point correlation function for 4 subsamples of the CfA catalogue. The galaxies are divided into intervals of absolute magnitude:

$$A: -18.5 \geq M \geq -19.5;$$
$$B: -19.5 \geq M \geq -20.5;$$
$$C: -20.5 \geq M \geq -21.5;$$
$$D: -21.5 \geq M \geq -22.5.$$

Typical $\pm\sigma$ error bars as derived from a randomly generated distribution are given. Straight lines are from linear regression analysis.

formation, where the environment (density enhancement) plays an important part.

3.3 Typical Scales

(see e.g. [61])

Nonbaryonic dark matter can be 'hot' or 'cold': Hot DM consists of light particles (such as a massive neutrino $m_\nu \leq 100$ eV) which are still relativistic when they decouple. Apart from statistical factorrs they are therefore as abundant as photons, $n \sim n_\gamma$.

Cold DM is made up of constituents which decouple very early or have small random velocities, such that they are nonrelativistic when they become important gravitationally.

(A) The 'Jeans' Mass. Collisionless nonbaryonic matter can resist the pull of gravity by its velocity dispersion. The balance of gravitation and effective pressure $p \sim \frac{1}{3}\langle v^2 \rangle$ defines the 'Jeans' mass (similar to the baryonic case). For a massive neutrino with mass m_ν, e.g. $M_{J,\nu} \sim M_{pl}^3 m_\nu^{-2}$ (M_{pl} is the Planck mass $\sim 10^{19}\,\mathrm{GeV}$).

Numerically

$$M_{J,\nu} = 3.2 \times 10^{15} \left(\frac{m_\nu}{30\,\mathrm{eV}}\right)^{-2} M_\odot . \tag{3.8}$$

(B) The Horizon Scale. An important quantity is the mass within the horizon ($\lambda_H = ct$) when the energy density in relativistic particles equals the energy density in nonrelativistic particles.

$$(1 + z_{eq}) = \frac{\Omega_{x,0}}{\Omega_{\gamma,0}} = 4.2 \times 10^4 (\Omega_{x,0} h_0^2)^{-2} \left(\frac{T_{\gamma 0}}{2.7}\right)^{-4} . \tag{3.9}$$

The corresponding mass

$$M_{Heq,x} = 2.2 \times 10^{15} (\Omega_{x,0} h_0^2)^{-2} M_\odot . \tag{3.10}$$

Scales with $M_J \leq M \leq M_{Heq}$ can grow. For hot DM one has $M_{Heq} \sim M_J$, for cold DM $M_J \ll M_{Heq}$.

(C) Damping Scales. Relativistic particles can remove density perturbations by travelling freely away from a region of compression, and also by the directional dispersion of velocities. This has the consequence that the first scale on which a perturbation can grow is the horizon scale at the epoch when the particles become nonrelativistic.

This free-streaming scale

$$\lambda_{FS} = ct(T = m_x) \approx (G\rho)^{-\frac{1}{2}} \sim M_{pl} m_x^{-2} g^{-\frac{1}{2}} \tag{3.11}$$

corresponds to a damping mass

$$M_{FS} \approx \rho \lambda_{FS}^3 \approx M_{pl}^{-2} g^{-\frac{1}{2}} .$$

For hot DM $M_{FS} \sim M_J$, as well as $M_J \sim M_{Heq}$, i.e. regardless of the initial spectrum, there is only one characteristic scale which gives growing fluctuations. Models of galaxy formation with massive neutrinos, e.g., have a characteristic length scale corresponding to the size of clusters. For cold DM the free streaming scale is of no importance.

Another damping scale is related to the damping of sound waves. The Jeans mass (Fig. 17) drops rapidly to small values after recombination. Fluctuations which oscillate as sound waves before can be above the Jeans mass after recombination and start to grow. Due to the friction between photons and matter there is, however, damping of these oscillations (Silk damping). The scale can be estimated quite simply from a comparison of the photon diffusion time for a length λ

$$t_{diff} = \frac{\lambda^2}{c^2} \tau_\gamma^{-1}$$

(τ_γ is the mean free time for Thomson scattering) with the expansion time $t_{exp} \approx t$. Damping is possible if $t_{diff} \leq t_{exp}$, i.e.

$$\lambda \leq (\ell_\gamma \ell_H)^{\frac{1}{2}} \equiv \ell_s \quad (\ell_\gamma \equiv c\tau_\gamma; \ell_H \equiv ct). \tag{3.12}$$

The Silk damping mass $M_s \sim \ell_s^3 \rho$ for the baryon dominated epoch ($t \propto \rho^{-\frac{1}{2}}$, $n_e \propto \rho$) is

$$M_s = 3 \times 10^{12} (\Omega_0 h_0^2)^{-\frac{5}{4}} M_\odot \,. \tag{3.13}$$

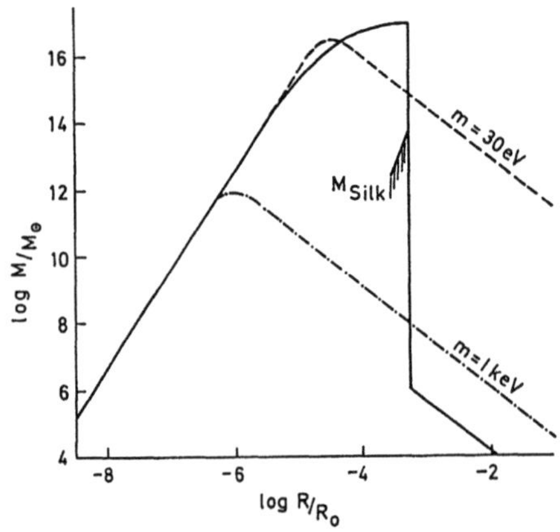

Fig. 17. The Jeans mass for baryonic matter, and for a nonbaryonic collisionless component ("hot" (m = 30eV) and "warm" (m = 1keV) DM). The steep fall-off at the time of recombination occurs for baryonic matter, because its sound speed changes drastically after decoupling of the photons. The damping mass M_{Silk} is also given.

(D) Conclusion. These characteristic scales are interesting constraints for models of galaxy formation, but they do not really explain why systems of a typical size appear, and why they are correlated in a typical way.

Numerical simulations with hot DM have the difficulty that the large scale structures collapse first, but too close to the present epoch, and thus there is not enough time to form galaxies by fragmentation. There is the hope that a more detailed investigation of nonlinear effects involved in pancake formation will show a way to overcome this difficulty.

In cold DM simulations the problem occurs to model the voids and the observed large scale structure. Here an appeal has been made to the so called model of biased galaxy formation, where – for some mysterious reason – galaxies only form in peaks of the density enhancement above a certain threshold. In such a model the formation of galaxies depends upon the environment. One could perhaps imagine that the luminous galaxies form preferentially in big clusters, and the fainter ones outside. Our recent analysis [60] seems to give support to such a picture.

A fundamental difficulty of all these models is their inability to derive a primeval spectrum of fluctuations. This problem has, in my opinion, not even seen an attractive suggestion of a solution. TUROK will say more about this.

3.4 The "Pancake Scenario" of Zel'dovich

There is a model by Zel'dovich ([62], [63]) which describes in an approximate, analytic way some features of the collapse of collisionless matter. It turns out that an approximate solution to the Newtonian equations of gravity can be found by considering the integral curves of the particle trajectories in Lagrange space.

The relation between Euler (x_i) and Lagrange (ξ_i) coordinates ($i = 1,2,3$) is

$$x_i = R(t)\xi_i + b(t)\frac{\partial\psi}{\partial\xi_i} \tag{3.14}$$

$R(t)$ is the expansion factor, and $b(t)$ is obtained from the linear evolution equations $b(t) \sim t^{\frac{4}{3}}$.

The function $\psi(\xi)$ describes the initial perturbation spectrum of the particle motion. Mass conservation

$$dm = \rho dx_1 dx_2 dx_3 = \rho_0 \; d\xi_1 d\xi_2 d\xi_3$$

leads to the relation

$$\rho = \rho_0 \left\{\det\left(\frac{dx_i}{d\xi_k}\right)\right\}^{-1} \tag{3.15}$$

between the homogeneous background density ρ_0, and the real density. Clearly something drastic must happen at the zeros of the determinant! Transformation to principal axes gives eigenvalues α_i ($i = 1,2,3$) with

$$\rho = \rho_0 \left(1 - \frac{b}{R}\alpha_1\right)^{-1} \left(1 - \frac{b}{R}\alpha_2\right)^{-1} \left(1 - \frac{b}{R}\alpha_3\right)^{-1} . \tag{3.16}$$

Let $\alpha_1 > \alpha_2 > \alpha_3$, then ρ becomes infinite for $(b/R)\alpha_1 \to 1$!

This approximation by Zel'dovich describes the formation of regions of very high density. The generic collapse seems to be preferentially one dimensional, i.e. into a plane. Therefore the name "pancake" scenario was coined for the model.

Fig. 18 compares a numerical simulation with a plot of the Zel'dovich solution. Which is which?

The mathematical background of this structure formation is the formation of caustics in particle trajectories (described and classified in the context of "catastrophe theory" by Arnold). We have been working on some aspects of this model for some time. Exact analytic solutions for the pancake picture have been found ([64], [65]). High resolution numerical simulations of particle trajectories have also been done ([65], [66]).

Fig. 19 shows a 2-dimensional simulation, where voids (black regions), the linear regime (faint), and high-density objects (strongly white) can be seen. Pancake structures penetrating each other can be discerned. The singularity classification given by Arnold can be applied to the complicated structures at the knots in this picture.

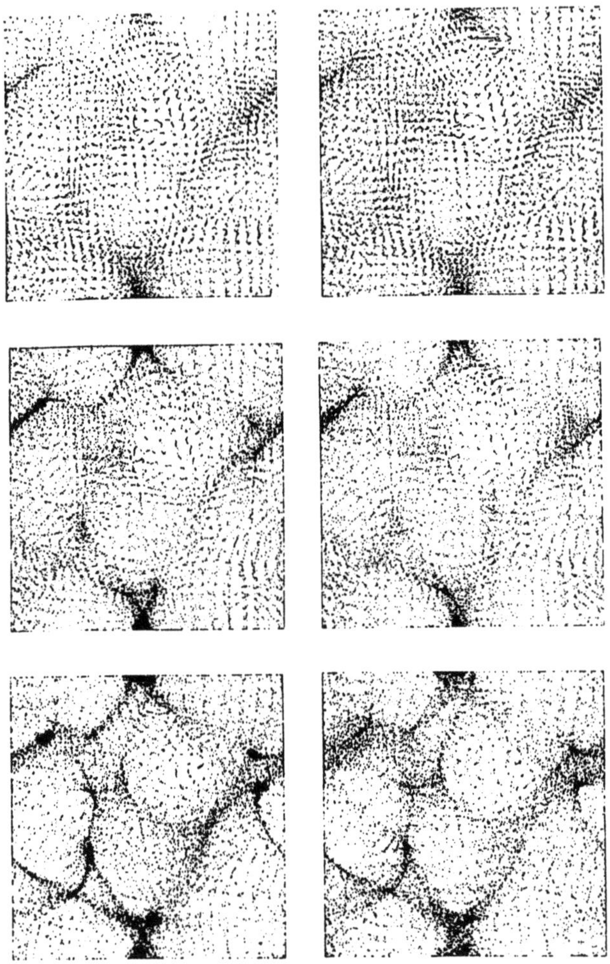

Fig. 18. Plot of a numerical solution and of the Zel'dovich solution for the same initial condition. Time increases from top to bottom. Which is which?

Fig. 20 gives an impression of a 3-dimensional simulation of particle trajectories. Pancakes have been determined by the magnitude of the eigenvalues of (det $\frac{dx_i}{d\xi_k}$). The function $\psi(\boldsymbol{\xi})$ is always taken as

$$\psi(\boldsymbol{\xi}) = \sum A(\boldsymbol{K}) \sin(\boldsymbol{K} \cdot \boldsymbol{\xi}) + B(\boldsymbol{K}) \cos(\boldsymbol{K} \cdot \boldsymbol{\xi})$$

with random numbers $A(\boldsymbol{K})$ and $B(\boldsymbol{K})$.

So these simulations give a general impression of nonlinear effects, but not yet a realistic scenario that can be compared to the observations. To do this, we shall have to implement initial conditions derived from the linear perturbation theory of a hot or cold DM model.

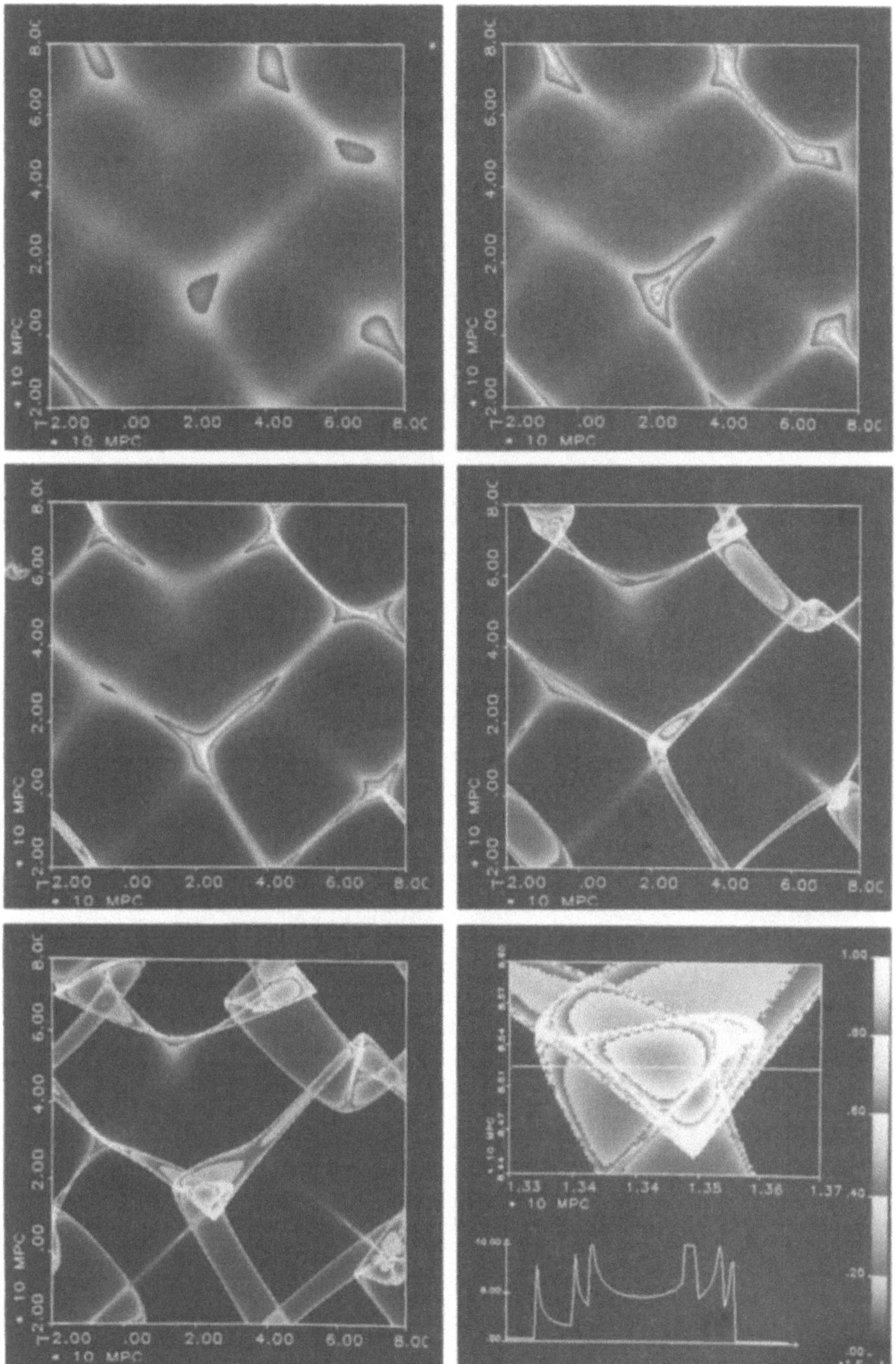

Fig. 19. High-resolution 2-dimensional simulation of Zel'dovich solution [65]. Different shades from black to white give the range from "voids" ($\frac{1}{5}$ of the mean density) to high density "caustics". Interesting singularity structures can be seen (e.g. pancakes penetrating each other). Initial data are Fourier spectra with Gaussian amplitudes. Density and velocity perturbations are prescribed independently with 25 modes each. Initial density amplitude $= 10^{-3}$, initial redshift 1000; $\Omega = 1$. 10^8 particle trajectories are collected in $(512)^2$ surface elements of a Eulerian plane $z =$ const. 5 pictures at successive redshifts $z = 10, 6, 4, 2, 1$. The last picture

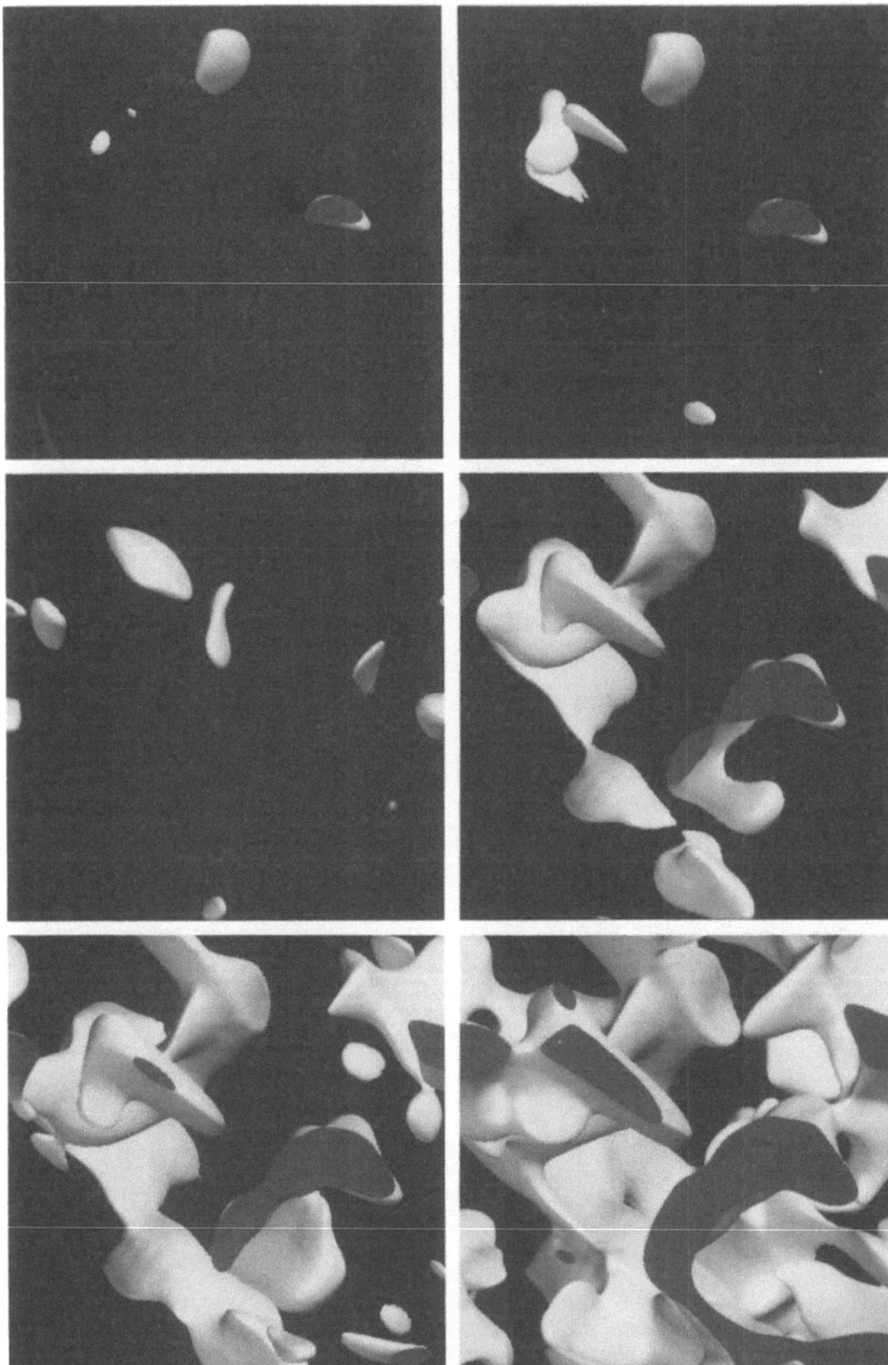

Fig. 20. 3-dimensional simulations of the Zel'dovich solution [66]. The pictures are projections of contours in Lagrange space. From left to right and top to bottom the surfaces surround regions with successively smaller positive eigenvalues of the matrix $\frac{\partial \psi}{\partial \xi_i \xi_k}$. Surfaces with smaller eigenvalues enclose those belonging to larger ones. Pancake structures develop inside these contours. Top right and middle left show two different views at right angles of the same structure.

References

1. Elsässer H., Scheffler H., 1974 "Physik der Sterne und der Sonne" (Bibl. Inst.)
2. Ritter H., 1987, Vorlesungsmanuskript "Sternentwicklung" (Inst. f. Astron. u. Astrophys., Univ. München, Scheinerstr. 1, D–8000 München 80)
3. Maeder A., 1987, ESO workshop on SN1987A (I.J. Danziger ed.)
4. Hillbrandt W., 1983, Enzykl. Naturw. und Technik (Verl. Mod. Ind. 1983), p. 81
5. Hillebrandt W., Höflich P., Truran J.W., Weiss A., 1987, Nature 327, 597
6. Woosley S.E., Pinto P.A., Ensman L., 1987, Ap.J. (in press)
   Chiosi C., Maeder A., 1986, Ann. Rev. Astron. Astrophys. 24, 329
7. Arnett W.D., Bowers R.L., 1977, Ap. J. Suppl. 33, 415
8. Witten E., 1984, Phys. Rev. D30, 272
   Farhi E., Jaffe R.L., 1984, Phys. Rev. D30, 2379
9. Alcock C., Farhi E., Olinto A., 1986, Ap. J. 310, 261
10. Kaplan D.B., 1985, Nucl. Phys. B260, 215
    Srednicki M., 1985, Nucl. Phys. B260, 689
11. Raffelt G., Dearborn D., 1987, Phys. Rev. D36, 2211
12. Börner G., 1980, Phys. Rep. 60, 151
    Hayakawa S., 1985, Phys. Rep. 121, 317
13. Taylor J.H., Weisberg J.M., 1984, Phys. Rev. Lett. 52, 1348
14. Rawley L.A., Taylor J.H., Davis M.M., Allan D.W., 1987, preprint
15. Lamb R.C., Weekes T.C., 1988, Science 238, 1528
16. Hubble E.P., 1929, Proc. NAS 15, 168
17. Penzias A.A., Wilson R.W., 1965, Ap. J. 142, 419
18. Friedmann A., 1922, Zs. f. Physik 10, 377
    Lemaître A., 1931, M.N.R.A.S. 91, 490
19. Hawking S.W., Ellis G.F.R., 1973, "The Large scale structure of spacetime" (Cambridge Univ. Press)
20. Sandage A., Tammann G.A., 1982, Ap. J. 207, L1
21. Sandage A., Tammann G.A., 1976, Ap. J. 256, 339
22. DeVaucouleurs G., 1981, N.Y. Ac. Sci., p. 90
23. Tully R.B., 1986, Ap. J. 303, 25
24. DeLapparent V., Geller M.J., Huchra J.P., 1986, Ap. J. 302, L1
25. Zel'dovich Ya.B., Einasto J., Shandarin S.F., 1982, Nature 300, 407
26. Peebles P.J.E., 1971, "Physical Cosmology" (Princeton Univ. Press),
    Peebles P.J.E., 1980, "The Large Scale Structure of the Universe" (Princeton Univ. Press)
27. Matsumoto T. et al, 1988, (Ap. J. in press)
    Woody D.P., Richards P.L., 1979, Phys. Rev. Lett. 42, 925
28. Weiss R., 1980, Ann. Rev. Astron. Astrophys. 18, 489
29. Sunyaev R.A., Zel'dovich Ya.B., 1980, Ann. Rev. Astron. Astrophys. 18, 537
30. Melchiorri F., Moreno G., Dall'Oglio G., Melchiorri B., 1986, Proc. 2nd ESO-CERN Symposion, (ed. by G. Setti, L. van Hovel)
31. Smoot G., Gorenstein F., Muller R.A., 1977, Phys. Rev. Lett. 39, 898
32. Lubin P.M., Epstein G.L., Smoot G.F., 1983, Phys. Rev. Lett. 50, 616
33. Jüttner F., 1928, Zs. f. Physik 47, 542
34. Weinberg S., 1972, "Gravitation and Cosmology" (Wiley, New York)
35. Steigman G., 1979, Ann. Rev. Nucl. Part. Sci. 29,313
36. Hayashi C., 1950, Prog. Theor. Phys. 5, 224
37. Yang J., Schramm D.N., Steigman G., Rood R.T., 1979, Ap. J. 227, 697
38. Boesgaard A.M., Steigman G., 1985, Ann. Rev. Astr. Astrophys. 23, 319
39. Rowan-Robinson M., Walker D., Yahil A., 1986, Ap. J. 301, L1
40. Sakharov A.D., 1967, Zh. Eksp. Teor. Fiz. Pisma 5, 32
41. Kolb E.W., Turner S.M., 1983, Ann. Rev. Nucl. Part. Sci. (33., 645
42. Fry J.N., Olive K.A., Turner S.M., 1980, Phys. Rev. Lett. 45, 2074
43. Ringwaldt A., 1988, Heidelberg Univ. preprint THEP 87-20
    Kuzmin V.A., Rubakov V.A., Shaposhnikov M.E., 1985, Phys. Lett. B155, 36
44. Brandenberger R.H., 1985, Rev. Mod. Phys. 57, 1
45. Jaffe A., Taubes C., 1980, "Vortices and Monopoles" (Birkhäuser)

46. Linde A.D., 1985, Comments Astrophys. 16, 229
47. Kennedy T., King C., 1985, Phys. Rev. Lett. 55, 776
Borgs C., Nill F., 1986, Phys. Lett. B171, 289
48. Guth H.A., 1981, Phys. Rev. D23, 347
Sato K., 1981, M.N.R.A.S. 195, 198, 487
49. Börner G., Götz G., 1988, in preparation
Barrow J.D., Turner M.S., 1982, Nature 292, 35
50. Misner C.W., 1968, Ap. J. 151, 431
51. Penrose R., 1979, "General Relativity" p. 581. An Einstein Centenary Survey (Hawking S.W., Israel W. (eds.)) (Cambridge Univ. Press)
52. Bond J.R., Efstathiou G., 1987, M.N.R.A.S. 226, 655
53. Uson J., Wilkinson D.T., 1984, Nature 312, 427
54. Readhead et al, 1988, Proc. IAU Symp. #130 (Reidel, in press) (J. Audouze, A. Szalay (eds.))
55. Davies R.D. et al, 1987, Nature 326, 462
56. Gouda N., Sasaki M., 1986, RIFP-preprint
57. Vishniac E.T., 1988, Proc. IAU Symp. #130
58. Huchra J.P. et al, 1983, Ap. J. Suppl. 53, 89
59. De Lapparent V., Geller M.J., Huchra J.P., 1986, Ap. J. 302, L1
60. Börner G., Deng Z., Xia X.Y., Zhou Y.Y., 1988, Astron. Astrophys. (to be published)
61. Efstathiou G., Silk J., 1983, Fund. Cosmic Phys. 9, 1
62. Zel'dovich Ya.B., 1970, Astron. Astrophys. 5, 84
63. Shandarin S.F., Zel'dovich Ya.B., 1988, MPA-preprint 350
64. Buchert T., Götz G., 1987, J. Math. Phys. 28, 2714
65. Buchert T., 1988, Ph. D. Thesis (MPI für Astrophysik – Garching)
66. Klaffl R., 1988, Diploma Thesis (in prep. – MPI für Astrophysik – Garching)

This article was processed by the author using the TEX Macropackage from Springer-Verlag.

Physics of Type II Supernova Explosions

N. Straumann

Institut für Theor. Physik, Universität Zürich, Schönberggasse 9,
CH–8001 Zürich, Switzerland

I. Introduction and Historical Remarks

Supernova explosions are so energetic that one of the following two basic mechanisms must be responsible for each such event: Either a star-size fusion bomb explodes more or less as a whole, or the central core of a massive star in an advanced evolutionary phase becomes unstable and collapses to nuclear densities, thereby releasing an enormous gravitational energy. In both cases the physics is clearly very complicated in detail and our present models are comparably still rather primitive. It will take generations of astrophysicists to come, in order to develop a satisfactory detailed understanding. However, some important aspects of the physics of supernovas are qualitatively and even semi-quantitatively understood.

Apart from very few remarks about nuclear detonation models I will discuss in these lectures only the core collapse scenario. The existence of the Chandrasekhar limit implies that sufficiently massive stars must sooner or later run into an instability – even if they loose a substantial fraction of their mass in the course of their evolution – and must eventually collapse catastrophically. A small fraction of the released gravitational energy can account for the observed energy of even the brightest supernova explosion.

It was truly a historical moment when a cosmic firework of this kind in the Large Magellanic Cloud became suddenly visible roughly a year ago. Perhaps the most remarkable aspect of this event was, so far, the detection of a neutrino pulse in underground laboratories, which provided a direct test of our theoretical ideas on stellar collapse and neutron star formation. The impact of this supernova on astrophysics and particle physics is already enormous. Other exciting observations are ahead of us. It is, for example, likely that a very fast pulsar will show up within a year or so.

Since this is a school of a fairly wide breath of interest I will proceed in a smooth fashion, without assuming specialized astrophysical knowledge. Hopefully, the lectures will be useful for nonexperts, in particular for high energy physicist (the majority at this school). I shall emphasize qualitative considerations and estimates based on simplified models. This should help you to digest the outcome of much more detailed numerical simulations.

Before we start, I would like to add a few historical remarks.

There is an ongoing investigation of the historical records on supernovae by Chinese astronomers. An interesting status report has recently been given at a workshop on Schloss Ringberg [1]. A fascinating discussion of the historical records – in particular of the 8 "new stars" of long duration of A.D. 185, 386, 393, 1006, 1054, 1181, 1572, and 1604 – can be found in Ref. [2].

Among these, the star which blazed forth in A.D. 1006 was by far the brightest of the new stars on record and was definitely a supernova. (With the Einstein-satellite one has obtained beautiful pictures of the Crab Nebula-like remnant.) It was not only observed in China and Japan, but because of its extreme brilliance, it was also recognized in the West. In particular, the new star was also noticed in St. Gallen. In the monastic chronicals *Annales Sangallenses Maiores* one finds [2]:

1006. A new star of unusual size appeared, glittering in aspect, and dazzling (verberans) the eyes, causing alarm. In a wonderful manner this was sometimes contracted, sometimes diffused, and moreover sometimes extinguished. It was seen likewise for three months in the inmost limits of the south (in intimis finibus austri), beyond all the constellations which are seen in the sky.

For those of you who know St. Gallen, it is clear that this is an interesting observation, because the southern horizon is dominated by the Alpstein with the famous mountain Säntis (2502 m). This implies, that no star south of declination -38.5^o would be visible from the neighborhood of the town [2]. On the other hand, the description implies that the new star cannot have been five degrees north of this.

The most famous event was, as you all know, the "guest star" in 1054, witnessed by Chinese, Japanese, Korean and Arab astronomers/astrologers, whose remnant is the Crab nebula. This nebula was independently discovered by Bevis in 1731 and Messier in 1758. The name Crab was given by Lord Rosse in 1844, and the nebula was first associated with the 1054 A.D. event by Hubble in 1928. This identification was conclusively established by Duyvendak, Oort, and Mayall in 1942. The Crab nebula was also the first radio source and one of the first X-ray sources to be indentified with an optical object (apart from the sun). Since the discovery of a pulsar by Staelin and Reifenstein in 1968, we know for sure that neutron stars can be formed in supernova explosions, as was suggested by Baade and Zwicky in 1934.

Further famous examples are SN 1572 observed by Tycho Brahe, and SN 1604 which was observed and described in great detail by Kepler. The following quotation shows the surprise, when Tycho Brahe first saw the "new star":

When I had satisfied myself that no star of that kind had ever shone forth before, I was led into such perplexity by the unbelievability of the thing that I began to doubt the faith of my own eyes, and so, turning to the servants who were accompanying me, I asked them whether they too could see a certain extremely bright star when I pointed out the place directly overhead. They immediately replied with one voice that they saw it completely and that it was extremely bright.

Tycho measured the position of the new star in relation to the principal stars of Cassiopeia and finally came to the conclusion:

Hence this new star is located neither in the region of the Element, below the Moon, or in the orbits of the seven wandering stars but in the eighth sphere, among the other fixed stars.

This was against one of the basic tenets of Aristoles' concept of the Universe, namely that the eighth sphere of the fixed stars was immutable.

On August 31, 1885, Hartwig discovered a "nova" near the center of the Andromeda nebula (M 31), which was visible for about 18 months. In 1919, Lundmark estimated the distance of M 31 to be 650 000 light years and concluded from this that some stars could flare up to luminosities thousands of times larger than a normal nova. Another such event was observed in 1895 in NGC 5253 ("nova" Z Centauri), which appeared to be around 5 times brighter than the entire galaxy. The word super-novae for a class of objects fundamentally different from ordinary novae was coined by Zwicky who began in 1934 organized searches for supernovae and discovered himself 122 extragalactic events. The visionary paper [3] by Baade and Zwicky in 1934 contains a brief paragraph, appended as an "additional remark", which says:

In addition, the new problem of developing a more detailed picture of the happenings in a supernova now confronts us. With all reserve we advance the view that a supernova represents the transition of an ordinary star into a neutron star, consisting mainly of neutrons. Such a star may possess a very small radius and an extremely high density. As neutrons can be packed much more closely than ordinary nuclei and electrons, the "gravitational packing" energy in a cold neutron star may become very large, and under certain circumstances, may far exceed the ordinary nuclear packing fractions. A neutron star would therefore represent the most stable configuration of matter as such. The consequences of this hypothesis will be developed in another place, where also will be mentioned some observations that tend to support the idea of stellar bodies made up mainly of neutrons.

We know now five Crab like supernova remnants which contain unresolved central X-ray objects which have been observed to pulse at radio and/or X-ray frequencies. There can be no doubt that these are neutron stars. Another five remnants have central unresolved X-ray sources, which are probably neutron stars, but pulsations have yet to be detected [4].

II. Some Observational Facts

Until recently, two major classes of supernovae, called Type I and Type II were distinguished. (Zwicky had already created other classes.)

The spectra of classical Type I supernovae (now SN Ia) show no spectroscopic evidence for hydrogen, which implies that the progenitor stars should be hydrogen-deficient stars. Type Ia events have been observed in elliptical galaxies as well as in spiral and irregular galaxies. They are not concentrated in spiral arms. Another characteristic of Type Ia supernovae is the exponential tail in their light curves (see Fig. II.1). Such a tail requires an additional energy source at late times. Radioactive decay models ($^{56}\mathrm{Ni} \xrightarrow{6\,\mathrm{d}} {}^{56}\mathrm{Co} \xrightarrow{80\,\mathrm{d}} {}^{56}\mathrm{Fe}$) have been quite successful in fitting the observed light curves. Type Ia events are undetectable at radio wavelength, and they show expansion velocities up to four times that of Type II events.

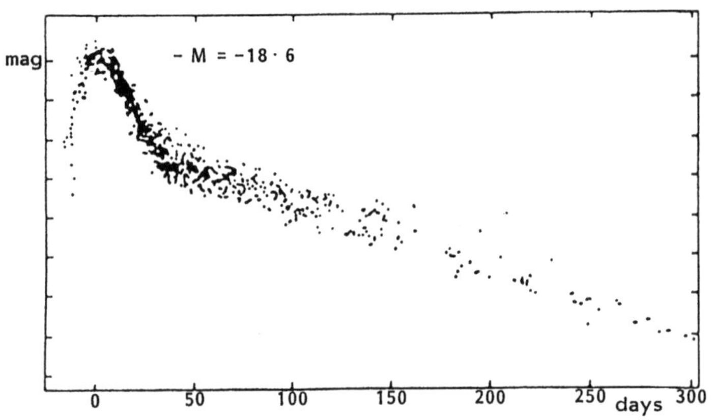

Fig. II.1. Composite blue light curves of 38 classical Type I supernovae. Time is in days from maximum light; brightness in blue magnitudes, adjusted to make peaks of all events coincide. (From Ref. [5].)

Type II supernovae appear only in the arms of spiral galaxies, usually in H II regions that represent the birth place of stars and include many young, massive stars. They are generally less bright and form a less homogeneous class (see Fig. II.2). The spectra show normal abundances, including hydrogen. Several months after the explosion they become luminous radio emitters at centimeter wavelengths.

Further observed properties of these classical types are listed in Table II.1.

All these features suggest that Type II supernovae arose from the explosion of short-lived massive stars whose cores had collapsed after they had exhausted their nuclear fuel.

Type Ia supernovae are probably powered by runaway thermonuclear reactions. Thermonuclear disruption of accreting white dwarfs in close binaries is the most popular scenario. In such a system the white dwarf draws in matter from the companion and this leads to an increase of the temperature and

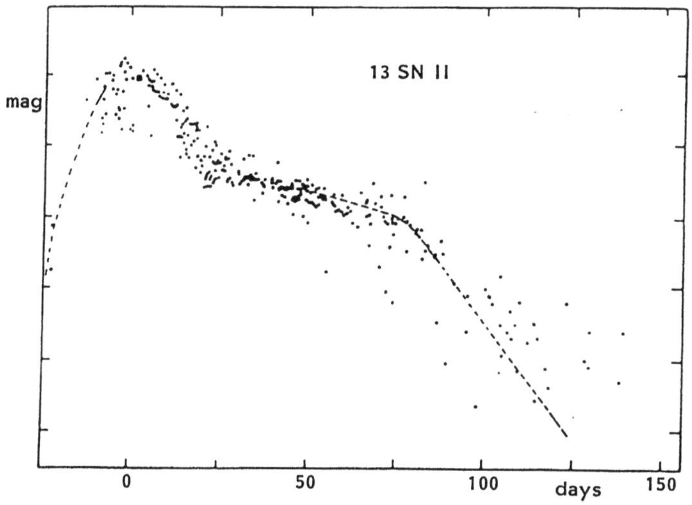

Fig. II.2. Composite light curve obtained from 13 Type II supernovae. (From Ref. [5].)

Table II.1. Some properties of supernovae

Energetics	**Light curves and spectra**		
		Type I	Type II
Maximum luminosity 10^{44} erg sec^{-1}	Near maximum light for	30 days	10 days
Energy in visual light 10^{48} – 10^{50} erg	Plateau	no	100 days
Total energy output 10^{51} – 10^{52} erg	Duration	2 years	1 year
Expansion velocity 10^4 km sec^{-1}	H-lines	no	strong
Temperature near maximum light 15 000 K	Abundances	Co, Fe ?	solar

density in the degenerate oxygen-carbon core. A carbon burning front may be triggered, which rips outward, completely destroying the star.

Several recent supernovae have complicated the classification scheme. A new class of events, now called Type Ib, has been identified, which showed no evidence of hydrogen, but were found to be radio emitters and had peak luminosities a factor roughly five below the nominal value of classical Type I supernovae. Furthermore, nearly all of these events occurred near star-forming regions, suggesting massive progenitor stars. It was also found that the expanding gases contain helium, an element never definitely detected in the spectrum of classical Type I and II supernovae. (For a very readable account, see Ref. [6].)

At the same time, evidence accumulated for a distinction, based on light-curve morphology, between two subclasses of Type II (II-p and II-ℓ). At least one Type II event showed exponential decay with a Type I's 55-day halflife. The light curves of the various types are shown in Fig. II.3.

It has now become clear that the physics of supernovae is only loosely connected to spectral classification schemes. Both mechanisms – runaway thermonuclear reactions and release of gravitational energy by an imploding core – can occur either in a star with a hydrogen-rich envelope or in one that has lost its envelope. A massive star may loose its outer envelope, via a strong stellar wind or through accretion onto a binary companion. It is now believed that Type Ib events originate from massive stars that have lost their hydrogen envelopes. The binary X-ray sources may perhaps be relics of Type Ib supernovae. SN 1987 A has added to this complexity (Fig. II.4). One gets the impression that the existing classification scheme is not really of great importance.

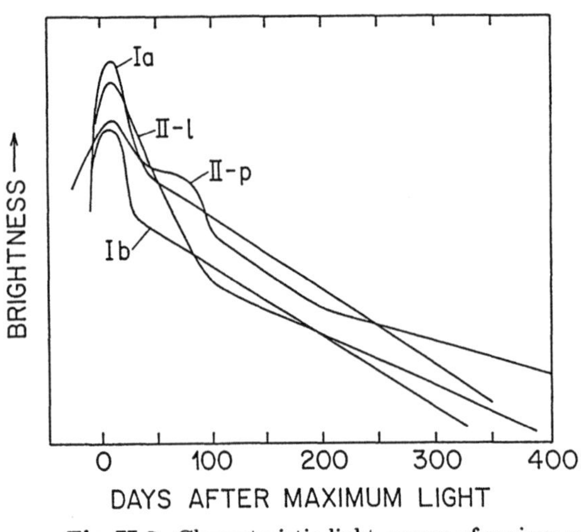

Fig. II.3. Characteristic light curves of various supernova types. (From Ref. [6].)

Let me add some remarks about the energetics. The integrated energies of a supernova outburst range from 10^{48} to 10^{50}erg in visual light, which is only of the order of a few percent of the total energy of the ejected material. Note, for instance, that $\frac{1}{2}M_{\odot}v^2 \sim 10^{51}$erg for $v/c \simeq 0.3 \times 10^{-2}$.

If thermonuclear reactions are responsible, then at least about 0.5 to 1 $M_{\odot}$ of He, C or O have to be fused explosively to iron group elements. (In explosive carbon burning about 0.5 MeV per nucleon are released. Note that the sun contains roughly 10^{57} nucleons.)

The binding energy of a neutron star is about

$$\frac{GM^2}{R} = 3 \times 10^{53}\text{erg} \left(\frac{M}{M_{\odot}}\right)^2 \left(\frac{R}{10\text{ km}}\right)^{-1} . \tag{1}$$

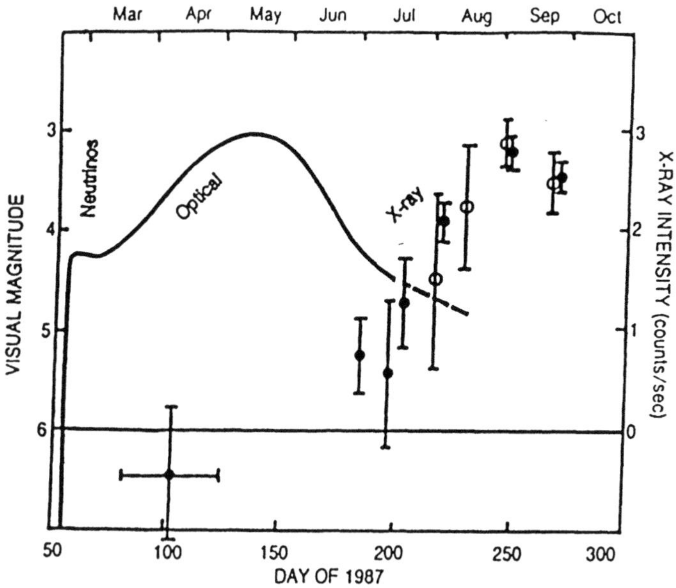

Fig. II.4. Optical light curve of SN 1987 A and X-ray intensity as observed by the large-area counter of the X-ray satellite Ginga (energy range 10–30 keV). (From Ref. [7].)

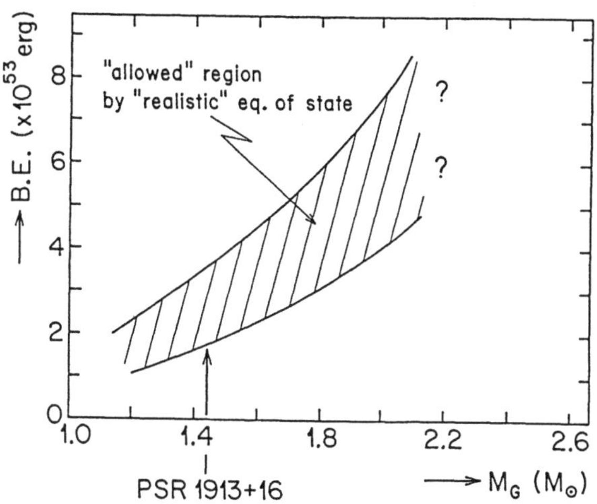

Fig. II.5. Binding energy versus gravitational mass. The shaded region is covered by various equations of state and indicates the theoretical uncertainties.

The range of binding energies covered by various "realistic" equations of state is shown in Fig. II.5.

This shows that only about 1% of the gravitational energy release in core collapse is needed for an explosion. As will be discussed in detail, the other 99% are radiated away in a gigantic neutrino burst, within less than about 10 seconds.

I discuss in these lectures only models triggered by core collapse and therefore describe next briefly the evolution of relatively massive stars ($\gtrsim 10 M_\odot$) and the structure of supernova progenitors.

III. Presupernova Evolution of Massive Stars

In this chapter we give a brief discussion of the structure and evolution of massive stars. This is a vast field and I can only emphasize some of the most important aspects, which are relevant for the formation of neutron stars and/or black holes by gravitational collapse and for Type II (Ib) supernova explosions.

III.1 Formation of Iron Cores

For sufficiently massive stars ($\gtrsim 10 M_\odot$) the central temperature and density increase more or less monotonically during their evolution (see Fig. III.1). The history of such a star consists of a continued contraction of its core, with halts during stages of nuclear energy generation. There is a succession of nuclear fuels which can be effective in producing such halts: These include fusion of hydrogen into helium, burning of ^{4}He, ^{12}C, ^{16}O, and ^{28}Si. These are marked along the track corresponding to $25\,M_\odot$ in Fig. III.1 and cause, together with shell burning, the excursions of the track. The last of these stages of nuclear energy generation produces the nuclear statistical equilibrium abundance peak centered around ^{56}Fe. At this stage, all the available sources of nuclear energy have been extracted from matter. Only stars more massive than about $11\,M_\odot$ go through this full sequence, whereby an onion structure is built up around the core of the iron peak elements, as a result of shell burning in various shells. (Details will be presented in Sect. 4.)

For these massive stars C-burning ignites under non-degenerate conditions of the electron gas. The degenerate and non-degenerate regions are separated by the dashed line in Fig. III.1. They correspond to the following expressions for the Fermi energy

$$\frac{\varepsilon_{\mathrm{F}}}{kT} = \begin{cases} 2.97 \times 10^5\, T^{-1} (\varrho Y_{\mathrm{e}})^{2/3} & \text{(nonrelativistic)} \\ 5.93 \times 10^7\, T^{-1} (\varrho Y_{\mathrm{e}})^{1/3} & \text{(relativistic)}\,. \end{cases} \tag{1}$$

Here the electron fraction Y_{e} is the number of electrons per baryon and the density ϱ is in g/cm^3 and the temperature T in K. Completely degenerate electrons become relativstic for $\varrho \geq 10^6$g/cm^3, since

$$\varrho = 0.97 \times 10^6 (\mathrm{g/cm}^3) \frac{A}{Z} \left(\frac{p_{\mathrm{F}}}{m_{\mathrm{e}}}\right)^3 , \tag{2}$$

where p_{F} is the Fermi momentum. Note also that $m_{\mathrm{e}}/k \simeq 6 \times 10^9$ K.

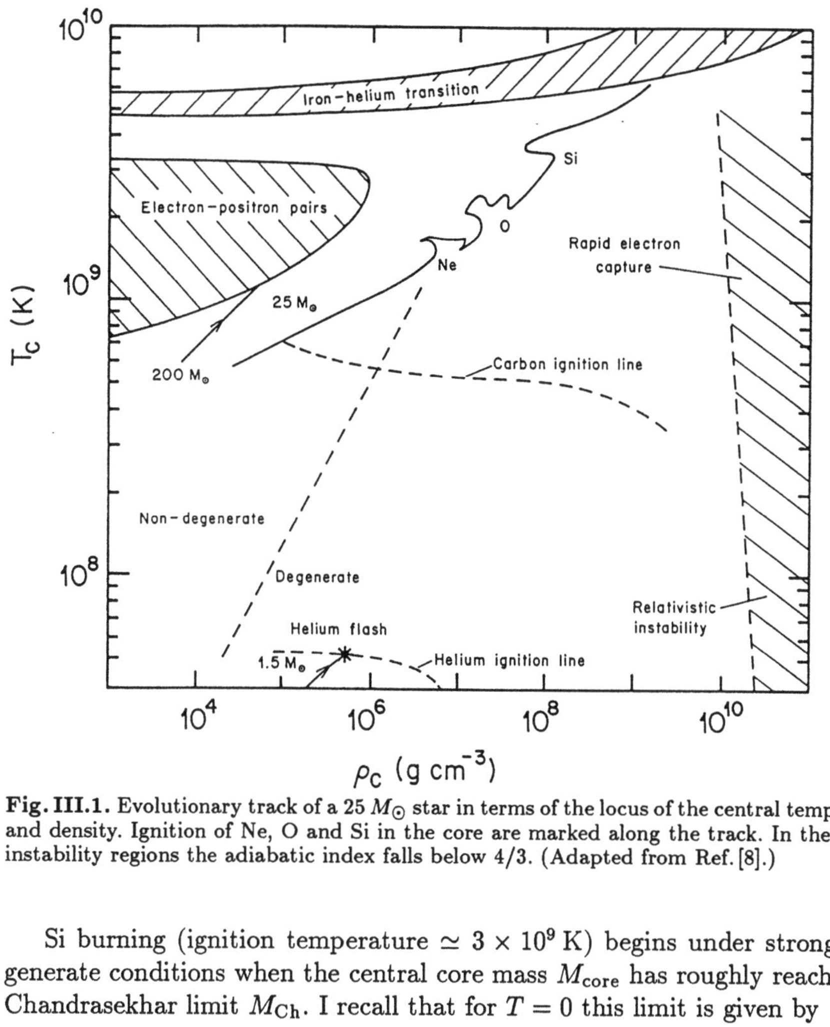

Fig. III.1. Evolutionary track of a 25 $M_\odot$ star in terms of the locus of the central temperature and density. Ignition of Ne, O and Si in the core are marked along the track. In the shaded instability regions the adiabatic index falls below 4/3. (Adapted from Ref. [8].)

Si burning (ignition temperature $\simeq 3 \times 10^9$ K) begins under strongly degenerate conditions when the central core mass $M_{\rm core}$ has roughly reached the Chandrasekhar limit $M_{\rm Ch}$. I recall that for $T = 0$ this limit is given by

$$M_{\rm Ch} = 3.1\, Y_{\rm e}^2 \,\frac{m_{\rm Pl}^3}{m_{\rm N}^2} = 5.84\, Y_{\rm e}^2\, M_\odot\ , \tag{3}$$

where $m_{\rm Pl}$ is the Planck mass and $m_{\rm N}$ the nucleon mass. Finite temperature effects increase this critical mass, because the electron pressure is for $T/\varepsilon_{\rm F} \ll 1$ given by

$$P_{\rm e} = P_{\rm e}(T=0)\left(1 + \frac{2}{3}\frac{\pi^2 T^2}{\varepsilon_{\rm F}^2}\right), \tag{4}$$

which gives the effective Chandrasekhar limit

$$M_{\rm eff} = M_{\rm Ch}\left[1 + \left(\frac{\pi T}{\varepsilon_{\rm F}}\right)^2\right]. \tag{5}$$

Si burning is completed within a day, or so, and an iron-nickel core develops with a mass close to $M_{\rm eff}$. Since the electrons are fully relativistic, the adiabatic index is close to the critical value 4/3, and thus the core is only marginally stable. (For the standard stability analysis see, e.g., Chapter 6 of [9].) We shall see shortly that it runs inevitably into an instability and begins to collapse.

Stars with mass $M \lesssim 8M_\odot$ will in their last phases become red giants that will eventually eject their outer envelopes due to pulsational instabilities to become white dwarfs. Fig. III.2 shows that for a certain mass range such stars converge to a common track and ignite carbon under strongly degenerate conditions [10], which leads to a carbon flash.

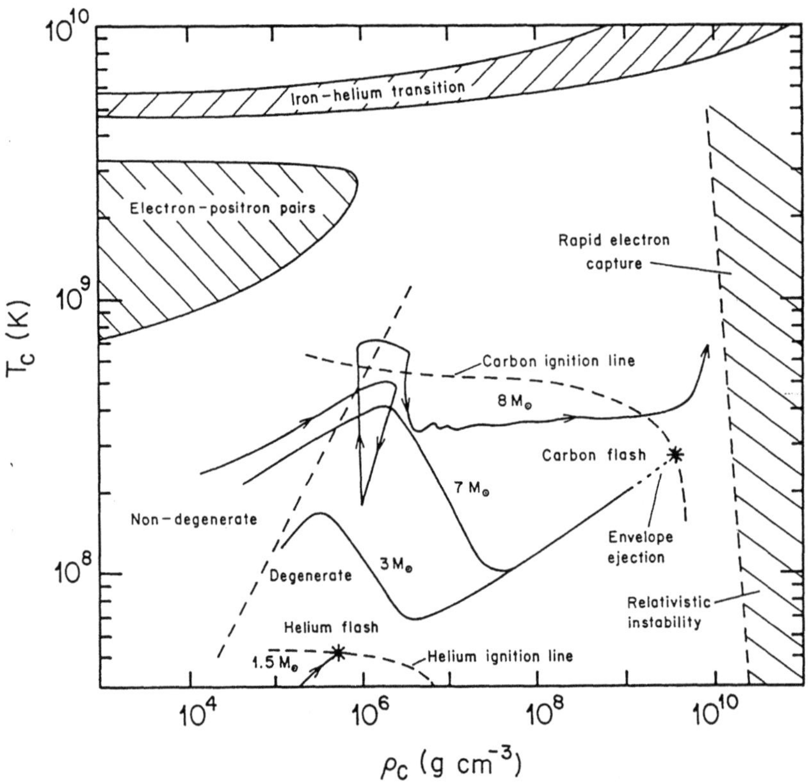

Fig. III.2. Evolutionary tracks for stars with mass $M \lesssim 8M_\odot$. Carbon ignites under strongly degenerate conditions.

III.2 Qualitative Considerations on Stellar Evolution

The qualitative difference between the evolution of massive stars and low-mass stars can be seen in rough terms as follows.

We start from the virial theorem

$$E_G + 3\int P dV = 0, \tag{6}$$

where E_G is the gravitational energy and the integral extends over the star. For the pressure we use the following interpolation between degenerate and nondegenerate behavior

$$P = P_e(T=0) + nkT, \tag{7}$$

where n is the number density of free particles of all kind. Now

$$P_e(T=0) = (\gamma - 1)\, u_e(T=0),$$

where $u_e(T=0)$ is the zero point energy of the electrons and $4/3 \leq \gamma \leq 5/3$. The lower limit in this inequality corresponds to the extreme relativistic limit and the upper one to the nonrelativistic limit. Furthermore,

$$u_e(T=0) = n_e m_e[(1+x^2)^{1/2} - 1] \tag{8}$$

with

$$x = \bar{p}_e/m_e \simeq \frac{1}{m_e} n_e^{1/3} \simeq \lambda\!\!\!^{-}_e \frac{N_e^{1/3}}{R}; \tag{9}$$

here N_e is the total number of electrons and R the radius of the star. In (9) we have used the uncertainty relation for the average momentum $\bar{p}_e$ of the electrons and the Pauli principle: At $T=0$ there is one electron in each de Broglie cube having volume $(\hbar/p)^3$. If N denotes the total number of particles (of all kind), then $N = N_e/Y_e\mu$, where μ is the average molecular weight, and we obtain from the virial theorem:

$$-\frac{GM^2}{R} + 3Nk\bar{T} + 3(\gamma-1)N_e m_e(\sqrt{x^2+1} - 1) \simeq 0\,. \tag{10}$$

Defining

$$N_0 = \frac{m_{\rm Pl}^3}{m_{\rm N}^3}, \quad N_{e0}^{1/3} = N_0^{1/3} Y_e = \frac{m_{\rm Pl}}{m_{\rm N}} Y_e, \tag{11}$$

we obtain from (9) and (10)

$$\frac{k\bar{T}}{m_e} \simeq \frac{1}{3}\mu Y_e \left[\left(\frac{N_e}{N_{e0}} \right)^{2/3} x + (1 - \sqrt{1+x^2}) \right]. \tag{12}$$

This equation describes very roughly the evolutionary track for a star or a stellar core with average temperature $\bar{T}$ and density determined by x. It allows us to draw the following interesting conclusions:

(i) With increasing x, i.e., with increasing density, $\bar{T}$ reaches a maximum provided $N_e < N_{e0}$, and then decreases monotonically. Note that the stellar mass belonging to N_{e0} is equal to $(N_{e0}/Y_e)m_{\rm N} = (m_{\rm Pl}^3/m_{\rm N}^2)Y_e^2$. Up to a numerical factor, this is nothing else than the Chandrasekhar mass (3).

(ii) If $M > M_{\rm Ch}$, the temperature increases without limit as $\bar{\varrho}^{1/3}$. We thus expect that the evolution of the star will encounter an instability.

(iii) If $M < M_{\rm Ch}$ and $x \ll 1$, the maximum temperature is given approximately by

$$\left(\frac{kT}{m_e}\right)_{\max} \simeq \left(\frac{M}{M_{Ch}}\right)^{4/3}. \tag{13}$$

This demonstrates once more that the Chandrasekhar limit sets a crucial scale for stellar structure and evolution.

III.3 Detailed Models

Detailed calculations show that stars between $0.08M_{\odot}$ and $0.25M_{\odot}$ will ignite hydrogen but not helium, and will leave helium white dwarfs. Stars between $0.25M_{\odot}$ and $4-8M_{\odot}$ will ignite helium but – depending on mass losses – perhaps not carbon, and may leave carbon and oxygen white dwarfs. It is, however, possible that some stars in this range do ignite carbon under very degenerate conditions (see Fig. III.2) and either flash or are completely disrupted in some sort of supernova. (This is a very delicate problem for numerical simulations.) Stars in the mass range $8-11M_{\odot}$ will burn carbon nondegenerately to leave O–Ne–Mg cores. Beyond this mass scale ($\simeq 11M_{\odot}$) nuclear fusion proceeds all the way up to iron.

There are considerable uncertainties in determining presupernova models, some of which should be mentioned at this point. First of all, there is no reliable theory of convection. Most workers still use the Schwarzschild criterion for convective stability. This criterion is an expression of the requirement that an isolated rising fluid element will experience a net force downwards. Thus at the boundary of a convective region, the acceleration imparted by the buoyancy force to convective elements vanishes. But the velocity of convective elements is not zero and thus they will penetrate (overshoot) into the formally stable radiative zones and increase the convective region. This phenomenon can have significant effects [11]. (Convective overshooting is usually formulated in the framework of the mixing length theory.) In particular, the carbon abundance after helium burning depends on overshooting, beside the reaction rate for $^{12}C\,(\alpha,\gamma)\,^{16}O$, which has recently been revised by a factor three.

The mass of the evolved core, which is close to the Chandrasekhar mass, is sensitive to Y_e and to corrections in the equation of state. Inclusion of Coulomb corrections in the equation of state decreases the Chandrasekhar mass and, hence, the core mass at each step by a few percent. This has been taken into account in a recent study of Nomoto and Hashimoto [12]. These authors have also included a better treatment of electron capture, which affects the value of Y_e. These and other improvements in the input physics led to final pre-collapse structures, which are substantially different from earlier results by Woosley and Weaver [13]. This is shown in Figs. III.3 and III.4, taken from Ref. [14].

There are also significant differences in the chemical structures. In Figs. III.5 and III.6 we show some results of Nomoto and Hashimoto.

It should be mentioned that convective overshooting has been neglected in these calculations.

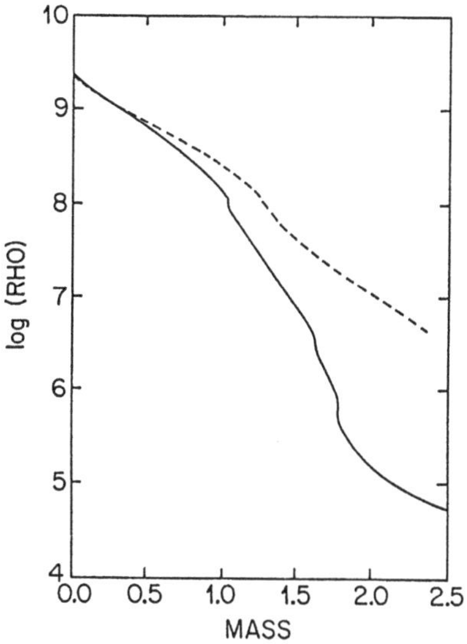

Fig. III.3. Density distribution of a $8M_\odot$ helium star near the onset of iron core collapse of Ref. [12] (solid line), compared with that of the $8M_\odot$ helium core of the $25M_\odot$ of Ref. [13] (dashed line).

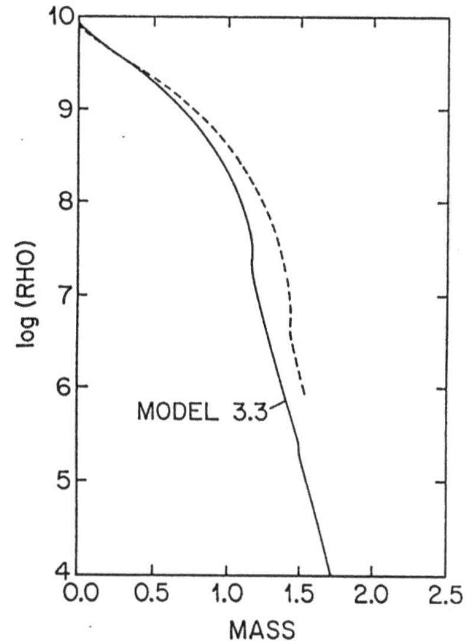

Fig. III.4. Density distribution of a $3.3M_\odot$ helium star at the onset of core collapse of Ref. [12] (solid line), compared with that of the $4M_\odot$ helium core of the $15M_\odot$ star of Ref. [13] (dashed line).

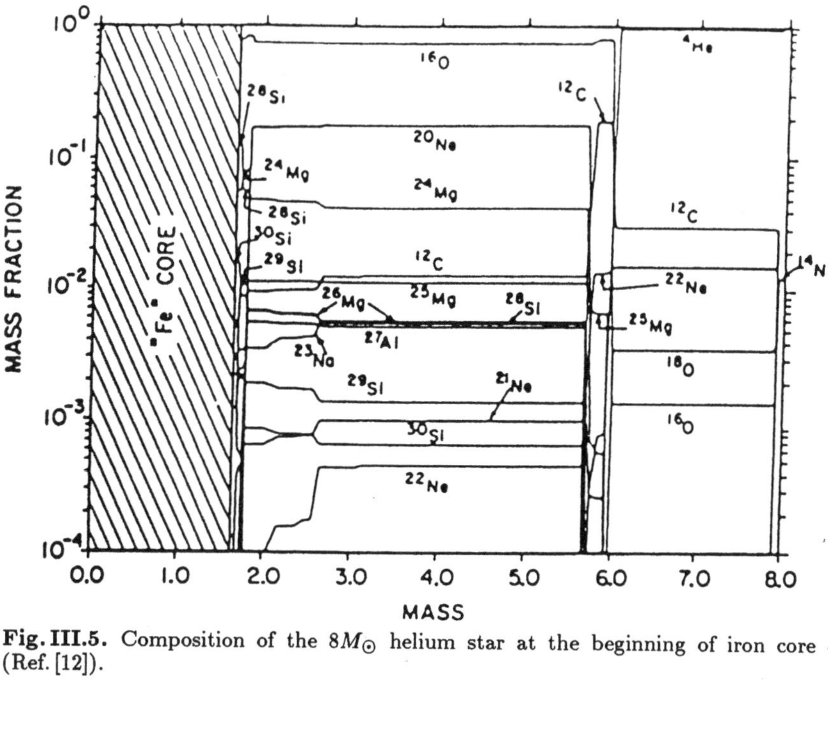

Fig. III.5. Composition of the $8M_\odot$ helium star at the beginning of iron core collapse (Ref. [12]).

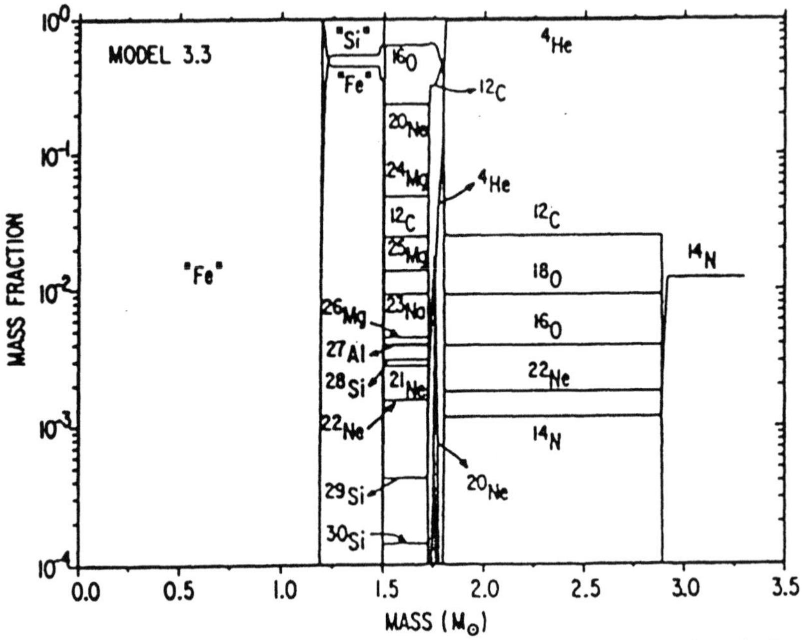

Fig. III.6. Same as Fig. III.5 for the $3.3M_\odot$ helium star (Ref. [12]).

III.4 Neutrino Losses in Advanced Evolutionary Phases

After the ignition of carbon, the core temperatures and densities are sufficiently high that neutrino radiation becomes the dominant sink of energy and entropy of a star. This accelerates the subsequent evolution very much. For instance, for a $20M_\odot$ star, carbon burning lasts only 450 yr and neon burning is completed after only 0.4 yr [14].

The most important neutrino processes, which will also play a crucial role after core collapse, are the following pair emission reactions:

pair annihilation:	$e^- + e^+ \longrightarrow \nu + \bar{\nu}$
photoneutrino process:	$\gamma + e^- \longrightarrow e^- + \nu + \bar{\nu}$
plasmon neutrino process:	plasmon $\longrightarrow \nu + \bar{\nu}$.

As an example, we calculate the loss rate by pair annihilation. (The plasma-neutrino process is treated in Appendix D.)

In late phases the electrons are fully relativistic (see Fig. III.1, 2). Then there are no interference terms between V−A and V+A contributions of the standard model Lagrangian:

$$\begin{aligned} L = -\frac{G_F}{\sqrt{2}}\{&\bar{\nu}_e\gamma^\alpha(1-\gamma_5)\nu_e\,\bar{e}\gamma_\alpha(C_V - C_A\gamma_5)e \\ &+ \bar{\nu}_\mu\gamma^\alpha(1-\gamma_5)\nu_\mu\,\bar{e}\gamma_\alpha(C'_V - C'_A\gamma_5)e + (\nu_\mu \to \nu_\tau)\} \end{aligned} \tag{14}$$

with

$$C_{\rm V} = \frac{1}{2} + 2\sin^2\Theta_{\rm W}, \quad C_{\rm A} = \frac{1}{2}$$
$$C'_{\rm V} = C_{\rm V} - 1, \quad C'_{\rm A} = C_{\rm A} - 1 \,. \tag{15}$$

Hence, we first work out the loss rate for a pure V−A interaction

$$\begin{aligned} L_{\rm V-A} &= -\frac{G_{\rm F}}{\sqrt{2}}\bar{\nu}\gamma^\alpha(1-\gamma_5)\nu\ \bar{e}\gamma_\alpha(1-\gamma_5)e \\ &= -\frac{8G_{\rm F}}{\sqrt{2}}e^{*\alpha}\nu^{\dot\beta}\ \nu^*_\alpha e_{\dot\beta}\,. \end{aligned}$$

In the second line we have used two-component Weyl spinors which simplify the calculation. The T-matrix element for $e^-(p_-)+e^+(p_+) \to \bar\nu(k)+\nu(k')$ with the indicated four momenta is

$$T = \frac{8G_{\rm F}}{\sqrt{2}}v^{*\alpha}(p_+)v^{\dot\beta}(k')\ v^*_\alpha(k)v_{\dot\beta}(p_-)\,.$$

The corresponding differential reaction rate is

$$\begin{aligned} v\cdot d\sigma^{\rm V-A} &= \left(\frac{8G_{\rm F}}{\sqrt{2}}\right)^2 \frac{1}{(2\pi)^2}\frac{1}{2p_-^0\,2p_+^0}\frac{1}{4}\sum_{\rm spins}|v^{*\alpha}(p_+)\ldots|^2 \\ &\times \delta^4(k+k'-p_--p_+)\frac{d^3k}{2k^0}\frac{d^3k'}{2k'^0}\,. \end{aligned}$$

With the use of

$$v^*_\alpha(p)v_{\dot\beta}(p) = p_{\alpha\dot\beta}$$

we obtain, without computing any traces,

$$v\cdot\sigma^{\rm V-A} = \frac{G_{\rm F}^2}{2\pi^2}\frac{1}{E_+E_-}\int(p_-\cdot k')(p_+\cdot k)\delta^4(k+k'-p_--p_+)\frac{d^3k}{2k^0}\frac{d^3k'}{2k'^0}\,.$$

Finally, we make use of the phase space integral

$$\int k^\mu k'^\nu\delta^4(P-k-k')\frac{d^3k}{2k^0}\frac{d^3k'}{2k'^0} = \frac{\pi}{6}[2P^\mu P^\nu + g^{\mu\nu}P^2]$$

and obtain (neglecting the electron mass)

$$v\cdot\sigma^{\rm V-A} = \frac{G_{\rm F}^2}{3\pi}\frac{1}{E_-E_+}(p_-\cdot p_+)^2\,. \tag{16}$$

The same expression holds, of course, for the V+A interaction. Hence, we conclude from (15) that the total rate for pair annihilation into all neutrino flavors is given by

$$v\cdot\sigma = v\cdot\sigma^{\rm V-A}\left\{\frac{1}{2}(C_{\rm V}^2+C_{\rm A}^2) + [(C_{\rm V}-1)^2+(C_{\rm A}-1)^2]\right\}\,. \tag{17}$$

The factor in the curly bracket is close to one:

$$\{\ldots\} = 0.84 \qquad \text{for} \quad \sin^2\Theta_\mathrm{W} = 0.23 \; .$$

Once the temperature approaches $m_\mathrm{e}/k \simeq 6\times 10^9$ K, electron–positron pairs are created in equilibrium with radiation, and the electron and positron number densities, $n_\pm(p)$, are given by Fermi distributions

$$n_\pm(p) = \frac{2}{(2\pi)^3} \frac{1}{\mathrm{e}^{(E(p)\mp\mu)/kT} + 1} \; . \tag{18}$$

(Recall that the chemical potentials of particles and antiparticles are opposite.) The chemical potential μ is related to the lepton number density.

The energy loss rate, $\varepsilon_\mathrm{pair}$, is obtained by multiplying Equation (17) with $(E_- + E_+)\, n_-(p_-)\, n_+(p_+)$ and integrating over phase space. For simplicity, we first average $v \cdot \sigma$ over angles:

$$\langle v \cdot \sigma \rangle \simeq \{\ldots\} \frac{4}{9\pi} G_\mathrm{F}^2 E_+ E_- \; .$$

This leads to the energy loss rate per unit volume

$$\begin{aligned} \varepsilon_\mathrm{pair} \simeq \frac{4}{9} &\left\{ \frac{1}{2}(C_\mathrm{V}^2 + C_\mathrm{A}^2) + (C_\mathrm{V} - 1)^2 + (C_\mathrm{A} - 1)^2 \right\} \\ &\times \frac{G_\mathrm{F}^2}{\pi^5} m_\mathrm{e}^9 \left(\frac{kT}{m_\mathrm{e}}\right)^9 F(\mu/kT) \end{aligned} \tag{19}$$

with

$$F(y) = \int_0^\infty \frac{u^4 du}{1 + \mathrm{e}^{u-y}} \int_0^\infty \frac{v^3 dv}{1 + \mathrm{e}^{v+y}} + \int_0^\infty \frac{u^3 du}{1 + \mathrm{e}^{u-y}} \int_0^\infty \frac{v^4 dv}{1 + \mathrm{e}^{v+y}} \; . \tag{20}$$

In the late phases of stellar evolution the electrons are degenerate. In this limit we find

$$F(y) \simeq \frac{6}{5} \left(y^5 + \frac{5}{y} \right) \mathrm{e}^{-y}$$

and thus in the extreme relativistic and degenerate limit:

$$\frac{1}{\varrho} \varepsilon_\mathrm{pair}^\mathrm{E.D.} \simeq 1.2 \times 10^{15} Y_\mathrm{e} \left(1 + \frac{5kT}{\varepsilon_\mathrm{F}} \right) \left(\frac{\varepsilon_\mathrm{F}}{m_\mathrm{e}} \right)^2 T_{10}^4 \mathrm{e}^{-\varepsilon_\mathrm{F}/kT} \mathrm{erg\ g^{-1} sec^{-1}} . \tag{21}$$

As an example we take $\varrho = 10^9\ \mathrm{g/cm^3}$, $kT = m_\mathrm{e}$. With the help of (1) we find $\varepsilon_\mathrm{pair}/\varrho \simeq 6 \times 10^{11}$erg/g-sec. If we multiply this with one solar mass, we obtain a few times $10^{11} L_\odot$. This is in rough agreement with the result of a more detailed calculation for a $15 M_\odot$ star shown in Fig. III.7, which includes all neutrino pair processes.

The total emissivity of all pair processes is plotted in Fig. III.8. (For comparison we note that the average energy loss rate of the sun is 2 erg/g-sec.)

It is instructive to compare the result (21) and Fig. III.8 with the radiation energy density

$$aT^4 = 7.6 \times 10^{17}\ T_{10}^4\ \mathrm{erg/cm^3},$$

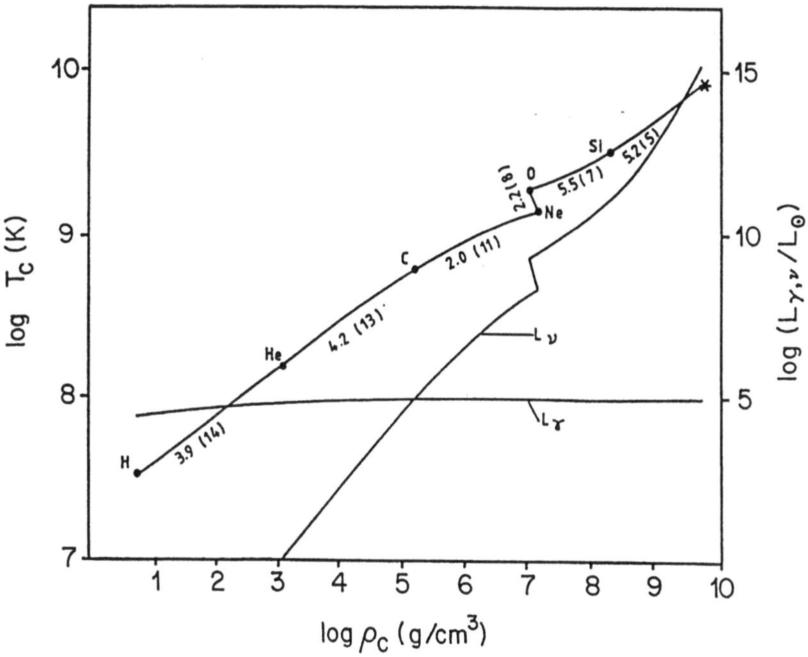

Fig. III.7. Total optical (L_γ) and neutrino (L_ν) luminosities for a $15M_\odot$ star, together with its evolutionary track.

which dominates the thermal energy density for advanced stages of massive stars ($T_9 \gtrsim 1$). One sees that the cooling time aT^4/ε becomes very short.

The neutrino luminosity of a $15M_\odot$ star in Fig. III.8 reaches at least 10^4 times the optical luminosity of a whole galaxy, just before the star becomes unstable.

III.5 Core Instability

As already remarked, the central core of a massive star is only marginally stable when Si-burning starts. The central region of the $25M_\odot$ star in Fig. III.1 becomes finally unstable through photodesintegration of the iron-peak elements to α-particles. The nuclear desintegration of iron to helium costs a lot of internal energy and reduces the adiabatic index (often called Γ_1) below the critical value 4/3 (see Appendix B).

For stars in the lower part of the considered mass range, electron capture, which also reduces the pressure support within the core, is a further destabilizing factor. General relativity has also a destabilizing effect, because it increases the critical value $\Gamma_{\rm crit}$ above 4/3 by $\mathcal{O}(R_{\rm S}/R)$, where $R_{\rm S}$ is the Schwarzschild radius and R the radius of the star (see, e.g., Sect. 6.9 of Ref. [9]). These regions of instability are shown in Fig. III.1.

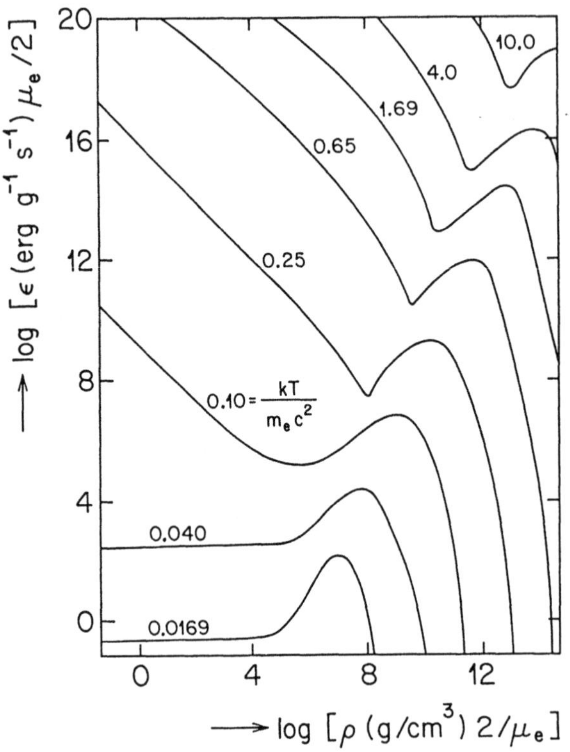

Fig. III.8. Neutrino emissivity ε in erg/g-sec, due to pair processes, as a function of density, in $\mathrm{g\,cm^{-3}}$, for a variety of values of temperature, T. The curves are marked by the value of $kT/m_e c$. Y_e is defined by $n(e^-) - n(e^+) = Y_e\, \varrho/m_N$. (Adapted from Ref. [15].)

Although there are differences in detail (as discussed in Sect. III.3), all the models computed up to this point give similar results. Thus the dynamic collapse of the iron core is unavoidable.

For readers not familiar with pulsation theory, we give here an oversimplified argument, which shows that $\Gamma > 4/3$ is necessary for stability. Let us consider a star of uniform density. Its total energy is $E = E_G + E_{\rm int}$, with $E_G = -(3/5)GM^2/R$. In equilibrium $\partial E/\partial R = 0$ and thus we obtain from

$$\partial E_G/\partial R = -E_G/R$$

and

$$\frac{\partial E_{\rm int}}{\partial R} = \frac{\partial E_{\rm int}}{\partial V}\frac{\partial V}{\partial R} = -3PV/R$$

($V = (4\pi/3)R^3$ is the volume of the star) the virial theorem $E_G + 3PV = 0$. Stability requires $\partial^2 E/\partial R^2 > 0$, i.e.,

$$\begin{aligned}\frac{\partial^2 E}{\partial R^2} &= 2\,\frac{E_G}{R^2} - 6\,\frac{PV}{R^2} - 3\,\frac{V}{R}\frac{\partial P}{\partial R}\\ &= -9\,\frac{PV}{R^2}\left(\frac{4}{3} + \frac{V}{P}\frac{\partial P}{\partial V}\right) > 0,\end{aligned}$$

which implies that

$$\Gamma := -\frac{V}{P}\frac{\partial P}{\partial V} > \frac{4}{3} .$$

From pulsation theory one can show that the frequency ω_0 of the fundamental mode for adiabatic pulsations satisfies $\omega_0^2 <$ const $(\bar{\Gamma}_1 - 4/3)$, where the constant is a positive number determined by the structure of the star, and

$$\bar{\Gamma}_1 = \int \Gamma_1 P dV \left(\int P dV \right)^{-1}, \qquad \Gamma_1 = \left(\frac{\partial \ln P}{\partial \ln \varrho} \right)_{\text{ad}} . \tag{22}$$

Thus the fundamental mode becomes unstable for $\bar{\Gamma}_1 < 4/3$. The necessary condition $\bar{\Gamma}_1 > 4/3$ for stability can also be derived with an energy argument.

III.6 Progenitor Star of SN 1987 A

It came as a great surprise to many astrophysicists, that the progenitor of the supernova, Sanduleak—69202, was a blue (B3I) supergiant with $\log T_{\text{eff}} \simeq 4.2$ and $\log L/L_\odot \simeq 5.1$. Since B supergiants were not believed to be evolved enough to undergo core collapse, it had been thought that only M giants or supergiants could be progenitors of supernovae. Soon after the night of 23/24 February 1987, it was suggested that the explosion of a blue supergiant may be attributable to the low metallicity of the Large Magellanic Cloud. (The metal content Z is only $(\frac{1}{3} - \frac{1}{4})Z_\odot$.) Some earlier evolutionary calculations (without mass loss!) had indeed shown that metal poor B stars may never become M supergiants prior to the onset of carbon burning [16].

Contrary to many claims that the explosion of SN 1987 A from a blue progenitor cannot be understood in the framework of "normal" stellar evolution, André Maeder has shown quite convincingly that this is not the case if substantial mass loss of massive stars is taken into account [17]. The progenitor star may have lost about half of its initial mass in a previous *red* supergiant stage.

The existence of very strong stellar winds of massive stars was shown by UV spectroscopy from the COPERNICUS and IUE satellites and the importance of mass loss in stellar evolution has been clearly demonstrated in recent years. In particular, the tracks in the HR diagram are substantially changed by mass loss (see Ref. [11]).

It was pointed out by Maeder that it is not enough to consider Sanduleak - 69202 in isolation. In the Large Magellanic Cloud there exist many bright red-supergiants corresponding to stars with main sequence masses up to $\sim 50 M_\odot$. The relative frequency of these red-supergiants is a strong constraint for evolutionary models. Maeder [17] has emphasized in particular that there are many red supergiant stars in LMC clusters with turnoff around $\log(L/L_\odot) = 5.1$. On the basis of this and other observations he constructed models for the progenitor of SN 1987 A which match the various observational constraints. Fig. III.9 shows the evolutionary track in the HR diagram for a star with an initial mass of $20 M_\odot$ with various mass loss rates (which depend on the metallicity). One

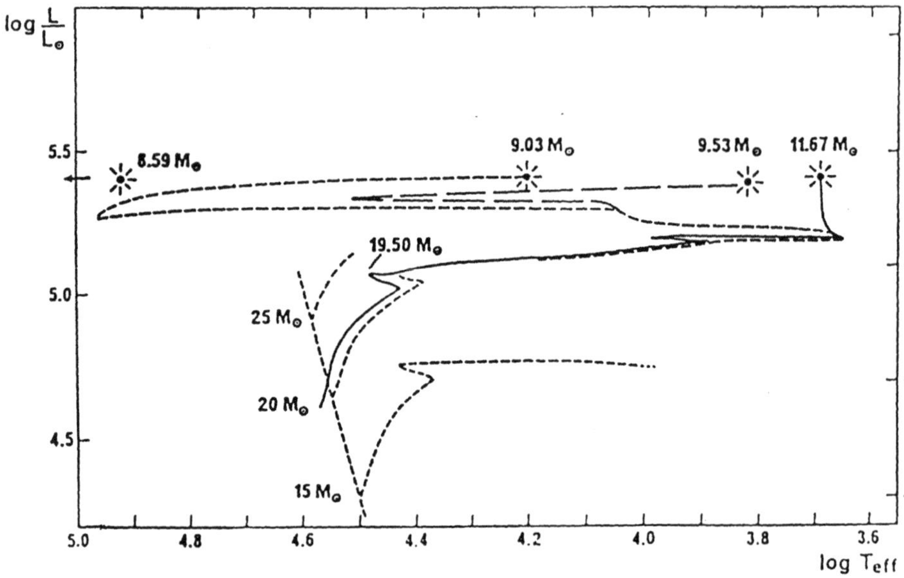

Fig. III.9. Evolutionary tracks in the HR diagram for a model with an initial mass of $20M_{\odot}$ and composition $X = 0.744$ and $Z = 0.006$ for various mass loss rates and corresponding final masses.

sees that over some interval of final masses a blue supergiant progenitor with $\log T_{\text{eff}} \simeq 4.2$ is produced. Thus the fact that the progenitor of SN 1987 A was blue may not be so mysterious. (For a more detailed discussion, see Ref. [17].)

IV. Physics of Stellar Collapse, Shock Energy

We discuss now in some detail the collapse of the central iron core, with

$$M_{\text{core}} \simeq 1.3 - 1.6 M_{\odot}, \qquad Y_{\text{e}} \simeq 0.45 \ ,$$

to nuclear densities and the generation of a shock wave by core bounce.

IV.1 Neutronization

Initially, when core collapse starts, the chemical potential μ_{e} of the electrons is about 6 MeV and rises rapidly according to

$$\mu_{\text{e}}/m_{\text{e}} \simeq 100\, Y_{\text{e}}^{1/3} \varrho_{12}^{1/3} . \tag{1}$$

For this reason neutrinos are not only generated by pair processes discussed in Sect. III.4, but soon neutronization reactions, i.e., electron capture on free protons and nuclei begin to play the dominant role. We shall see later that most of the nucleons remain bound in heavy nuclei and thus absorption on free protons is limited by their small abundance Y_{p}. On the other hand, shell blocking effects reduce the capture rates of the most abundant nuclei and for this reason, the capture on free protons should give the order of magnitude of the total capture rate. (More detailed calculations have been carried out in Refs. [18 – 20].)

A very similar calculation as for neutron decay shows that the capture rate $\langle v \cdot \sigma \rangle$ for $\mathrm{e}^- + \mathrm{p} \to \mathrm{n} + \nu_{\mathrm{e}}$ is, after taking spin averages,

$$\langle v \cdot \sigma \rangle = \frac{1}{2\pi} G_{\mathrm{F}}^2 m_{\mathrm{e}}^2 \left[1 + 3 \left(\frac{g_{\mathrm{A}}}{g_{\mathrm{V}}} \right)^2 \right] \left(\frac{E}{m_{\mathrm{e}}} \right)^2 , \tag{2}$$

where E is the electron energy. (We neglect, for simplicity, the difference between E_{e} and E_{ν} and ignore the mass difference $m_{\mathrm{n}} - m_{\mathrm{p}}$.) Thus the total capture rate Γ per proton in the extreme relativistic limit is given by

$$\begin{aligned} \Gamma &= \int \langle v \cdot \sigma \rangle n_{\mathrm{e}}(p) d^3 p \\ &= \frac{1}{2\pi^3} G_{\mathrm{F}}^2 \left[1 + 3 (g_{\mathrm{A}}/g_{\mathrm{V}})^2 \right] \int_0^{\infty} \frac{E^4}{\mathrm{e}^{(E-\mu_{\mathrm{e}})/kT} + 1} dE, \end{aligned} \tag{3}$$

as long as the neutrinos escape freely. We can use the strongly degenerate approximation and find with (1) and $g_{\mathrm{A}}/g_{\mathrm{V}} \simeq 1.25$

$$\begin{aligned} \Gamma &= 1.3 \times 10^{-4} \left(\frac{\mu_{\mathrm{e}}}{m_{\mathrm{e}}} \right)^5 \mathrm{sec}^{-1} \mathrm{proton}^{-1} \\ &= 1.3 \times 10^6 Y_{\mathrm{e}}^{5/3} \varrho_{12}^{5/3} \ \mathrm{sec}^{-1} \mathrm{proton}^{-1} \ . \end{aligned} \tag{4}$$

For comparison, we note that the free fall time is given by

$$\tau_{\mathrm{ff}} = \left(\frac{8\pi}{3} G \varrho \right)^{-1/2} = 1.3 \times 10^{-3} \varrho_{12}^{-1/2} \ \mathrm{sec} \ . \tag{5}$$

We shall show in the next Section that neutrinos are trapped when the core reaches a density $\varrho_{\mathrm{trap}} \simeq 3 \times 10^{11} \mathrm{g/cm}^3$ at which point the Pauli principle forbids further electron capture. We estimate now the change ΔY_{e} until this density is reached. Considering only capture on free protons we have $\dot{Y}_{\mathrm{e}} = -\Gamma Y_{\mathrm{p}}$ with Γ given by (4). Setting $Y_{\mathrm{e}} = Y_{\mathrm{p}}$ and using (5) we obtain roughly

$$\begin{aligned} Y_{\mathrm{e},f} - Y_{\mathrm{e},i} &= \int \dot{Y}_{\mathrm{e}} dt \sim \left[\dot{Y}_{\mathrm{e}}(\varrho) \, \tau_{\mathrm{ff}}(\varrho) \right] \Big|_{\varrho = \varrho_{\mathrm{trap}}} \\ &\sim -2 \times 10^{-3} \, Y_{\mathrm{e},f}^{8/3} \, \varrho_{12,\mathrm{trap}}^{7/6} \end{aligned}$$

and from this we find $Y_{\mathrm{e},f} \sim 0.07$. In reality, matter is not so strongly neutronized. A more careful analysis, properly allowing for neutronization on heavy nuclei, gives $\Delta Y_{\mathrm{e}} \simeq -0.1$, corresponding to $Y_{\mathrm{e},f} \simeq 0.35$ [18–20]. This value is rather insensitive to variations of the underlying physics [21].

IV.2 Neutrino Trapping

Neutrinos are produced by electron capture until the core becomes sufficiently opaque that the time for a neutrino to diffuse out of the core becomes large compared with the dynamical time scale (5). Let us investigate when this happens.

Various processes contribute to the neutrino opacity. Scattering from free nucleons and from nuclei through neutral current interactions are the dominant sources.

The effective neutrino-hadron interaction in the standard model is

$$L = -\frac{G_{\mathrm{F}}}{\sqrt{2}}\, \bar{\nu}\gamma^{\alpha}(1-\gamma_5)\nu\; J^{\mathrm{Z}}_{\alpha}, \tag{6}$$

where the neutral hadron current J^{Z}_{μ} has the following form

$$J^{\mathrm{Z}}_{\mu} = (V^3_{\mu} \;-\; A^3_{\mu}) \;-\; 2\sin^2\Theta_{\mathrm{W}}\, J^{\mathrm{elm}}_{\mu}. \tag{7}$$

For nucleons V^3_{μ} and A^3_{μ} are the isospin partners of the charged vector and axial currents which govern β-decay.

Coherent Neutrino-Nucleus Scattering. The axial term is not important for heavy nuclei because it gives spin-dependent contributions which are largely averaged out.

For the vector contribution of J^{Z}_{μ} the matrix elements for nuclei are proportional to $I_3 - 2\sin^2\Theta_{\mathrm{W}} \cdot Q$ (I_3 = third component of the isospin, Q = electric charge). For a nucleus (Z,N) this is equal to $\frac{1}{2}(Z-N) - 2\sin^2\Theta_{\mathrm{W}} Z \simeq -\frac{1}{2}N$, where N is the number of neutrons. Thus the current matrix element for a heavy nucleus is in good approximation:

$$\langle f \mid J^{\mathrm{Z}}_{\mu} \mid i\rangle = -\frac{N}{2} F(q^2)(P_f + P_i)_{\mu}, \tag{8}$$

where $F(q^2)$ is the form factor for the neutron distribution, with $F(0) = 1$. The momentum transfer is sufficiently small that we have coherence: $F(q^2) \simeq F(0) = 1$. Thus the T-matrix element for neutrino scattering on (Z,N) is

$$T = \frac{G_{\mathrm{F}}}{\sqrt{2}}\frac{N}{2} \bar{u}(k')\gamma^{\mu}(1-\gamma_5)u(k)(P_f + P_i)_{\mu}. \tag{9}$$

A very short calculation gives for the differential cross section

$$\frac{d\sigma^{\nu \mathrm{A}}}{d\Omega} = \frac{1}{4\pi}\frac{N^2}{16}\sigma_0 \left(\frac{\nu}{m_{\mathrm{e}}}\right)^2 (1+\cos\theta), \tag{10}$$

where ν is the neutrino energy, θ the scattering angle and

$$\sigma_0 = \frac{4}{\pi} G_{\mathrm{F}}^2 m_{\mathrm{e}}^2 = 1.7 \times 10^{-44}\mathrm{cm}^2. \tag{11}$$

The quantity which is relevant for neutrino transport is the cross section Σ for momentum transfer:

$$\Sigma_{\nu A} = \int (1-\cos\theta)\frac{d\sigma^{\nu A}}{d\Omega}d\Omega = \frac{1}{24}N^2\sigma_0\left(\frac{\nu}{m_e}\right)^2 . \tag{12}$$

(The reader is invited to estimate the non-coherence corrections.)

Neutrino-Nucleon Scattering. Similarly, the T-matrix element for neutrino-neutron scattering is

$$T = \frac{1}{2}\sqrt{2}G_F\left[v^*(k')v(k)\langle 1\rangle - \frac{g_A}{g_V}\langle\boldsymbol{\sigma}\rangle v^*(k')\boldsymbol{\sigma}v(k)\right],$$

where $v(k)$ and $v(k')$ are the 2-component left-handed Weyl spinors for the initial and final neutrons with energy-momenta k and k' and $\langle 1\rangle, \langle\boldsymbol{\sigma}\rangle$ are the Fermi and Gamow-Teller matrix elements. A standard calculation gives

$$\frac{d\sigma^{\nu n}}{d\Omega} = \frac{G_F^2}{(2\pi)^2}\frac{1}{4}\nu^2\left[(1+\cos\theta) + 3\left(\frac{g_A}{g_V}\right)^2\left(1-\frac{1}{3}\cos\theta\right)\right]. \tag{13}$$

For protons only the axial current in (7) contributes in the approximation $\sin^2\Theta_W = 1/4$. Thus we have

$$\frac{d\sigma^{\nu p}}{d\Omega} = \frac{G_F^2}{(2\pi)^2}\frac{1}{4}\nu^2\cdot 3\left(\frac{g_A}{g_V}\right)^2\left(1-\frac{1}{3}\cos\theta\right). \tag{14}$$

For the cross section for momentum transfer this gives

$$\Sigma_{\nu p} \simeq \Sigma_{\nu n} = \frac{1}{24}\left[1+5\left(\frac{g_A}{g_V}\right)^2\right]\sigma_0\left(\frac{\nu}{m_e}\right)^2 . \tag{15}$$

From (12) and (15) we obtain for the mean free path due to neutrino scattering on nucleons and nuclei

$$\lambda_\nu = \frac{m_N}{\varrho\Sigma_{\nu N}}\left(X_N + \frac{\bar{N}^2}{A[1+5(g_A/g_V)^2]}X_A\right)^{-1} . \tag{16}$$

Here X_A and X_N are the mass fractions of heavy nuclei and "free" nucleons and $\bar{N}$ is the average number of neutrons in each heavy nucleus.

Numerically we obtain

$$\lambda_\nu \simeq (1.0\times 10^6\text{cm})\,\varrho_{12}^{-1}\left(\frac{\langle\nu\rangle}{10\text{MeV}}\right)^{-2}\left(X_N + \frac{\bar{N}^2}{A(1+5\times 1.25^2)}X_A\right)^{-1} . \tag{17}$$

For the average neutrino energy $\langle\nu\rangle$ we use $\langle\nu\rangle = (5/6)\mu_e$, because of the E-dependence in the capture rate (2). Together with (1) and $Y_e \simeq 0.4$ we expect

$$\langle\nu\rangle \simeq 33\,\varrho_{12}^{1/3}\text{MeV} . \tag{18}$$

As an example we take $X_N = 0$, (Z,A)=^{56}Fe, $X_A = 1$ for which

$$\lambda_\nu \simeq 0.3\times 10^5\text{cm}\,\varrho_{12}^{1/3}\text{ MeV}. \tag{19}$$

Take, for example, $\varrho_{12} = 1$. The radius at which $\varrho_{12} = 1$ is about 40 km and this is much larger than the mean free path. The corresponding diffusion time is

$$\tau_{\text{diff}} \simeq \frac{R^2(\varrho)}{\pi^2 c \lambda_\nu(\varrho)/3} \simeq 0.02\, \varrho_{12} \text{ sec} \tag{20}$$

and this is larger than the free fall time (5) for $\varrho > \varrho_{\text{trap}}$,

$$\varrho_{\text{trap}} \simeq 1.5 \times 10^{11} \text{g/cm}^3 . \tag{21}$$

From this, we conclude that for ϱ larger than a few times 10^{11} g/cm^3 neutrinos will be *trapped* in the star for the duration of the collapse. The density of the neutrinos increases very fast and they become degenerate. The Pauli principle forbids further electron capture and matter stays close to beta equilibrium.

IV.3 Entropy and Equation of State During Collapse

After neutrino trapping the weak interactions are also essentially in equilibrium and little entropy is, therefore, generated (the electromagnetic and strong interactions are always in equilibrium). Thus the collapse is almost adiabatic, shortly after neutrino trapping, with a fixed lepton fraction Y_ℓ. It is important to know the specific entropy of matter during the collapse.

At the onset of collapse, the entropy per nucleon is close to one in units of k_{B}, as we now show. In units with $\hbar = c = k_{\text{B}} = 1$ we have for relativistic electrons for the entropy per electron

$$s_{\text{e}} = \frac{4}{3} \frac{F_3(\mu_{\text{e}}/T)}{F_2(\mu_{\text{e}}/T)} - \frac{\mu_{\text{e}}}{T} , \tag{22}$$

where $F_n(\eta)$ are the Fermi integrals

$$F_n(\eta) = \int_0^\infty \frac{x^n}{e^{x-\eta}+1} dx . \tag{23}$$

This follows, for instance, from the general homogeneity relation

$$S = \frac{1}{T}(PV + U - \mu N), \quad \text{and} \quad PV = \frac{1}{3}U$$

for a relativistic ideal gas. For the strongly degenerate case, one has the well-known expansion of Fermi integrals

$$\int_0^\infty \frac{\varphi'(x)}{e^{x-\eta}+1} dx = \varphi(\eta) + 2\left[\frac{\pi^2}{12}\varphi''(\eta) + \frac{7\pi^4}{720}\varphi^{(iv)}(\eta) + \cdots\right] . \tag{24}$$

With this we find

$$s_{\text{e}} = \pi^2 \frac{T}{\mu_{\text{e}}} \quad \text{(relativistic)}. \tag{25}$$

Below we need also the entropy of an ideal Fermi gas in the nonrelativistic limit

$$s = \frac{\pi^2}{2} \frac{T}{\mu} \qquad \text{(nonrelativistic)}. \tag{26}$$

The entropy of a nucleus due to translational motion is given by the Sackur-Tetrode formula

$$(s_{\rm Fe})_{\text{per nucleus}} = \frac{5}{2} + \ln \left[\frac{1}{n_{\rm Fe}} \left(\frac{m_{\rm Fe} T}{2\pi} \right)^{3/2} \right] . \tag{27}$$

Typical values at the onset of collapse are (see Fig. III.1):

$$\varrho_{\rm c} = 3.7 \times 10^9 {\rm g/cm}^3, \qquad T_{\rm c} = 8 \times 10^9 {\rm K} .$$

For these values we obtain for the sum of (25) and (27) per nucleon ($Y_{\rm e} = 26/56 = 0.46$):

$$\left(s_{\rm Fe}^{\rm trans}\right)_{\rm nucleon} + \left(s_{\rm e}\right)_{\rm nucleon} = \frac{1}{56} \times 16.7 + 0.46 \times 1.1 \;=\; 0.81 . \tag{28}$$

For nuclei we should also include the entropy of the excited states. Let us estimate this contribution with the Fermi gas model of a nucleus. From (26) we have

$$\left(s_{\rm exc}\right)_{\rm nucleon} = \frac{\pi^2}{2} \frac{T}{T_{\rm F}} , \tag{29}$$

where $T_{\rm F}$ is the Fermi temperature corresponding to the Fermi energy (excluding the rest mass)

$$\varepsilon_{\rm F} = \frac{(3\pi^2 \varrho)^{2/3}}{(2 m_{\rm N})^{5/3}} = 39 \text{ MeV} \tag{30}$$

for $\varrho = \varrho_{\rm nucl} \simeq 2.8 \times 10^{14} {\rm g/cm}^3$. (Note that each momentum state can contain *four* particles.) For $T = T_{\rm c} = 8 \times 10^9$ K Eq. (29) gives the contribution 0.1 and thus we find for the total initial entropy s_i per nucleon

$$s_i = 0.9 \qquad \text{(at onset of collapse)}. \tag{31}$$

This agrees with more detailed calculations which take also the partial dissociation of ^{56}Fe into α-particles and neutrons into account [18]. (It is instructive to compare (30) with the entropy of interstellar matter, which is typically ~ 100 per nucleon. Thus the central core is in a much more ordered state.)

It turns out that the entropy does not increase much until the neutrinos are trapped. There are two competing effects. We expect some increase of entropy, because the weak interactions are out of β-equilibrium. On the other hand, neutrino losses will generally lead to a decrease of entropy. Several calculations [18, 21] show that the final entropy is about 1.5 when neutrinos are trapped and a β-equilibrium is established. At this point, net neutronization ceases and the lepton fraction is, as already discussed, $Y_\ell \simeq 0.35$. These results appear to be rather insensitive to various variations of the underlying physics, as was discussed in Ref. [21].

Fig. IV.1 shows the range of low-entropy adiabates during collapse.

The crucial point is now that nuclei survive all the way to nuclear densities. This can already be seen with the Fermi gas model. According to (29) and (30) $s = 1.5$ corresponds to a temperature of about 10 MeV. At such low temperatures, in comparison to T_F, thermal dissociation of the nuclei does not take place until the nuclei fuse into a stellar sized nucleus at nuclear densities, at which point a phase transition to a nucleon Fermi liquid takes place. Fig. IV.1 shows also the transition region where the nuclei are desintegrated. This lies, as expected, clearly above the low-entropy adiabates.

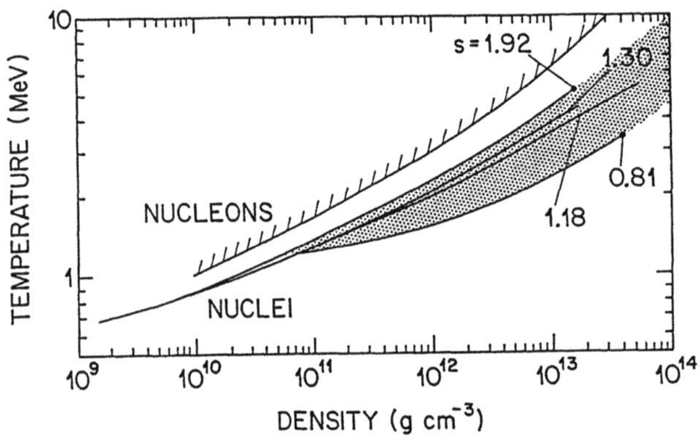

Fig. IV.1. Typical adiabates that represent the actual trajectories of mass zones in stellar collapse. The transition region from nuclei to nucleons is also shown. (Adapted from Ref. [22].)

These considerations show that the pressure is dominated by relativistic degenerate electrons up to nuclear densities. Thus we have

$$P \propto Y_e^{4/3} \varrho^{4/3} , \tag{32}$$

and the adiabatic index Γ remains always close to 4/3. Note that $(Y_{e,f}/Y_{e,i})^{4/3} \simeq (0.42/0.35)^{4/3} \simeq 0.7 - 0.8$.

IV.4 Adiabatic Collapse

In the previous section we have shown that the core collapse can be described by approximating the equation of state by a polytropic law

$$P = \kappa \, \varrho^\gamma , \tag{33}$$

where γ is the effective adiabatic index which is close to 4/3. Some useful insight into the collapse problem can be obtained by studying self-similar solutions for such an equation of state [23].

Such solutions exist because there are only two dimensional parameters (κ, G) in the problem, and these solutions are relevant to the collapse problem, if other parameters, such as the initial central density, affect only transients. Comparison with numerical calculations shows indeed that the memory of these additional parameters is lost when the density greatly exceeds the initial central density.

The unique similarity variable is

$$\chi = r(-t)^{\gamma-2}\kappa^{-1/2}G^{(\gamma-1)/2} , \tag{34}$$

where the origin of time is chosen to be the catastrophic moment when the central density becomes infinite. (Note that $[G] = L^3T^{-2}M^{-1}$, $[\kappa] = M^{1-\gamma}T^{-2}L^{3\gamma-1}$.)

For self-similar solutions we can express all the hydrodynamic variables by dimensionless functions of χ. Specifically

$$\begin{aligned} \varrho &= G^{-1}(-t)^{-2}D(\chi) \\ v &= \kappa^{1/2}G^{(1-\gamma)/2}(-t)^{1-\gamma}V(\chi) \\ m &= \kappa^{3/2}G^{(1-3\gamma)/2}(-t)^{4-3\gamma}M(\chi), \end{aligned} \tag{35}$$

where $m(r)$ is the enclosed mass inside the radius r. $M(\chi)$ and $D(\chi)$ are related by

$$M(\chi) = 4\pi \int_0^{\chi} x^2 D(x)dx . \tag{36}$$

If one substitutes these expressions into the continuity and Euler equations, one obtains the following ordinary differential equations for D and V:

$$\begin{aligned} [V-(\gamma-2)\chi]\,\frac{D'}{D} + V' &= -2 - \frac{2}{\chi}V , \\ \gamma D^{\gamma-2}D' + [V+(2-\gamma)\chi]\,V' &= -\frac{M}{\chi^2} - (\gamma-2)V . \end{aligned} \tag{37}$$

We note that for the special case $\gamma = 4/3$ the first equation in (37) has the homologous solution $V = (\gamma-2)\chi$. The second equation can then be reduced with the help of (36) to an ordinary differential equation for D. This case is studied in detail in Sect. 6.9 of [24] (see also Ref. [25]). It turns out that the homologously collapsing core can not have a mass exceeding the Chandrasekhar mass belonging to (33) (i.e., 4.56 $(\kappa/G)^{3/2}$) by more than 4.5 %.

The general case has been investigated by Yahil [23]. Fig. IV.2 shows the resulting dimensionless infall velocity $-V$, together with the speed of sound for $\gamma = 1.30$, as a function of χ. The velocity profile is clearly seen to be homologous ($v \propto r$) in the inner core, and a power law in the outer core. From the previous equations one can indeed easily deduce the asymptotic behavior

$$\begin{aligned} V(\chi) &\simeq -2\chi/3 \qquad && (\chi \ll 1) , \\ V(\chi) &\propto \chi^{(1-\gamma)/(2-\gamma)} \qquad && (\chi \gg 1) . \end{aligned} \tag{38}$$

It is important that most of the inner homologous core is subsonic. The sonic point is at $\chi = 1.39$ and the maximal infall velocity at $\chi = 2.96$ with $(-V/A) = 2.1$. The Mach number is almost constant in the outer core with the asymptotic value 2.94. (For further details we refer to [23].)

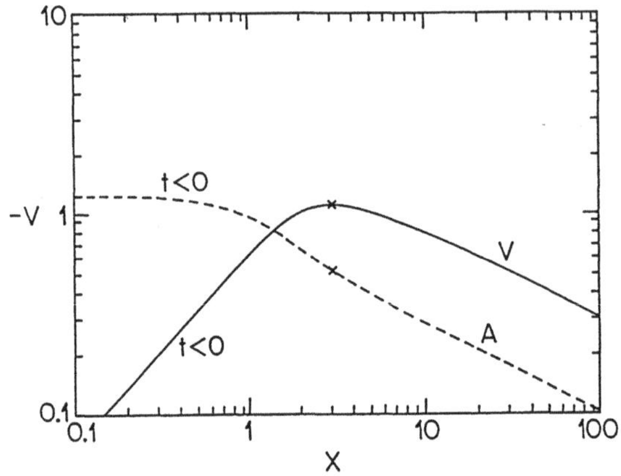

Fig. IV.2. Dimensionless infall velocity $-V$ and sound speed A for $\gamma = 1.30$ and $t < 0$. (Adapted from Ref. [23].)

These results are in agreement with numerical calculations, as is shown in Fig. IV.3. Note that the maximum speed becomes somewhat larger than $0.1\,c$.

IV.5 Shock Formation and Shock Energy

Collapse is finally halted when nuclear densities are achieved. At these densities the nuclei overlap and undergo a (phase) transition into a degenerate nucleon Fermi liquid with strong repulsive forces at small distances. The equation of state stiffens suddenly and within tenths of milliseconds the subsonic inner core rebounds as a unit. The supersonic mantle cannot communicate with the homologous core by means of pressure waves and continues its infall. With other words, the homologous core is oblivious to the supersonic mantle; no signals can travel upstream to cause the infalling material to adjust to the central obstacle. As always in such cases, the fluid adjusts suddenly and dramatically to the obstacle by forming a shock wave. The shock which is generated by the two colliding flows is formed at the interface where the Mach number is about 1. The boundary between the shocked and unshocked matter encloses roughly the mass $M_{\rm m}$ interior to the point, where the infall velocity is maximal. The shock energy is determined by the expansion work the inner core does in coming to rest.

This picture enables us to estimate the shock energy. Since the initial energy of the collapsing core was nearly zero, it is expected that the initial shock energy is equal to the negative of the binding energy of the unshocked inner core. (There is also some inner core heating which reduces the effective energy of the shock as it begins to propagate.) The mass, $M_{\rm m}$, of the inner core is established during infall and can be estimated from the self-similar solutions discussed in the last Section.

In order to estimate the binding energy of the unshocked inner core, we model this residue as a composite of two nested polytropes [27]. In a high

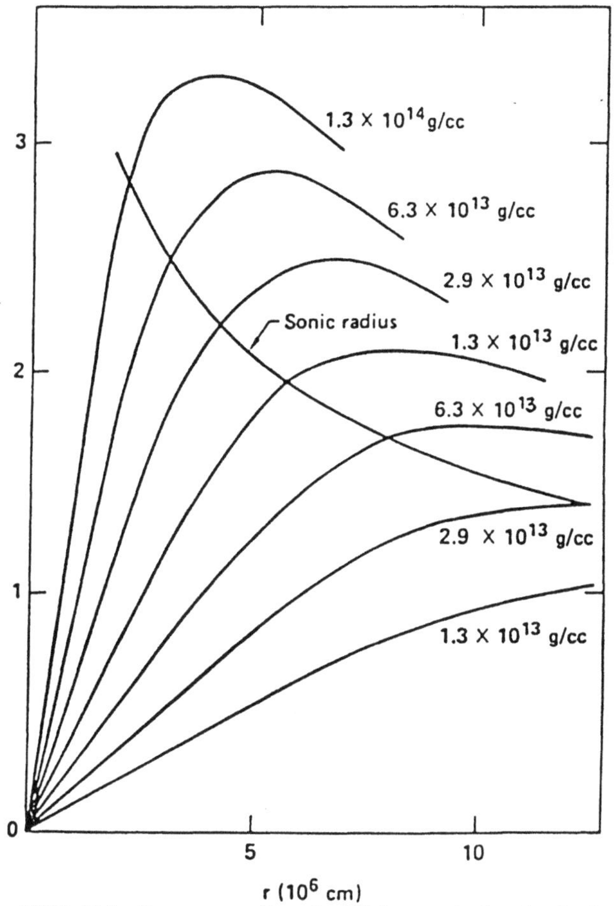

Fig. IV.3. Velocity versus radius of a $15 M_\odot$ star during the final 0.001 seconds before bounce. (From Ref. [26].)

density region $\varrho > \varrho_t \simeq (\frac{1}{2} - \frac{2}{3})\varrho_{\rm nucl}$ the nuclei are broken up and matter consists of a rather stiff nucleon-electron plasma which we approximate by a polytrop with inner index $n_1 \simeq 0$ to $n_1 \simeq 1$. ($n_1 = 0$ corresponds to an incompressible fluid; $n_1 < 1$ is plausible from baryon incompressibility at saturation [27].) The hard core is surrounded by an envelope of polytropic index n which should be slightly larger than three, because lepton pressure dominates. Figs. IV.4 and IV.5 show the binding energy of such a structure for various choices of parameters.

Fig. IV.5 shows that the binding energy depends only weakly on n_1 for realistic values of n ($n < 3.5$). From this we conclude that the binding energy is insensitive to the poorly known equation of state above nuclear densities. Furthermore, we see that the binding energy is about 5×10^{51} erg and that the unshocked inner core has a mass of $0.7 - 0.8\ M_\odot$. These results are also borne out by numerical calculations.

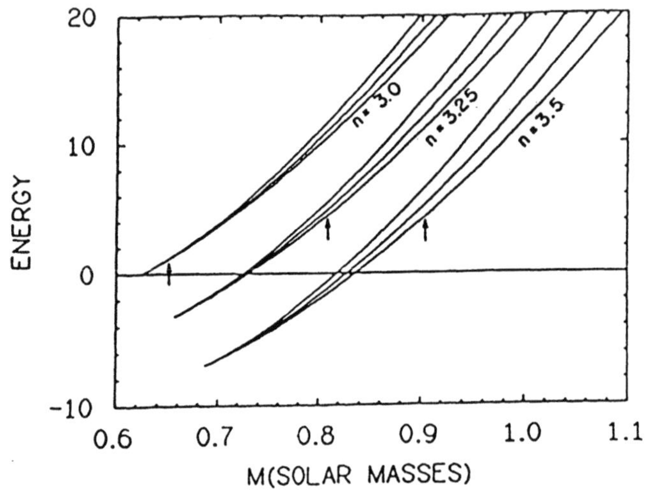

Fig. IV.4. Binding energy (in units of 10^{51} erg) versus total mass for the composite polytrope with $Y_\ell = 0.4$ for $n = 3.0, 3.25, 3.5$, $n_1 = 0, 0.5, 1$. Arrows mark M_m, the maximum velocity mass. (From Ref. [27].)

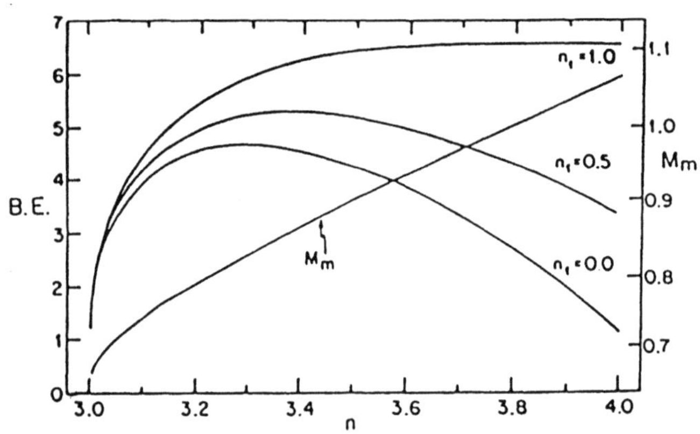

Fig. IV.5. Binding energy (in units of 10^{51} erg, left hand scale) of the composite polytrope with total mass M_m, plotted against n, for $n_1 = 0, 0.5, 1$. Also plotted is the maximal velocity mass M_m (in solar masses, right hand scale).

Summarizing, we expect that a shock is formed at a radius encompassing about $0.7 M_\odot$ with an initial energy of roughly 5×10^{51} erg. This is the supernova explosion in its infancy. The propagation and damping of this shock through the outer parts of the core, and most especially whether such a shock wave is capable of producing mass ejection in the overlying mantle and envelope, will be discussed later (Chap. VII).

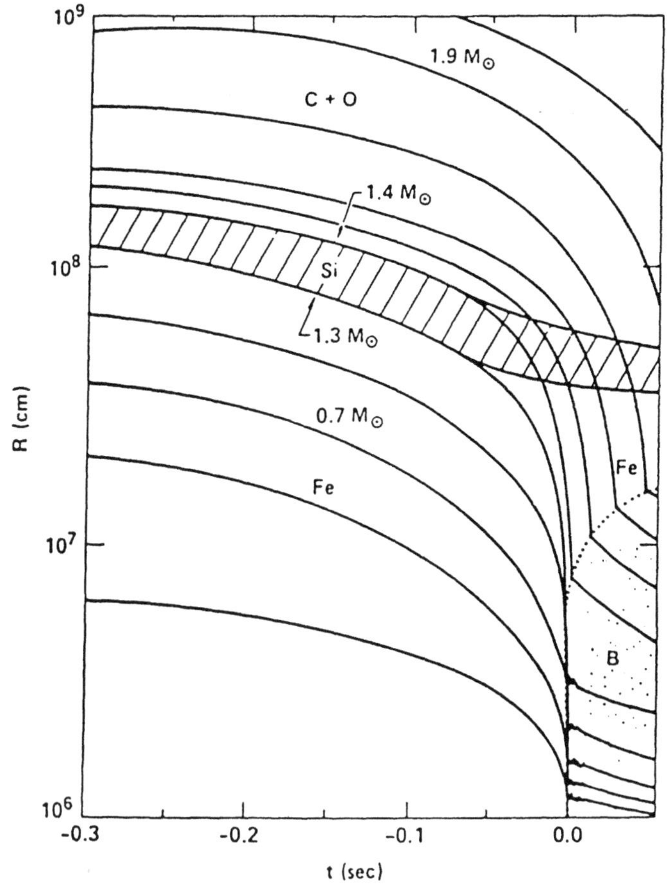

Fig. IV.6. Collapse and bounce of a $15 M_\odot$ star. (From Ref. [26].) The symbol B indicates free baryons.

V. Neutrino Radiation

The prominent role of neutrinos in massive star evolution and neutron star formation has already been emphasized repeatedly. In particular, most of the (negative) binding energy of a cold neutron star is carried away by neutrinos within a few seconds.

It was a great moment when the neutrino burst from the supernova SN 1987 A in the Large Magellanic Cloud was registered [28] in underground detectors last February (see Fig. V.1). We shall see that these observations are fully consistent with our standard model of neutron star formation in core collapse.

In this chapter we shall discuss the properties of the initial neutrino radiation in some detail. I begin with simple estimates and then present the results of detailed numerical calculations. We shall analyse the Kamiokande II and IMB data for SN 1987 A shown in Fig. V.1. In particular, we will discuss mass

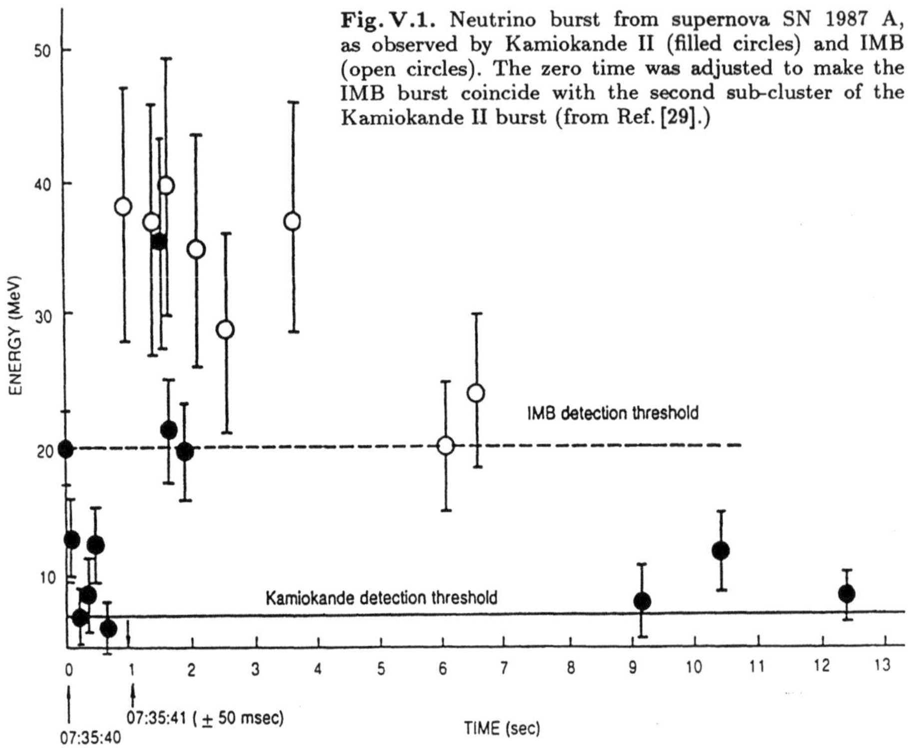

Fig. V.1. Neutrino burst from supernova SN 1987 A, as observed by Kamiokande II (filled circles) and IMB (open circles). The zero time was adjusted to make the IMB burst coincide with the second sub-cluster of the Kamiokande II burst (from Ref. [29].)

bounds for the electron neutrino which can be inferred from the energy versus time of arrival pattern of the detected neutrinos.

V.1 Estimates of the Properties of the Neutrino Burst

The initially formed proto-neutron star does not resemble the final cold neutron star with which you are familiar. It takes tens of seconds of quasi-hydrostatic thermal, structural, and compositional readjustment until the residue becomes a neutron star. Let me first describe qualitatively what happens during this period. (Neutrino emission dominates the further cooling of the neutron star, until the interior temperature falls to about 10^8 K [24].)

Initially the proto-neutron star has the following structure:

(i) A low entropy *inner core*, which has not been shocked, with
$M \simeq 0.6 - 0.8 M_{\odot}$, $\varrho = 3 \times 10^{13} - 4 \times 10^{14} \text{g/cm}^3$
$Y \simeq 0.35$.

(ii) An *outer core*, with $M \simeq 0.6 - 0.8 M_{\odot}$, $\varrho \geq 10^{11} \text{g/cm}^3$, through which a shock propagates outward, with a resulting jump in entropy.

A. Deleptonization. Let us first consider the deleptonization neutrino pulse from the outer core, which turns out to be too short and too weak to be observed with present detectors.

By the same argument as for the neutrino trapping during core collapse, the neutrinos in the hot, shocked outer core remain confined until the shock reaches the neutrinosphere (where the optical depth is one) at a density around $3 \times 10^{11}\mathrm{g/cm^3}$, corresponding to a radius of about 70 km [1]. Furthermore, the capture of electrons by protons behind the shock is limited by the neutrinos which are still present there, or which come from the inside. Once the shock front has passed the neutrinosphere, deleptonization is very fast, as we now show.

The lepton number flux F can be estimated by assuming a thermal spectrum characterized by a temperature T and a chemical potential μ_ν:

$$F = \frac{1}{4}\left(n_{\nu_e} - n_{\bar{\nu}_e}\right) = \frac{1}{4}\,\frac{1}{2\pi^2}T^3\left[F_2(\mu_\nu/k_\mathrm{B}T) - F_2(-\mu_\nu/k_\mathrm{B}T)\right], \qquad (1)$$

where $F_k(\eta)$ are the Fermi integrals (IV.23).

The difference in (1) and similar combinations, which will be needed later, can be computed exactly in the following manner. For $k = 0$ we have

$$F_0(\eta) = \ln(1 + \mathrm{e}^\eta) = \eta + \ln(1 + \mathrm{e}^{-\eta})$$

and thus

$$F_0(\eta) - F_0(-\eta) = \eta\ . \qquad (2)$$

One now uses the recursion relation

$$F_k' = kF_{k-1} \qquad (3)$$

and finds by integration

$$\begin{aligned} F_1(\eta) + F_1(-\eta) &= 2F_1(0) + \int_0^\eta [F_0(\eta') - F_0(-\eta')]d\eta' \\ &= 2F_1(0) + \frac{1}{2}\eta^2\ . \end{aligned}$$

For $F_k(0)$ we have

$$F_k(0) = (1 - 2^{-k})\ k!\zeta(k+1)\ , \qquad (4)$$

i.e.

$$F_1(0) = \frac{\pi^2}{12}, \quad F_2(0) = \frac{3}{2} \times 1.202, \quad F_3(0) = \frac{7\pi^4}{120}\ .$$

Thus

$$F_1(\eta) + F_1(-\eta) = \frac{1}{6}(\pi^2 + 3\eta^2)\ . \qquad (5)$$

Integration of this equation gives

$$F_2(\eta) - F_2(-\eta) = \frac{1}{3}\eta(\pi^2 + \eta^2)\ . \qquad (6)$$

1 The notion of a neutrinosphere is, of course, not very well defined, because the cross sections of neutrino reactions are energy dependent.

Continuing in this manner, we obtain

$$F_3(\eta) + F_3(-\eta) = \frac{7\pi^4}{60} + \frac{1}{2}\eta^2(\pi^2 + \frac{1}{2}\eta^2) , \tag{7}$$

etc. From (1) and (6) we find

$$F = \frac{1}{24\pi^2}\mu_\nu^3 \left[1 + \left(\frac{\pi T}{\mu_\nu}\right)^2\right] . \tag{8}$$

The time scale for deleptonization is $\tau_\nu = MN_\mathrm{A}\Delta Y_\ell/4\pi R_\nu^2 F$, where M is the mass of the shocked region, R_ν the radius of the neutrinosphere, ΔY_ℓ the average change of lepton fraction, and N_A is Avogadro's number. With (8) we find (with $\pi T/\mu_\nu \simeq 1$)

$$\tau_\nu \simeq 1.5 \text{ msec} \left(\frac{\Delta Y_\ell}{0.2}\right)\left(\frac{M}{3\times 10^{32}\mathrm{g}}\right)\left(\frac{R_\nu}{70 \text{ km}}\right)^{-2}\left(\frac{\mu_\nu}{10 \text{ MeV}}\right)^3 . \tag{9}$$

For typical parameters all dimensionless factors in (9) are close to unity. Thus we see that the deleptonization time scale is of the order of a few milliseconds.

Let us also estimate the luminosity of this very short pulse. The energy flux is with (7)

$$\begin{aligned} S &= \frac{1}{4}\,\frac{1}{2\pi^2}T^4\left[F_3(\mu_\nu/k_\mathrm{B}T) + F_3(-\mu_\nu/k_\mathrm{B}T)\right] \\ &= \frac{1}{4}\,\frac{1}{8\pi^2}\mu_\nu^4\left[1 + 2\left(\frac{\pi T}{\mu_\nu}\right)^2 + \frac{7}{15}\left(\frac{\pi T}{\mu_\nu}\right)^4\right] . \end{aligned} \tag{10}$$

This gives for the average neutrino energy in terms of $y := (\pi T/\mu_\nu)^2$:

$$\langle\varepsilon_\nu\rangle = \frac{S}{F} = \frac{3\mu_\nu}{4}\,\frac{1 + 2y + \frac{7}{15}y^2}{1+y} \simeq 10 \text{ MeV} \tag{11}$$

for typical neutrinospherical parameters $\mu_\nu \simeq 10$ MeV, $T_\nu \simeq 4$ MeV. For the luminosity we find

$$L_\nu = \frac{MN_\mathrm{A}\Delta Y_\ell\langle\varepsilon_\nu\rangle}{\tau_\nu} \sim \frac{10^{51}\mathrm{erg}}{1.5 \text{ msec}} \simeq 6\times 10^{53}\mathrm{erg/sec} . \tag{12}$$

After deleptonization ($\mu_\nu \simeq 0$) we have roughly blackbody neutrino emission

$$\begin{aligned} L_\nu &= N_\nu \frac{7}{8}\,\sigma\, T_\nu^4\, 4\pi R_\nu^2 \\ &= (1.4\times 10^{53}\mathrm{erg/sec})\left(\frac{N_\nu}{3}\right)\left(\frac{T}{4 \text{ MeV}}\right)^4\left(\frac{R_\nu}{40 \text{ km}}\right)^2 . \end{aligned} \tag{13}$$

Here N_ν is the number of neutrino generations. (Later we shall see that R_ν decreases from about 70 km to 40 km within less than one second.) Eq. (13) shows that the total binding energy (II.1) (see also Fig. II.5) is radiated in a *few seconds*, which is much longer than the typical hydrodynamical time scale (9). This long time scale is due to the slow diffusion of neutrinos out from the inner core. It is instructive to estimate this diffusion directly.

B. Diffusion Time. For the mean free path we can again use (IV.17) which represents the effect of neutrino scattering from nucleons and nuclei. (In an accurate treatment, one should include other processes; see Sect. V.4.) In the interior part of the unshocked region, the nuclei are broken up and hence (IV.17) reduces to

$$\lambda_\nu \simeq 1.0 \times 10^6 \mathrm{cm}\ \varrho_{12}^{-1} \left(\frac{\langle\nu\rangle}{10\ \mathrm{MeV}}\right)^{-2} . \tag{14}$$

The chemical potential μ_ν for β-equilibrium is close to μ_e and for degenerate neutrinos we have $\langle\nu\rangle = \frac{3}{4}\mu_\nu$. Since

$$\mu_e = 51.6(Y_e \varrho_{12})^{1/3}\ \mathrm{MeV} \tag{15}$$

we take $\langle\nu\rangle \simeq (150 - 200)$ MeV as a reasonable initial value in the central region. (This will be confirmed by the results of numerical simulations.) This gives for the diffusion time through the inner core

$$\tau_{\mathrm{diff}} \simeq \frac{R^2}{\pi^2 c\lambda/3} \simeq 3\ \mathrm{sec} \left[\frac{R_{\mathrm{inner\ core}}}{10\ \mathrm{km}}\right]^2 \left(\frac{\langle\nu\rangle}{150\ \mathrm{MeV}}\right)^2 \varrho_{14} ,$$

which is typically a few seconds. Thus we expect a luminosity

$$L_\nu = \frac{\mathrm{B.E.\ of\ neutron\ star}}{\tau_{\mathrm{diff}}} \simeq 10^{53} \mathrm{erg/sec} , \tag{16}$$

which compares favorably with Eq. (13).

V.2 Estimate of the Total Neutrino Loss of SN 1987 A

Let us now estimate the total neutrino loss of SN 1987 A from the Kamiokande II experiment [28], which has produced the richest data set and has a relatively low energy threshold. The detectors of Kamioka and Cleveland are similar in design: large volumes of ultrapure water surrounded by thousands of phototubes and located in deep mines to minimize the background from cosmic rays. The neutrino burst from SN 1987 A was detected via Cerenkov light from positrons created in the reaction $\bar{\nu}_e + p_{(\mathrm{free})} \to n + e^+$. This conclusion is based on the near isotropy of the events and the fact that the cross section for the $\bar{\nu}_e\ p \to n\ e^+$ reaction at ~ 10 MeV is about 100 times that for ν_e - e^- scattering and even about 600 times that for ν_μ - e^- scattering [2]. Electron neutrino scattering events would be forward-peaked. Since the angles of the first two events with respect to LMC are relatively small (18 ± 18^0, 15 ± 27^0), it has been suggested that they are ν_e events. Clearly, the statistics are such that this can

[2] More precisely, the relevant cross sections are:

$$\begin{aligned}
\sigma(\bar{\nu}_e p \to n\ e^+) &= 0.977 \times 10^{-41}\ \mathrm{cm}^2\ (E/10\ \mathrm{MeV})^2 \\
\sigma(\nu_e e^- \to \nu_e e^-) &= 0.933 \times 10^{-43}\ \mathrm{cm}^2\ (E/10\ \mathrm{MeV}) \\
\sigma(\bar{\nu}_e e^- \to \bar{\nu}_e e^-) &= 0.388 \times 10^{-43}\ \mathrm{cm}^2\ (E/10\ \mathrm{MeV}) \\
\sigma(\nu_\mu e^- \to \nu_\mu e^-) &= 0.159 \times 10^{-43}\ \mathrm{cm}^2\ (E/10\ \mathrm{MeV}) \\
\sigma(\bar{\nu}_\mu e^- \to \bar{\nu}_\mu e^-) &= 0.130 \times 10^{-43}\ \mathrm{cm}^2\ (E/10\ \mathrm{MeV}) .
\end{aligned}$$

never be proven and we discard this possibility. We note also that the Kamioka detection threshold is $\simeq 7$ MeV, and that the electron energies as inferred from the emitted Cerenkov light have 20 % – 25 % errors (see Fig. V.1).

Since the observed events are due to $\bar{\nu}_e$–absorption, we consider first the total energy loss $U_{\bar{\nu}_e}$ of the anti-electron neutrinos. The time integrated *number* flux has the spectral distribution

$$\Phi(E_\nu)dE_\nu = \frac{U_{\bar{\nu}_e}}{4\pi D^2} \cdot \frac{f(E_\nu)E_\nu^2 dE_\nu}{\int_0^\infty f(E_\nu)E_\nu^3 dE_\nu} , \tag{17}$$

where f is approximated by a thermal distribution with an effective temperature T and zero chemical potential,

$$f(E_\nu) = \left(\mathrm{e}^{E_\nu/T} + 1\right)^{-1} ; \tag{18}$$

D is the distance to the LMC. We assume that the spectrum does not change appreciably with time. Hence, the total number of detected neutrinos is

$$N_\mathrm{d} = \int_{E_\mathrm{th}}^\infty N_\mathrm{p}\sigma(E_\mathrm{e})\Phi(E_\nu)\varepsilon(E_\mathrm{e})dE_\nu ,$$

where N_p is the total number of free protons, $\varepsilon(E_\mathrm{e})$ the position detection and reconstruction efficiency of the detector, E_th the e^+–threshold detection energy and $\sigma(E_\mathrm{e})$ the cross section for $\bar{\nu}_e + \mathrm{p} \to \mathrm{n} + \mathrm{e}^+$, given by

$$\sigma(E_\mathrm{e}) = \frac{1}{4}\sigma_0[1 + 3(g_\mathrm{A}/g_\mathrm{V})^2]\frac{p_\mathrm{e}E_\mathrm{e}}{m_\mathrm{e}^2} \tag{19}$$

with σ_0 given by (IV.11). Neglecting the electron mass and the difference of E_ν and E_e in $E_\nu \simeq E_\mathrm{e} + m_\mathrm{n} - m_\mathrm{p}$, we obtain

$$N_\mathrm{d} = N_\mathrm{p}\frac{9.7 \times 10^{-42}\mathrm{cm}^2}{4\pi D^2(10\ \mathrm{MeV})^2}\, U_{\bar{\nu}_e}\, T\, \frac{G_4(E_\mathrm{th}/T)}{F_3(0)} , \tag{20}$$

where F_n are the standard Fermi integrals (IV.23) and

$$G_n(E_\mathrm{th}/T) = \int_{E_\mathrm{th}}^\infty E^n \varepsilon(E)f(E)dE . \tag{21}$$

We rewrite this by introducing the average energy of the detected neutrinos

$$\langle E_\nu\rangle_\mathrm{d} = T\,\frac{G_5(E_\mathrm{th}/T)}{G_4(E_\mathrm{th}/T)} . \tag{22}$$

After solving Eq. (20) for $U_{\bar{\nu}_e}$ we find [30]

$$U_{\bar{\nu}_e} = 0.78 \times 10^{52}\mathrm{erg}\left(\frac{D}{50\ \mathrm{kpc}}\right)^2\left(\frac{1\ \mathrm{kt}}{M}\right)\left[\frac{F_3(0)G_5(E_\mathrm{th}/T)}{G_4^2(E_\mathrm{th}/T)}\right]\left(\frac{10\ \mathrm{MeV}}{\langle E_\nu\rangle_\mathrm{d}}\right)N_\mathrm{d} , \tag{23}$$

where M is the fiducial mass of the detector. Using published thresholds and efficiencies, Burrows and Lattimer [30] find by comparing (22) and (23) with the data of Kamiokande II

$$\begin{aligned} &T = 2.72 \pm 0.26\ \text{MeV}, \qquad \langle E_\nu \rangle_\text{d} = 8.57 \pm 0.82\ \text{MeV}, \\ &U_{\bar{\nu}_e} = 75\ {}^{-14}_{+36} \times 10^{51}\text{erg}. \end{aligned} \tag{24a}$$

The total neutrino emission of all neutrino species should be roughly six times larger. With the help of a second method, which involves predominantly sums instead of integrals, the errors are somewhat smaller [30]:

$$\begin{aligned} &T = 2.81 \pm 0.07\ \text{MeV}, \qquad \langle E_\nu \rangle_\text{d} = 8.86 \pm 0.22\ \text{MeV}, \\ &U_{\bar{\nu}_e} = 57\ {}^{-3}_{+2} \times 10^{51}\text{erg}. \end{aligned} \tag{24b}$$

Similar results have been obtained by many others (see, e.g., Ref. [31], and contributions in [32]). These estimates give a total neutrino loss which is close to the binding energy of a neutron star (Fig. II.5). This is very satisfactory and confirms the standard model of neutron star formation. Conversely, we can conclude that the number of light ($< 1\,\text{MeV}$) neutrino flavors cannot be substantially larger than three. Otherwise we would run into troubles with the supernova energetics.

Before we discuss detailed model calculations for the neutrino burst and their comparision with the data (Sect. V.4), we turn now to the question whether the observation of neutrinos from SN 1987 A can place interesting upper limits on the mass of the electron's neutrino, m_{ν_e}.

V.3 Bounds on the Electron Neutrino Mass

When the news spread around the world that a neutrino burst from the LMC supernova had been observed, everybody interested in neutrino masses made his private investigation on possible constraints from the arrival pattern of neutrinos as a function of their energy. A small fraction of easy calculations was immediately turned into a flood of circulating preprints. Unfortunately, only a few of these investigations where useful. In the meantime some authors have analysed neutrino mass limits more carefully, paying due attention to the question of statistical significance of small data samples, etc. [33, 34]. Clearly, upper mass limits should be pessimistic relative to all sorts of uncertainties. The published upper limit from the tritium decay by the Zürich group [35] has been established in this spirit and it would be unfair to beat this result from supernova data with improper means.

The velocity of a massive neutrino with energy E_ν is $((E_\nu^2 - m_\nu^2)^{\frac{1}{2}}/E_\nu) \simeq 1 - \frac{1}{2}(m_\nu/E_\nu)^2$. Hence, the time delay compared to the light travel time at a detector a distance D away from the supernova is

$$\Delta t \simeq \frac{D}{c}\frac{1}{2}\left(\frac{m_\nu c^2}{E_\nu}\right)^2 = 2.6 \text{ sec}\left(\frac{D}{50 \text{ kpc}}\right)\left(\frac{m_\nu c^2}{10 \text{ eV}}\right)^2\left(\frac{10 \text{ MeV}}{E_\nu}\right)^2 . \quad (25)$$

Using this formula, one can translate the observed arrival pattern for any hypothetical mass m_ν into an emission pattern. Constraints on m_ν can only be obtained by requiring that the emission pattern is not totally unreasonable from the point of view of supernova theory.

On should, for instance, not impose the condition that the average emitted neutrino energy $\langle E_\nu \rangle$ decreases with time, because the cross sections for neutral current interactions (Sect. IV.2) increase as E_ν^2 and hence the diffusion time increases also. On the other hand, there are, of course, also cooling rethermalization processes (dominated by neutrino electron scattering) and we will see that some numerical simulations show a slowly decreasing average energy after a second or so. However, the existing calculations do not yet treat the neutrino transport with sufficient accuracy (see Sect. V.4). In view of this, it is not even safe to require that the time delay (25) should not be larger than the duration of the burst (12.4 sec for Kamiokande II), which would imply the mass limit $m_\nu < 25$ MeV.

Let me now describe briefly the careful analysis of neutrino mass constraints in Refs. [33, 34], which is based on generally accepted theoretical ideas of neutrino emission by a supernova.

In [33] the authors assume basically only that some diffusion process on the time scale of seconds is at work, in order to constrain the time distribution $f(E_\nu, t_{\mathrm{SN}})$ of emitted neutrinos at fixed energy E_ν. In terms of this function, the expected normalized distribution of events in energy and earthy time, t, is proportional to $f(E_\nu, t_{\mathrm{SN}}(t))$, where the function $t_{\mathrm{SN}}(t)$ is determined by (25): $t_{\mathrm{SN}}(t) = t - \Delta t + \text{const}$. The proportionality factor is the expected time-integrated positron energy distribution dN/NdE_{e}, which is estimated as in Sect. V.2, by assuming a thermal spectrum for the emitted neutrinos characterized by an effective temperature T.

In Ref. [33] two types of extreme diffusion models are considered. In the first "hot spot" model the neutrino source is a small sphere from which the neutrinos escape and diffuse through relatively cooler material out to the much larger neutrino sphere with radius R_ν. In this model f is proportional to the time derivative of the Green function $(4\pi D^3 t^3)^{-\frac{1}{2}} \exp(-r^2/4Dt)$ of the diffusion equation (D = diffusion constant). If we normalize the time integral of f to one, we then have (ignoring contributions which decay faster)

$$f(E_\nu, t_{\mathrm{SN}}) = \frac{2}{\sqrt{\pi}} \frac{\tau^{3/2}}{t_{\mathrm{SN}}^{5/2}} \mathrm{e}^{-\tau/t_{\mathrm{SN}}} \quad (26\mathrm{a})$$

with $\tau = R_\nu^2/4D$. Diffusion theory gives $D = c\lambda/3$ where $\lambda = 1/\sigma\varrho$ is the mean free path (see Appendix C).

The authors set

$$\tau = 1 \text{ sec} \left(\frac{E_\nu}{10 \text{ MeV}} \right)^n \frac{1}{\gamma} \tag{27}$$

and fix γ (for various n) by the data. Note that in this model $t_{\rm SN}$ and E_ν appear for $n = 2$ (neutral current interactions) only in the combination $t_{\rm SN}/E_\nu^2$. This time-energy correlation is a crucial ingredient of the model.

As an other extreme a "hot ball" model is introduced for comparison. Here it is assumed that initially all the material within R_ν is uniformly hot. Then the time dependence is soon dominated by the principal mode of the diffusion equation of a ball, giving

$$f(E_\nu, t_{\rm SN}) = \frac{1}{\tau} e^{-t_{\rm SN}/\tau}, \tag{26b}$$

with $\tau = R_\nu^2/\pi^2 D$, which is again parametrized in the form (27).

These models are now fitted with a likelihood method to the data. The "free" parameters are m_ν, T, n, γ, and $t_{\rm SN}^1$ (= time between core collapse and the emergence of the first neutrino to be observed). The goodness of the fits is determined with the Kolmogorov-Smirnov test. Clearly, the hot spot model with $n = 2$ is conservative with respect to upper limits on the neutrino mass. Actually, the different functional forms of (26a) and (26b) are in practice indistinguishable for a fixed value for n. What really matters is this power in the energy dependence of τ in (27).

The most stringent limits come from the Kamiokande II (K2) experiment. I summarize now the conclusions of the analysis in [33]. (Event 6, which has been eliminated by the Kamiokande authors as a likely low-energy background, is always ignored.)

(i) The hot-ball model with $n = 0$ (time-independent shape of the energy spectrum, decreasing in magnitude exponentially with time) fits the 11 K2 events with a vanishing neutrino mass.

(ii) Keeping all events (except 6) the hot-spot model with $n = 2$ gives the following pessimistic upper limits on m_ν: 9 eV (90 %), 11 eV (95 %), 15 eV (99 %). This competes favorably with the present Tritium β-decay upper limit [35].

(iii) If event 3, which is close to the threshold energy cut ($\simeq$ 7.5 MeV), is interpreted as a background event and is eliminated, then the conservative upper limits on m_ν move strongly upwards: 27 eV (90 %), 30 eV (95 %), 33 eV (99 %). These constraints are worse than the accepted laboratory limits.

(iv) This upper move does not happen, if the last three K2 events are eliminated, because they might be attributed to a late non-diffusive supernova "hickup".

For further details, we refer to the original paper. As a summary, one should keep in mind that the strong conclusions in (ii) depend crucially on the reality of the most suspicious event in the data set. Its elimination considerably weakens the upper limits on the electron (anti-) neutrino mass.

The analysis of Abott et al. [33] assumes that the neutrino temperature does not increase with time. In their analysis, Spergel and Bachall [34] have relaxed this assumption and considered a more general class of models, allowing for time changes in the temperature and also for the radius of the neutrino sphere. This introduces more parameters in the expected distribution function of events in energy and time. The model parameters are again determined by maximizing a natural likelihood function. The goodness of a fit was also measured by a (two-dimensional) Kolmogorov-Smirnov test, using Monte Carlo procedures to simulate synthetic data from the various models. These simulations are used to assign statistical significance to the mass limits.

From the combined Kamiokande II and IMB data sets, neutrino masses greater than 16 eV are rejected at the 5% significance level.

V.4 Detailed Models

In a more detailed treatment of the neutrino emission, one must solve the quasi-static evolutionary equations that govern the proto-neutron star. This is a problem very much like standard stellar evolution theory, but with the following differences. Photon transport is completely negligible at the high densities of the core and the energy transport is dominated by neutrinos. In addition, we have to include lepton number (L_e) transport. Furthermore, an accurate calculation should be done in the framework of GR.

A. Basic Equations. The basic general relativistic equations will be derived in detail in Chap. VI. For simplicity, we ignore in the following discussion general relativistic effects.

a) Hydrostatic Equilibrium. First of all we have the (quasi-) hydrostatic equations:

$$\frac{dP}{dr} = -\varrho \frac{Gm}{r^2} , \qquad (28)$$

$$\frac{dm}{dr} = 4\pi r^2 \varrho , \qquad (29)$$

where $m(r)$ denotes the enclosed rest mass. P is the sum of the matter pressure P_{m} and the radial component of the neutrino pressure P_ν.

b) Electron Lepton Number Conservation. L_e-conservation can be written as follows. Define $\bar{Z}$ by

$$n_{e^-} - n_{e^+} = \frac{1}{m_N} \varrho \bar{Z} . \qquad (30)$$

Then we have

$$\frac{1}{m_N} \partial_t \bar{Z} = -\mathcal{S} , \qquad (31)$$

$$\partial_t \left(\frac{n_{\nu_e} - n_{\bar{\nu}_e}}{\varrho} \right) + \frac{\partial}{\partial m} \left(4\pi r^2 F \right) = \mathcal{S} , \qquad (32)$$

where F is the lepton number flux (flux of ν_e minus flux of $\bar{\nu}_e$) and $\mathcal{S}$ the net rate of production of ν_e minus $\bar{\nu}_e$.

c) Energy Equation. The energy equation can be written in the form (nonrelativistic approximation of (VI.80)):

$$\partial_t(\varepsilon_{\rm m} + u_\nu/\varrho) + P\partial_t(1/\varrho) = -\frac{\partial}{\partial m}(4\pi r^2 S). \tag{33}$$

Here $\varepsilon_{\rm m}$ is the specific internal energy of the matter part and u_ν denotes the energy density of all neutrinos. (See also Eq. (69).)

d) Boltzmann Equation. Finally, we discuss the Boltzmann equations for the neutrino distribution functions $f_{\nu_i}(\boldsymbol{x}, \boldsymbol{k})$, which are normalized such that $f_{\nu_i}(\boldsymbol{x}, \boldsymbol{k})d^3x d^3k(2\pi)^{-3}$ is the number of neutrinos of species ν_i in the phase element $d^3x d^3k$ around $(\boldsymbol{x}, \boldsymbol{k})$.

The Boltzmann equation for each f has the following general form:

$$D_t f = -\Lambda(\omega)f + \Gamma(\omega)(1-f) + \left(\frac{df}{dt}\right)_{\rm scattering}. \tag{34}$$

($D_t := \partial_t + \hat{\boldsymbol{k}} \cdot \nabla_x$, $\omega = |\boldsymbol{k}|$). The first term on the right hand side describes true absorption and the second spontaneous as well as induced emission, with relative weights reflecting one of the Einstein relations (for fermions). For the appearance of the factor $1 - f$ the normalization of f is, of course, crucial. A second Einstein relation between $\Lambda(\omega)$ and $\Gamma(\omega)$ follows from T-invariance (microscopic reversibility). If we assume that matter is in local thermal equilibrium (LTE), we obtain, as for photons,

$$\frac{\Gamma(\omega)}{\Lambda(\omega)} = {\rm e}^{-(\omega-\mu_\nu)/kT}, \tag{35}$$

where μ_ν is the chemical potential

$$\mu_\nu = \mu_{\rm e} + \mu_{\rm p} - \mu_{\rm n}. \tag{36}$$

This enables us to write (34) in the form

$$D_t f = -\Lambda(\omega)\left[1 + {\rm e}^{-(\omega-\mu_\nu)/kT}\right](f - f^{(0)}) + \left(\frac{df}{dt}\right)_{\rm scatt.}, \tag{37}$$

where $f^{(0)}$ is the equilibrium distribution function

$$f^{(0)} = \frac{1}{{\rm e}^{(\omega-\mu_\nu)/kT} + 1}. \tag{38}$$

Let us now consider the scattering contributions. To be specific, we consider first neutrino-electron scattering. Let $f_{\rm e}(\boldsymbol{p})$ be the distribution function for the electrons, which in our applications will be an equilibrium distribution. Taking occupancy and inoccupancy of initial and final states into account and using microscopic reversibility, one obtains with the usual arguments:

$$\left(\frac{df}{dt}\right)_{\text{e-scatt.}} = -\int d^3p d^3k' \{(1-f(\boldsymbol{k}'))(1-f_{\text{e}}(\boldsymbol{p}'))f(\boldsymbol{k})f_{\text{e}}(\boldsymbol{p}) \\ - (1-f(\boldsymbol{k}))(1-f_{\text{e}}(\boldsymbol{p}))f(\boldsymbol{k}')f_{\text{e}}(\boldsymbol{p}')\}\, v\frac{d^3\sigma}{d^3k'}(\boldsymbol{k},\boldsymbol{p}\to\boldsymbol{k}',\boldsymbol{p}'). \tag{39}$$

Here $v d\sigma(\boldsymbol{k},\boldsymbol{p}\to\boldsymbol{k}',\boldsymbol{p}')$ is the differential scattering rate (v = relative velocity) for ν–e scattering with the indicated momenta. In thermodynamic equilibrium the curly bracket in (39) vanishes.

Similar expressions hold for the scattering on nucleons and nuclei. It is easy to see that the effects of neutrino Fermi statistics cancel for elastic scattering (on nucleons and nuclei). This is not true for inelastic neutrino-electron scattering.

For LTE we have

$$\frac{f_{\text{e}}(p')(1-f_{\text{e}}(p))}{f_{\text{e}}(p)(1-f_{\text{e}}(p'))} = \text{e}^{(\omega'-\omega)/kT}$$

which can be used in (39):

$$\left(\frac{df}{dt}\right)_{\text{e-scatt.}} = -\int d^3k'\,\Sigma_{\text{es}}(\boldsymbol{k}\to\boldsymbol{k}')[(1-f(\boldsymbol{k}'))f(\boldsymbol{k}) \\ - \text{e}^{(\omega'-\omega)/kT}(1-f(\boldsymbol{k}))f(\boldsymbol{k}')], \tag{40}$$

where the scattering rate $\Sigma_{\text{es}}(\boldsymbol{k}\to\boldsymbol{k}')$ is given by

$$\Sigma_{\text{es}}(\boldsymbol{k}\to\boldsymbol{k}') = \int (1-f_{\text{e}}(\boldsymbol{p}'))f_{\text{e}}(\boldsymbol{p})\, v\frac{d^3\sigma}{d^3k'}\,(\boldsymbol{k},\boldsymbol{p}\to\boldsymbol{k}',\boldsymbol{p}')d^3p\,. \tag{41}$$

We use these results to write the Boltzmann equation in terms of the intensity I, given by

$$I(\omega,\boldsymbol{n}) = \frac{\omega^3}{(2\pi)^3}f(\boldsymbol{k}) \qquad (\boldsymbol{n}=\hat{\boldsymbol{k}}). \tag{42}$$

The basic neutrino radiation transport equation becomes now

$$D_t I(\omega,\boldsymbol{n}) = -\kappa^{\text{a}}(\omega)[I(\omega,\boldsymbol{n}) - B_\omega(T)] \\ - \int d\omega' d\Omega' \kappa^{\text{s}}(\omega,\boldsymbol{n}\to\omega',\boldsymbol{n}')\left\{I(\omega,\boldsymbol{n})\left[1-\frac{(2\pi)^3}{\omega^3}I(\omega',\boldsymbol{n}')\right]\right. \\ \left. - \left(\frac{\omega}{\omega'}\right)^3 I(\omega',\boldsymbol{n}')\left[1-\frac{(2\pi)^3}{\omega^3}I(\omega,\boldsymbol{n})\right]\right\}. \tag{43}$$

Here we have introduced the *effective absorption coefficient* (including stimulated emission)

$$\kappa^{\text{a}}(\omega) = \Lambda(\omega)\{1+\exp[(\mu_{\text{e}}+\mu_{\text{p}}-\mu_{\text{n}}-\omega)/kT]\} \tag{44}$$

and the *scattering coefficient* for all scattering processes

$$\kappa^{\mathrm{s}}(\omega, \boldsymbol{n} \to \omega', \boldsymbol{n}') = \sum_{i=\mathrm{e,n,p},\ldots} \int (1 - f_i(\boldsymbol{p}')) f_i(\boldsymbol{p})\, v \frac{d^2\sigma_i(\boldsymbol{k}, \boldsymbol{p} \to \boldsymbol{k}', \boldsymbol{p}')}{d\omega' d\Omega'} d^3p \,. \tag{45}$$

Furthermore, $B_\omega(T)$ denotes the Fermi-Dirac distribution for two-component neutrinos of chemical potential $\mu_{\mathrm{e}} + \mu_{\mathrm{p}} - \mu_{\mathrm{n}}$,

$$B_\omega(T) = \frac{1}{(2\pi)^3} \frac{\omega^3}{\mathrm{e}^{(\omega - \mu_{\mathrm{e}} - \mu_{\mathrm{p}} + \mu_{\mathrm{n}})/kT} + 1} \,. \tag{46}$$

For the μ- and τ-neutrinos the absorption coefficient vanishes and only neutral currents contribute to the elastic scattering.

The quantities $F, S, \mathcal{S}, u_\nu, P_\nu, n_\nu$ appearing in the basic equations (30–33) can all be expressed in terms of the neutrinos distribution functions (and reaction rates in the case of $\mathcal{S}$). Together with an equation of state (to be discussed later), we have in principle a complete system of equations. In practice drastic simplifications are made. In particular, the neutrino transport has, so far, been mainly treated in the so-called flux limited diffusion approximation, to which we now turn.

B. Diffusion Approximation and Flux Limited Diffusion. Deep inside the core of the proto-neutron star the intensity I is close to (46), where T is the local matter temperature and we can use the diffusion approximation.

As a starting point for the diffusion approximation and an approximation called flux limited diffusion, one takes the first two moments of (43). In the diffusion limit, one neglects certain anisotropies and the time derivative $\partial_t \boldsymbol{S}$ of the energy current $\boldsymbol{S}$ in the first moment equation. As shown in detail in Appendix C, one then obtains *Fick's law* for the spectral energy flux

$$\boldsymbol{S}(\omega) = -\frac{1}{3\kappa(\omega)} \nabla u(\omega) \tag{47}$$

with

$$\kappa(\omega) = \kappa^{\mathrm{a}}(\omega) + \kappa^{\mathrm{s}}(\omega) \langle 1 - \cos\theta \rangle \,, \tag{48}$$

where $\langle \cos\theta \rangle$ is the average of the cosine of the scattering angle, θ, taken with $\kappa^{\mathrm{s}}(\omega, \boldsymbol{n} \to \omega', \boldsymbol{n}')$, and

$$\kappa^{\mathrm{s}}(\omega) = \int \kappa^{\mathrm{s}}(\omega, \boldsymbol{n} \to \omega', \boldsymbol{n}') d\omega' d\Omega'. \tag{49}$$

In the central region, we can take for the energy density u in (47) the equilibrium distribution, which is $4\pi B_\omega(T)$. Adding the contribution of antineutrinos, we obtain from (47) for the radial spectral energy flux:

$$S(\omega) = -\frac{4\pi}{3} \left[\frac{1}{\kappa_\nu(\omega)} \frac{dB_\omega(T, \eta)}{dr} + \frac{1}{\kappa_{\bar{\nu}}(\omega)} \frac{dB_\omega(T, -\eta)}{dr} \right], \tag{50}$$

where $\eta = (\mu_{\mathrm{e}} + \mu_{\mathrm{p}} - \mu_{\mathrm{n}})/kT$. If we use

$$\frac{dB_\omega}{dr} = \frac{\partial B_\omega}{\partial T} \frac{dT}{dr} + \frac{\partial B_\omega}{\partial \eta} \frac{d\eta}{dr},$$

we can write the frequency integrated flux in the form [36]:

$$S = -\frac{1}{3}\left(\frac{7}{8}\frac{4\pi^2}{15}T^3\right)\left[\ell_T\frac{dT}{dr} + T\ell_\eta\frac{d\eta}{dr}\right]. \tag{51}$$

Here $\ell_T = \ell_{T\nu} + \ell_{T\bar{\nu}}$, $\ell_\eta = \ell_{\eta\nu} - \ell_{\eta\bar{\nu}}$, with ($x := \omega/T$)

$$\begin{aligned} \ell_{T\nu}(\eta) &= \frac{15}{7\pi^4}\int_0^\infty \frac{x^4 e^{x-\eta}dx}{\kappa_\nu(\omega)(e^{x-\eta}+1)^2}, \\ \ell_{\eta\nu}(\eta) &= \frac{15}{7\pi^4}\int_0^\infty \frac{x^3 e^{x-\eta}dx}{\kappa_\nu(\omega)(e^{x-\eta}+1)^2}, \end{aligned} \tag{52}$$

and corresponding expressions for $\bar{\nu}$. Similarly, the lepton number flux is

$$F = -\frac{4\pi}{3}\int\left[\frac{1}{\kappa_\nu(\omega)}\frac{dB_\omega}{dr} - (\nu \rightarrow \bar{\nu})\right]\frac{1}{\omega}d\omega$$

or

$$F = -\frac{1}{3}\left(\frac{7}{8}\frac{4\pi^2}{15}T^2\right)\left[\lambda_T\frac{dT}{dr} + T\lambda_\eta\frac{d\eta}{dr}\right] \tag{53}$$

with $\lambda_T = \lambda_{T\nu} - \lambda_{T\bar{\nu}}$, $\lambda_\eta = \lambda_{\eta\nu} + \lambda_{\eta\bar{\nu}}$, where

$$\lambda_{T\nu} = \ell_{\eta\nu} \tag{54}$$

and

$$\lambda_{\eta\nu}(\eta) = \frac{15}{7\pi^4}\int_0^\infty \frac{x^2 e^{x-\eta}}{\kappa(\omega)(e^{x-\eta}+1)^2}dx, \tag{55}$$

etc. Note that the transport coefficients ℓ, λ are averages of $1/\kappa(\omega)$ with the *unnormalized weight* functions

$$\frac{15}{7\pi^4}\frac{x^k e^{x\mp\eta}}{(1+e^{x\mp\eta})^2}, \qquad k = 2,3,4.$$

For numerical results, see for instance, Ref. [36].

Beyond the neutrino sphere there is free streaming. The difficult region for handling the neutrino transport is that between the diffusive core and the neutrino sphere. Most workers have used there the so-called flux limited diffusion approximation, which is a kind of rather ad hoc interpolation between diffusion and free streaming [37]. More precisely, one proceeds as follows, with the goal to close the moment equations. First, one assumes Fick's law in the form

$$\boldsymbol{S}(\omega) = -D(\omega)\nabla u(\omega), \tag{56}$$

with some diffusion constant D which will be specified shortly. This should be considered as a rough substitute of the first moment equation. This is now used in the zeroth moment of Eq. (43), giving the following energy transport equation

$$\begin{aligned} \partial_t u =& \nabla\cdot(D\nabla u) + \kappa^a[4\pi B_\omega - u] \\ &- \int d\omega' d\Omega' d\Omega \kappa^s(\omega, \boldsymbol{n} \rightarrow \omega', \boldsymbol{n}')\{\cdots\}, \end{aligned} \tag{57}$$

where the curly bracket is the same as in (43). Note that only inelastic scattering contributes to the integral. This contribution is often treated in the Fokker-Planck approximation.

Eq. (47) shows that in the diffusion limit

$$D(\omega) = \frac{1}{3}\lambda(\omega), \quad \lambda(\omega) = \frac{1}{\kappa(\omega)}. \tag{58}$$

The relation (56) holds also in the free streaming limit, where $S(\omega) = u(\omega)$ and thus

$$D = -\left(\frac{d\ln u}{dr}\right)^{-1}. \tag{59}$$

In the intermediate regime D is quite arbitrary. Most authors use an interpolation of the following form

$$D = \frac{\lambda(\omega)}{3 + \lambda(\omega) \mid d\ln u(\omega) \,/\, dr \mid \xi}. \tag{60}$$

The dimensionless quantity ξ is chosen as some function of $x = \lambda\ |d\ln u/dr|$, such that the limits (58) and (59) are guaranteed for $x \ll 1$, resp. $x \gg 1$. The form of this function is sometimes determined by fits to more accurate treatments. (For further details, see [37], [38], [26], and references therein.)

C. Source Function $\mathcal{S}$. We need also an expression for the source term $\mathcal{S}$ in (31) and (32), to which the following reactions contribute as sources and sinks:

$$\mathrm{e}^- + \mathrm{p} \rightleftharpoons \mathrm{n} + \nu_\mathrm{e}, \qquad \mathrm{e}^+ + \mathrm{n} \rightleftharpoons \mathrm{p} + \bar{\nu}_\mathrm{e}\ .$$

In β-equilibrium ($\mu_\nu = \mu_\mathrm{e} + \mu_\mathrm{p} - \mu_\mathrm{n}$) $\mathcal{S}$ vanishes, of course. If the neutrino distribution is thermal (with chemical potential μ_ν) and the nucleons are treated as free particles, it is not difficult to derive an explicit expression for $\mathcal{S}$.

Let us first consider the loss rate per baryon due to $\mathrm{e}^- + \mathrm{p} \to \mathrm{n} + \nu_\mathrm{e}$. Taking into account the Pauli-blocking in the final state, we obtain from (IV.2) and (IV.11) for the differential capture rate of relativistic electrons

$$d\Gamma = \frac{\sigma_0}{8\pi^2}\left[1 + 3(g_\mathrm{A}/g_\mathrm{V})^2\right](1 - f_{\nu_\mathrm{e}})\left(\frac{E_\nu}{m_\mathrm{e}}\right)^2 E_\mathrm{e}^2 f_\mathrm{e}\ . \tag{61}$$

This gives the following loss contribution to $\mathcal{S}$ per baryon

$$-Y_\mathrm{p}\frac{\sigma_0}{8\pi^2}\left[1 + 3(g_\mathrm{A}/g_\mathrm{V})^2\right]\frac{1}{m_\mathrm{e}^2}\int E_\nu^2(1 - f_{\nu_\mathrm{e}})E_\mathrm{e}^2 f_\mathrm{e} dE_\mathrm{e}\ .$$

Similarly (using $Y_\mathrm{n}/Y_\mathrm{p} = \mathrm{e}^{(\mu_\mathrm{n} - \mu_\mathrm{p})/T}$), one finds the other contributions, with the result:

$$\mathcal{S}/\ \text{baryon} = Y_\mathrm{p}\frac{\sigma_0}{8\pi^2}\left[1 + 3(g_\mathrm{A}/g_\mathrm{V})^2\right]\frac{T^5}{m_\mathrm{e}^2}\int_0^\infty dx\ x^4\{\cdots\},$$

where

$$\{\cdots\} = (1 - f(\eta_\nu))f(\eta_e) - e^{\eta_n - \eta_p}(1 - f(-\eta_\nu))f(-\eta_e) \\ - e^{\eta_n - \eta_p}(1 - f(\eta_e))f(\eta_\nu) - (1 - f(-\eta_e))f(-\eta_\nu) .$$

Here we have used the notation $\eta_e = \mu_e/kT$, etc, and

$$f(\eta) = \left(e^{x-\eta} + 1\right)^{-1} .$$

After some rearrangements one finds

$$\mathcal{S}/\text{baryon} = Y_p \frac{\sigma_0}{8\pi^2}\left[1 + 3(g_A/g_V)^2\right] \frac{T^5}{m_e^2}\left[S_4(\eta_e) - S_4(-\eta_e)\right] \\ \cdot \frac{1 - e^{-(\eta_e + \eta_p - \eta_n - \eta_\nu)}}{1 - e^{-(\eta_e - \eta_\nu)}} \tag{62}$$

with

$$S_n(\eta) = F_n(\eta) - F_n(-\eta) , \tag{63}$$

where F_n are the Fermi integrals (IV.23). S_4 can be obtained from (7)

$$S_4(\eta) = \frac{1}{5}\eta^5 + \frac{2\pi^2}{3}\eta^3 + \frac{7\pi^4}{15}\eta . \tag{64}$$

The last factor in (62) drives the matter into β-equilibrium.

In the transparent regime of the outer mantle, a similiar formula for the source function can be obtained [39]. [We shall consider gains and losses in the transparent region in connection with the possibility of delayed explosions by neutrino heating in Chap. VII.]

D. Equation of State and Other Constitutive Relations. The equations discussed so far must be supplemented by various constitutive relations, in particular by an equation of state $P = P(\varrho, T, Y_e)$ and an expression $\varepsilon = \varepsilon(\varrho, T, Y_e)$ for the specific internal energy. The neutrino opacities depend on the abundances of the various species. For these and also in (62), we need in particular the important quantity $\mu_n - \mu_p$, again as a function of ϱ, T, and Y_e.

Realistic calculations of these constitutive relations are clearly difficult and it would be out of place to go into a detailed discussion of what has been done so far. In obtaining the numerical results which I will discuss in the next subsection, a low level of sophistication has been adopted. In particular, a simple phenomenological equation for the zero-temperature component of the baryonic pressure at supranuclear densities has been used. This contribution will be varied, since no satisfactory neutron star equation of state exists today. (We have to deal not only with a difficult many-body problem, but also with the fundamental difficulty that our description of the strong interaction is not complete.) For detailed calculations of the properties of hot, dense matter below nuclear densities I refer again to [18] and [21].

E. Numerical Results. In an important paper [39] Burrows and Lattimer followed numerically the evolution of the residue of collapse to the cold neutron star state. The equations of stellar structure in this context are (28), (29), (31), (32), and the energy equation in a form which we derive now (Eq. (69) below).

We use the first law of thermodynamics

$$\frac{1}{m_{\mathrm{N}}} T \partial_t s = \partial_t(\varepsilon_{\mathrm{m}} + u_\nu/\varrho) + P\partial_t \left(\frac{1}{\varrho}\right) - \frac{1}{m_{\mathrm{N}}} \sum_i \mu_i \partial_t Y_i \tag{65}$$

and write (33) as

$$\frac{1}{m_{\mathrm{N}}} \left[T\partial_t s + \sum \mu_i \partial_t Y_i\right] = -\frac{\partial}{\partial m}(4\pi r^2 S). \tag{66}$$

If we add (31) and (32) and use (30) we obtain

$$\frac{1}{m_{\mathrm{N}}} \left[\partial_t Y_{\mathrm{e}} + \partial_t Y_\nu\right] + \frac{\partial}{\partial m}(4\pi r^2 F) = 0. \tag{67}$$

(Note that Y_{e} is the fraction of electrons minus the fraction of positrons and Y_ν is defined correspondingly.) From charge and baryon conservation and using the nuclear statistical equilibrium conditions (if nuclei are present), one derives easily

$$\sum \mu_i \partial_t Y_i = (\mu_{\mathrm{e}} + \mu_{\mathrm{p}} - \mu_{\mathrm{n}})\partial_t Y_{\mathrm{e}} + \mu_\nu \partial_t Y_\nu. \tag{68}$$

If we use this in (66) and subtract μ_ν times Eq. (67) we obtain

$$\begin{aligned} \frac{1}{m_{\mathrm{N}}} T\partial_t s = &- \frac{\partial}{\partial m}(4\pi r^2 S) + \mu_\nu \frac{\partial}{\partial m}(4\pi r^2 F) \\ &- \frac{1}{m_{\mathrm{N}}}(\mu_{\mathrm{e}} + \mu_{\mathrm{p}} - \mu_{\mathrm{n}} - \mu_\nu)\partial_t Y_{\mathrm{e}} \end{aligned}$$

or

$$\begin{aligned} \frac{1}{m_{\mathrm{N}}} T\partial_t s = &- \frac{\partial}{\partial m}(L_{\mathrm{e}} - \mu_\nu L_{\mathrm{n}}) - L_{\mathrm{n}} \frac{\partial \mu_\nu}{\partial m} \\ &- \frac{1}{m_{\mathrm{N}}}(\mu_{\mathrm{e}} + \mu_{\mathrm{p}} - \mu_{\mathrm{n}} - \mu_\nu)\partial_t Y_{\mathrm{e}} \ , \end{aligned} \tag{69}$$

where

$$L_{\mathrm{e}} = 4\pi r^2 S, \quad L_{\mathrm{n}} = 4\pi r^2 F \tag{70}$$

are the energy and lepton number luminosities.

The last term in the energy equation (69) gives the entropy that is generated when matter is not in β-equilibrium. The second term on the right-hand side is always positive and increases the entropy. In the first term $L_{\mathrm{e}} - \mu_\nu L_{\mathrm{n}}$ behaves like a heat flux and can be thought of as a cooling term, decreasing the total entropy.

Burrows and Lattimer solve the basic equations (with general relativistic modifications included) with a method that closely resembles the standard Henyey relaxation technique used in stellar evolution theory. I describe now some of their results which were fortunately available before the appearance of SN 1987 A. The reader should always be aware that the neutrino transport is treated rather crudely.

In the first paper [39], the baryon mass was taken to be 1.4 $M_\odot$ and no accretion or convection was considered. The initial conditions for entropy and lepton fraction correspond roughly to postbounce structures arrived at by fully hydrodynamical calculations (discussed in later chapters). Their choice induces some transient behavior for half a second, or so, which should not be taken seriously.

Figure V.2 shows the evolution of the temperature profile in the residue versus time. Each line correponds to a snapshot, and the entire figure covers 20 sec. In the first 0.5 sec, the high entropy hot shocked mantle radiates neutrinos and, hence, loses entropy. Subsequent compression (clearly seen in Fig. V.4) increases the exterior temperatures resulting in a temperature inversion which drives a heat flux into the interior. The effect of this on the entropy is seen

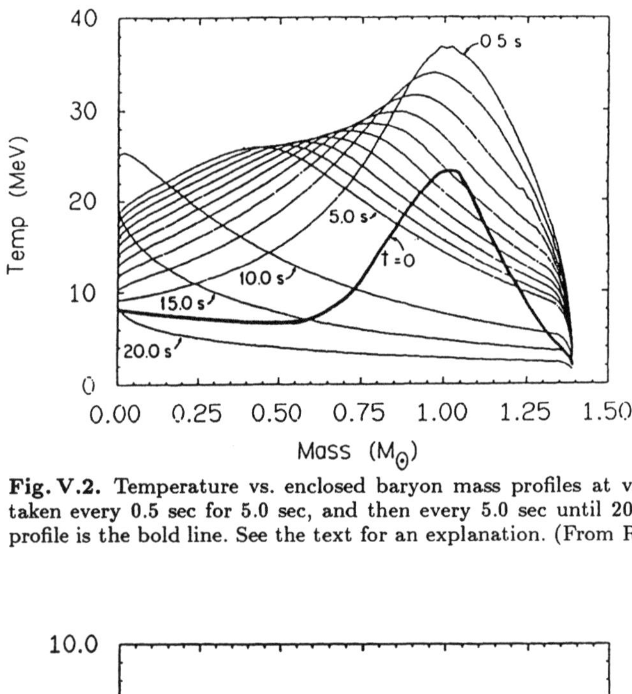

Fig. V.2. Temperature vs. enclosed baryon mass profiles at various times. Snapshots are taken every 0.5 sec for 5.0 sec, and then every 5.0 sec until 20.0 sec is reached. The $t = 0$ profile is the bold line. See the text for an explanation. (From Ref. [39].)

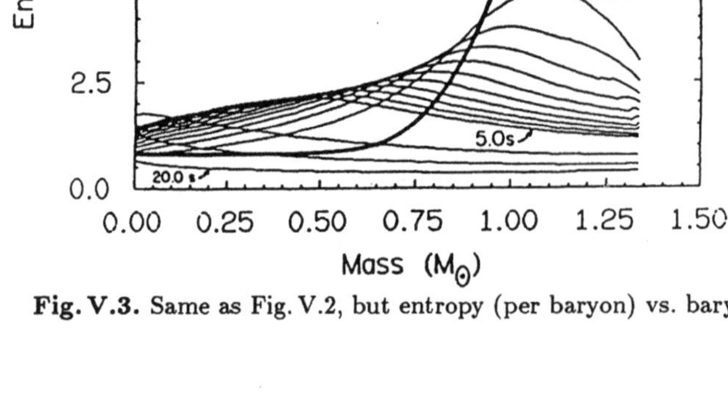

Fig. V.3. Same as Fig. V.2, but entropy (per baryon) vs. baryon mass. (From Ref. [39].)

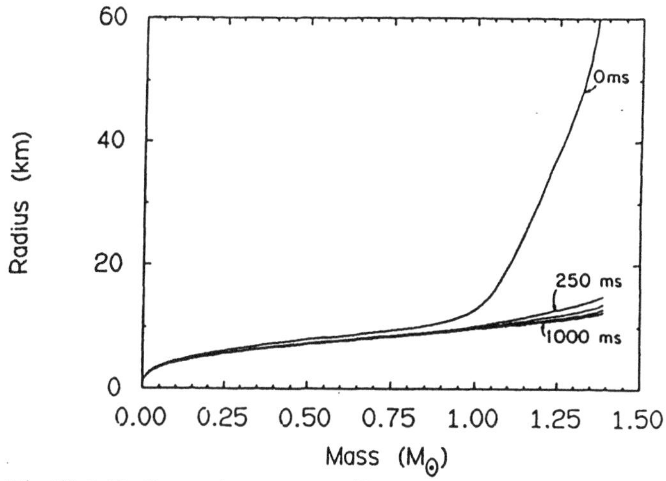

Fig. V.4. Radius vs. baryon mass. Dumps are every 250 msec for 1.0 sec. After ~ 1 sec the radius has already reached the final neutron star radius ~ 10 km. (From Ref. [39].)

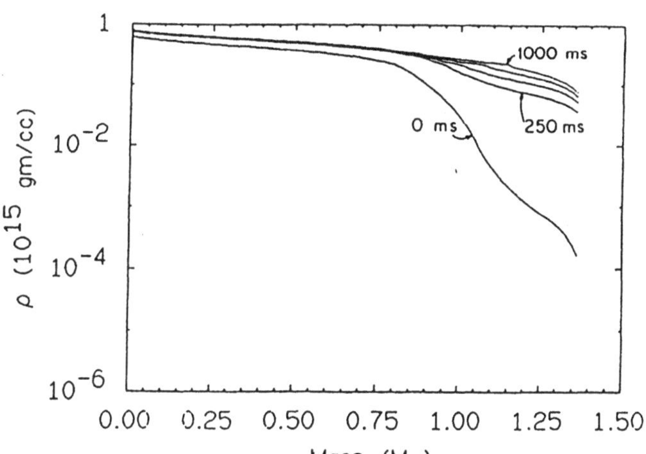

Fig. V.5. Rest mass density vs. baryon mass. (From Ref. [39].)

in Fig. V.3. This explains why temperatures and entropies can increase during the formation phase, despite the neutrino losses.

The dramatic compression of the outer layers is also seen in Fig. V.5.

The fast deleptonization of the exterior zones is seen in Fig. V.6, which shows also that the interior lepton fractions remain high for many seconds. It should be emphasized at this point that the uncertainties in the neutrinos opacities are typically a factor of two.

Fig. V.7 shows that after about 10 sec Y_ν is everywhere close to zero.

The behavior of the chemical potential μ_ν is shown in Fig. V.8. Again one sees

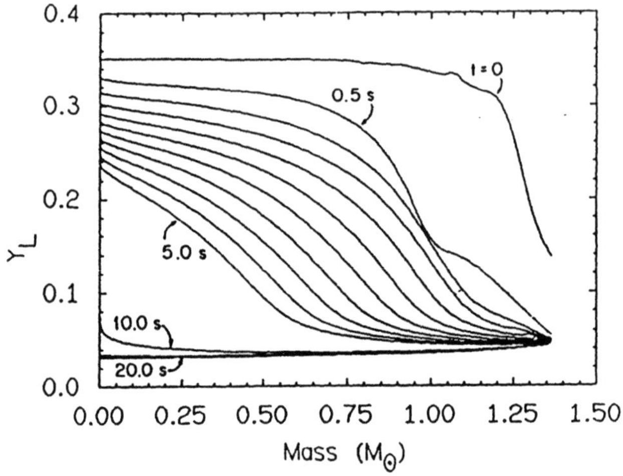

Fig. V.6. Lepton fraction $Y_L = Y_e + Y_\nu$, vs. baryon mass. Snapshots are at the same intervals as in Fig. V.2. (From Ref. [39].)

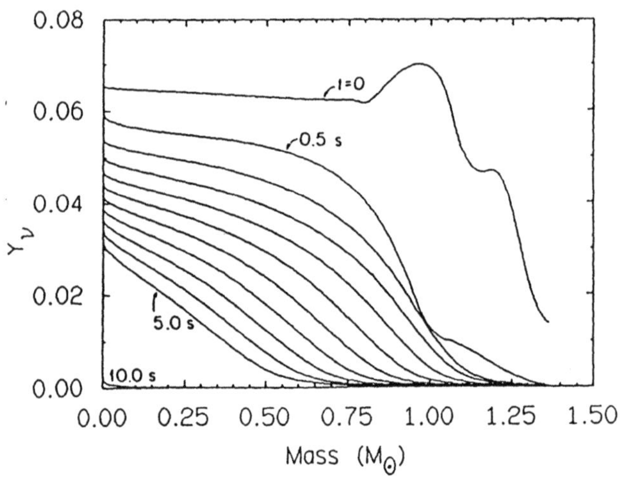

Fig. V.7. Same as Fig. V.6, but Y_ν vs. baryon mass. (from Ref. [39].)

that μ_ν remains high in the center for $\sim$ 5 sec and the ν_e's remain strongly degenerate. The gradient of μ_ν drives the lepton flux (see Eq. (53)).

Fig. V.9 shows that half the energy is radiated in the first 1.5 sec and three quarters after 5 sec.

Fig. V.10 should be considered as a representative example of the possible behavior of the average neutrino energy. (The results of other calculations will be shown below.) This and other results are not very reliable due to the poor neutrino transport treatment.

In Fig. V.11 the evolution of the average anti-electron neutrino energy during the first 12 seconds is shown. B 11 and B 17 are variations of model calculations

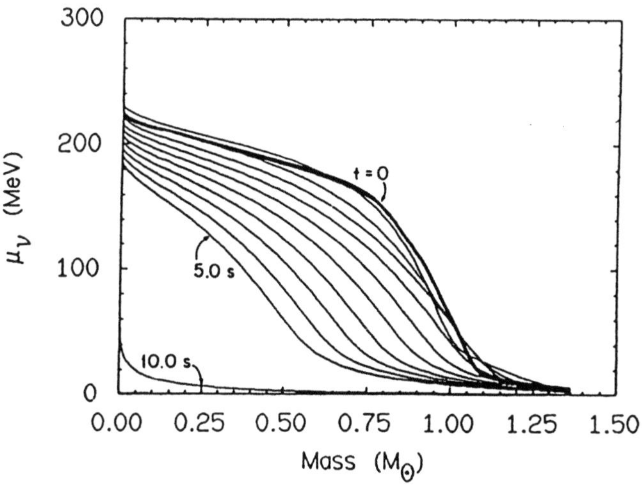

Fig. V.8. Chemical potential μ_ν (in MeV) vs. baryon mass. Snapshots are taken as in Fig. V.2. (From Ref. [39].)

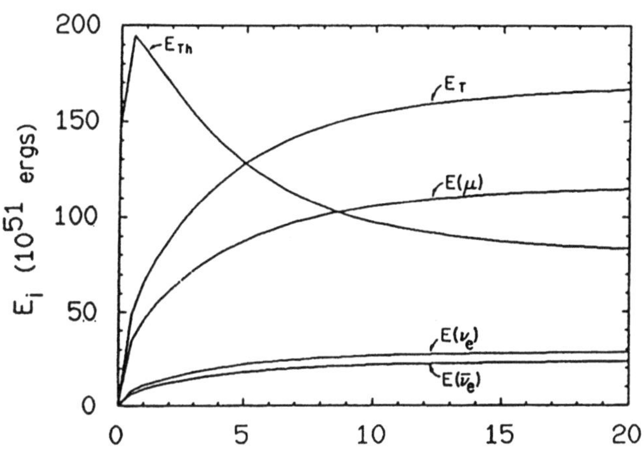

Fig. V.9. Energies E_i vs. time, where E_i is the total internal energy ($E_{\rm th}$), the total energy lost ($E_{\rm T}$), the energy lost in ν_μ's *and* ν_τ's [$E(\mu)$], the energy lost in ν_e [$E(\nu_e)$], or the energy lost in $\bar{\nu}_e$'s [$E(\bar{\nu}_e)$]. (From Ref. [39].)

discussed above [40]. B 17 incorporates a more realistic anti-neutrino cross section. The other theoretical curves are taken from [41] and [42].

Fig. V.12 depicts the anti-electron neutrino luminosity versus time for the first five seconds after bounce, as calculated by several authors. These model calculations should be considered as merely representative. Variations or ambiguities in the initial core mass, the neutron star equation of state, the amount of accretion after bounce, the degree of convection, the neutrino cross sections at supranuclear densities, and in particular the treatment of the neutrino transport imply rather large uncertainties in the behavior.

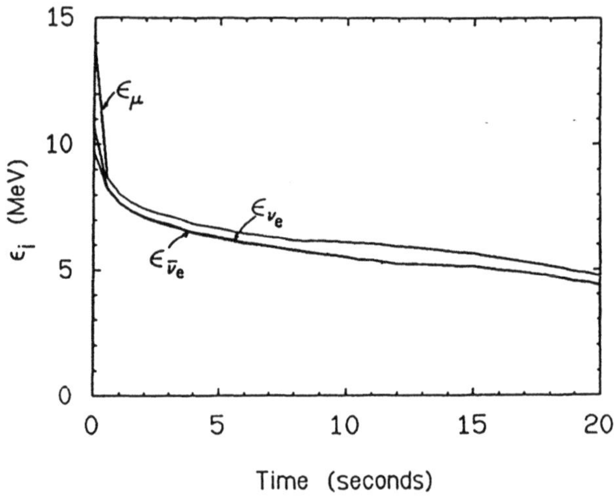

Fig. V.10. Average neutrino energies ε_i ($i = \nu_\mu, \nu_e, \bar{\nu}_e$) vs. time. (From Ref. [39].)

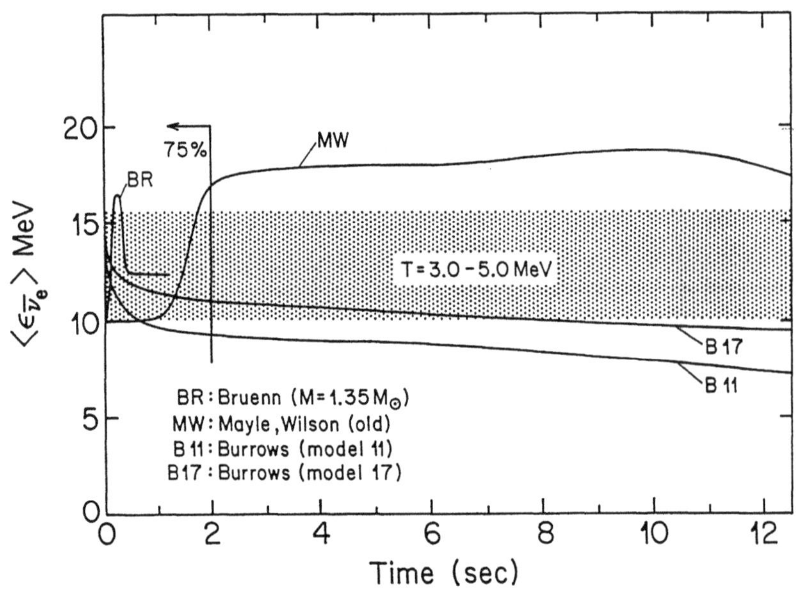

Fig. V.11. Average anti-electron neutrino energy vs. time for a sample of model calculations. The dark region fits the data best. (From Ref. [40].)

Burrows performed a whole series of model calculations which take into account some of the uncertainties just mentioned. Fig. V.13 shows his predictions for the first 18 seconds for a subset of this series. The detection efficiencies, thresholds, and fiducial masses for each detector were, of course, folded into the calculation and a distance of 50 kpc was used. *No normalization was imposed.* Some of the models incorporate accretion at the level indicated in the figure legend. The total $\bar{\nu}_e$ energy ($E_{\bar{\nu}_e}$) radiated in each model during the first

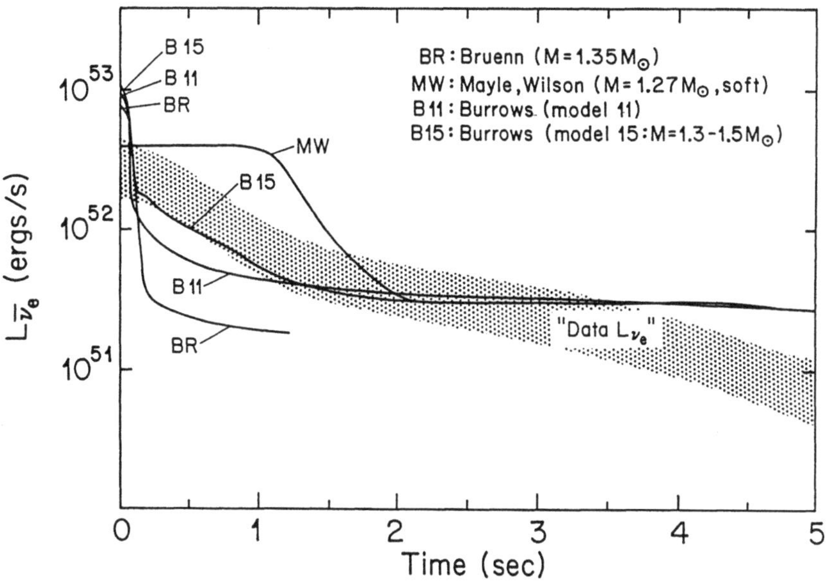

Fig. V.12. Anti-electron neutrino luminosity ($L_{\bar{\nu}_e}$) vs. time. (From Ref. [40].)

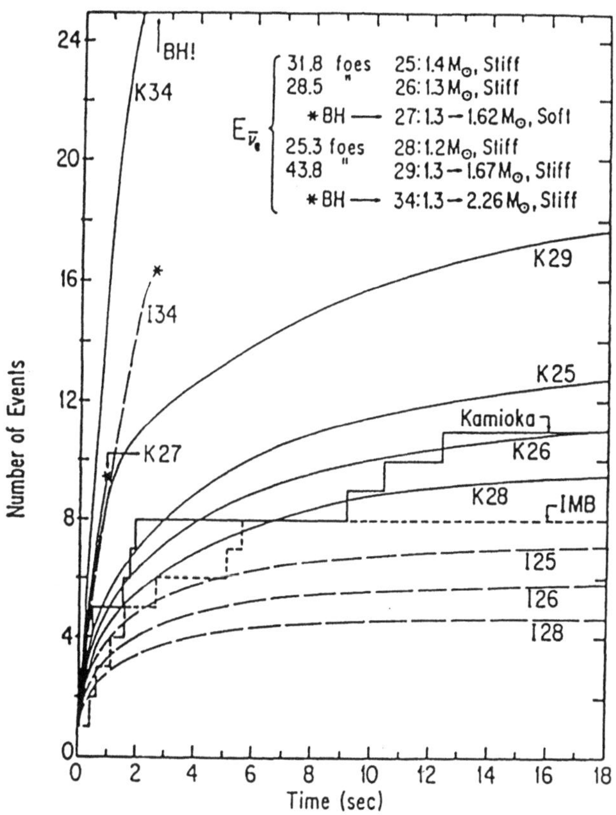

Fig. V.13. Total integrated number of events in IMB and Kamiokande II versus time for the first 18 seconds after bounce for various models. Superposed is the actual data (solid: Kamiokande II; dashed: IMB). Accretion rates, equation of state and total $\bar{\nu}_e$ energy loss is indicated in the legend. A more detailed discussion is given in the text. (From Ref. [40].)

20 seconds is also shown in the legend (1 "foe" = 10^{51}erg). "Stiff" means that the cold neutron star equation of state is similar to Bethe-Johnson I, and "soft" means that it is similar to Bethe-Johnson V (see Ref. [43]).

Fig. V.13 shows that some of the models do not fit the data. In particular, the formation of a black hole seems to be excluded. When a black hole is formed, the neutrino emission is abruptly (< 1 msec) stopped. Models with baryon core masses between $1.2M_\odot$ and $1.7M_\odot$ fit the data reasonably well. It is nice to see that the rapid rise during the first 1 – 2 seconds and the subsequent slow long-term cooling predicted by these models is borne out by the data.

VI. Basic Equations for Gravitational Collapse

In this section we discuss the basic equations which have to be solved for gravitational collapse problems. We begin with a general discussion and then specialize to spherical symmetry, since in most of the existing numerical simulations this symmetry is assumed (see, however, Ref. [44]).

A. General Discussion

We decompose the energy-momentum tensor into a "matter" part $T^{\alpha\beta}_{(\mathrm{m})}$ and a neutrino contribution $T^{\alpha\beta}_{(\nu)}$. First of all we have **Einstein's field equations**

$$G^{\alpha\beta} = 8\pi G(T^{\alpha\beta}_{(\mathrm{m})} + T^{\alpha\beta}_{(\nu)}) \tag{1}$$

and as a consequence the **energy-momentum "conservation"**

$$\nabla_\beta(T^{\alpha\beta}_{(\mathrm{m})} + T^{\alpha\beta}_{(\nu)}) = 0. \tag{2}$$

Let n_B be the proper baryon number density (measured in a frame for which matter is locally at rest), then the **conservation of baryon number** reads

$$\nabla_\alpha(n_\mathrm{B} U^\alpha) = 0, \tag{3a}$$

where U^α is the four-velocity of baryon transport. Neglecting the electron mass, we can write this as

$$\nabla_\alpha(\varrho_0 U^\alpha) = 0, \tag{3b}$$

where ϱ_0 is the matter rest mass density: $\varrho_0 \simeq m_\mathrm{B}\, n_\mathrm{B}$, m_B = average baryon mass.

Conservation of the lepton number L_e can be written as

$$\nabla_\alpha[(n_{\mathrm{e}^-} - n_{\mathrm{e}^+})U^\alpha] = -\nabla_\alpha(N^\alpha_{(\nu_\mathrm{e})} - N^\alpha_{(\bar\nu_\mathrm{e})}), \tag{4}$$

where n_{e^-}, n_{e^+} are the proper number densities of electrons and positrons and $N^\alpha_{(\nu)}$ denotes the number flux of neutrinos of species ν $(\nu_\mathrm{e}, \bar\nu_\mathrm{e}, \nu_\mu, \bar\nu_\mu, \nu_\tau, \bar\nu_\tau)$.

Conservation of electric charge can be expressed as

$$\nabla_\alpha[(n_{e^-} - n_{e^+})U^\alpha] = \nabla_\alpha(n_p U^\alpha), \tag{5}$$

where n_p denotes the proper density of protons.

The various neutrino species are described by one-particle distribution functions $f_{(\nu)}$ obeying Boltzmann equations. Before writing them down, we recall some generalities.

The metric g of the spacetime manifold M defines a natural diffeomorphism between the tangent bundle TM and the cotangent bundle T^*M, which can be used to pull back the natural symplectic form on T^*M. In terms of natural bundle coordinates, the diffeomorphism is given by $(x^\mu, p^\mu) \mapsto (x^\mu, p_\mu = g_{\mu\nu}p^\nu)$ and thus the induced symplectic 2-form on TM is

$$\omega = dx^\mu \wedge d(g_{\mu\nu}p^\nu). \tag{6}$$

The Lagrangian $L = \frac{1}{2}g_{\mu\nu}p^\mu p^\nu$ on TM defines a Hamiltonian vector field X_g on TM, determined by

$$i_{X_g}\omega = dL.$$

In terms of natural bundle coordinates the geodesic spray X_g is given by

$$X_g = (p^\mu, -\Gamma^\mu_{\alpha\beta}p^\alpha p^\beta), \tag{7}$$

where $\Gamma^\mu_{\alpha\beta}$ are the Christoffel symbols for (M, g). (For further details see [45].)

The one-particle phase space for particles of mass m, i.e., the submanifold $\{v \in TM : g(v,v) = m^2\}$, is invariant under the geodesic flow and we denote the restriction of X_g to the one-particle phase space also by X_g.

Let f be a distribution function on the one-particle phase space. The Vlasov and Boltzmann equatons for f involve the Lie derivative $L_{X_g}f$. If we consider the spatial components p^i relative to an orthonormal tetrad $\{e_\mu\}$ as independent variables of f, then the Liouville operator L_{X_g} can be written as

$$L_{X_g}f = p^\mu e_\mu(f) - \omega^i_\alpha(p)p^\alpha \frac{\partial f}{\partial p^i} \quad (i = 1,2,3), \tag{8}$$

where ω^α_β are the connection forms relative to the dual basis $\{\theta^\mu\}$.

In terms of $f_{(\nu)}$ we can express the energy-momentum tensor $T^{\alpha\beta}_{(\nu)}$ and the number flux $N^\alpha_{(\nu)}$ for each species as moments of $f_{(\nu)}$ in momentum space

$$T^{\alpha\beta}_{(\nu_i)} = \int p^\alpha p^\beta f_{(\nu_i)} d\Omega(p), \tag{9}$$

$$N^\alpha_{(\nu_i)} = \int p^\alpha f_{(\nu_i)} d\Omega(p), \tag{10}$$

where $d\Omega(p) = d^3p/p^0$ is the momentum phase space measure.

Assuming local thermodynamic equilibrium (LTE) for matter, the **Boltzmann equations** have the form

$$L_{X_g} f_{(\nu_i)} = -\kappa_{(\nu_i)} \varrho_0 \left[f_{(\nu_i)} - f^{(0)}_{(\nu_i)} \right] + \text{scattering terms}, \tag{11}$$

where $f^{(0)}_{(\nu_i)}$ is the equilibrium distribution with chemical potential $\mu_{(\nu_i)}$

$$f^{(0)}_{(\nu_i)}(p) = \left(e^{(p^0 - \mu_{(\nu_i)})/kT} + 1 \right)^{-1} \tag{12}$$

and $\kappa_{(\nu_i)}$ is the effective absorption coefficient including stimulated absorption (see Sect.V.4). In relativistic kinetic theory one shows [46] that the divergences of the moments (9), (10) can be written as

$$\nabla_\beta T^{\alpha\beta}_{(\nu_i)} = \int p^\alpha L_{X_g} f_{(\nu_i)} d\Omega(p) = \int [-\kappa_{(\nu_i)} \varrho_0 (f_{(\nu_i)} - f^{(0)}_{(\nu_i)}) + \cdots]\, p^\alpha d\Omega(p), \tag{13}$$

$$\nabla_\alpha N^\alpha_{(\nu_i)} = \int L_{X_g} f_{(\nu_i)} d\Omega(p) = \int [-\kappa_{(\nu_i)} \varrho_0 (f_{(\nu_i)} - f^{(0)}_{(\nu_i)}) + \cdots] d\Omega(p). \tag{14}$$

For simplicity, we assume that the matter part of the energy-momentum tensor can be taken in the ideal fluid approximation

$$T^{\alpha\beta}_{(\mathrm{m})} = [\varrho_0(1+\varepsilon_\mathrm{m}) + P_\mathrm{m}] U^\alpha U^\beta - P_\mathrm{m}\, g^{\alpha\beta}, \tag{15}$$

where ε_m is the specific internal energy density and P_m is the pressure. (Generalizations are discussed in [24]; see also Appendix A.)

Taking the scalar product of (2) with U^α, we obtain with (3b) the **energy conservation**

$$\varrho_0 [\nabla_U \varepsilon_\mathrm{m} + P_\mathrm{m} \nabla_U (1/\varrho_0)] = -U \cdot q, \tag{16}$$

where

$$q^\alpha := \sum_i \nabla_\beta T^{\alpha\beta}_{(\nu_i)}. \tag{17}$$

Projecting (2) on the space orthogonal to U, we find as an expression for **momentum conservation**

$$[\varrho_0(1+\varepsilon_\mathrm{m}) + P_\mathrm{m}] \nabla_U U - \nabla P_\mathrm{m} + (\nabla_U P_\mathrm{m}) U = -q + (q \cdot U) U. \tag{18}$$

These equations have to be supplemented by an equation of state $P_\mathrm{m} = P_\mathrm{m}(\varrho_0, \varepsilon_\mathrm{m})$ and by expressions for the opacities of various neutrino reactions.

B. Specialization to Spherical Symmetry in the Schwarzschild Gauge

For spherically symmetric configurations the gravitational field is not dynamical (no radiation). It is therefore possible to find a gauge in which the metric variables are determined instantaneously in terms of the matter variables [47]. Most of the numerical simulations have, however, been carried out in the

"Schwarzschild gauge" for which the metric takes the form [48]

$$g = e^{2\phi(r,t)}dt^2 - [e^{2\Lambda(r,t)}dr^2 + R^2(r,t)(d\vartheta^2 + \sin^2\vartheta d\varphi^2)] \tag{19}$$

and the matter four-velocity U is given by

$$U = e^{-\phi}\partial_t. \tag{20}$$

Connection and curvature forms, etc., relative to the orthonormal basis

$$\theta^0 = e^{\phi}dt,\ \theta^1 = e^{\Lambda}dr,\ \theta^2 = Rd\vartheta,\ \theta^3 = R\sin\vartheta d\varphi \tag{21}$$

are given in Ref. [48].

We parametrize the energy-momentum tensor of the neutrinos relative to (21) and its dual basis as

$$\left(T^{\alpha\beta}_{(\nu)}\right) = \left(\begin{array}{cc|cc} \varrho_\nu & S & 0 & 0 \\ S & P_\nu & 0 & 0 \\ \hline 0 & 0 & P_\nu + H & 0 \\ 0 & 0 & 0 & P_\nu + H \end{array}\right). \tag{22}$$

Neglecting the neutrino mass we see from (9) that the trace of $T^{\alpha}_{(\nu)\beta}$ vanishes and thus

$$H = \frac{1}{2}(\varrho_\nu - 3P_\nu)\ . \tag{23}$$

We set

$$\varrho = \varrho_0(1+\varepsilon_{\mathrm{m}}) + \varrho_\nu\ , \qquad P = P_{\mathrm{m}} + P_\nu \tag{24}$$

and thus the total energy-momentum tensor has the form

$$\left(T^{\alpha\beta}_{(\mathrm{m})} + T^{\alpha\beta}_{(\nu)}\right) = \left(\begin{array}{cc|cc} \varrho & S & 0 & 0 \\ S & P & 0 & 0 \\ \hline 0 & 0 & P + H & 0 \\ 0 & 0 & 0 & P + H \end{array}\right). \tag{25}$$

a) Baryon Conservation. Let us first consider the baryon number conservation (3b), which reads with Eq. (19)

$$\partial_t(\varrho_0 e^{\Lambda} R^2) = 0. \tag{26}$$

We can fix the gauge freedom in the radial coordinate r such that

$$e^{\Lambda} = \frac{1}{4\pi\varrho_0 R^2}\ . \tag{27}$$

Then the total rest mass enclosed by a sphere with radius r is equal to r, because

$$\int_0^r \varrho_0 4\pi R^2 e^{\Lambda} dr = r \ .$$

Carrying out the differentiation in (26) we obtain

$$\frac{1}{\varrho_0} D_t \varrho_0 = -D_t \Lambda - 2\frac{V}{R}, \tag{28}$$

where

$$D_t = e^{-\phi}\partial_t, \ V = D_t R. \tag{29}$$

b) Energy Equation. The energy equation can be written in the form (16), i.e.,

$$\varrho_0 [D_t \varepsilon_{\rm m} + P_{\rm m} D_t (1/\varrho_0)] = -q^0. \tag{30}$$

Alternatively, the 0-component (always relative to (21)) of Eq. (2) for the parametrization (25) leads to (for details see Appendix A):

$$\begin{aligned} D_t\varrho - \frac{\varrho + P}{\varrho_0} D_t \varrho_0 &= -D_r S - 2(D_r\phi + \frac{\Gamma}{R})S - 2\frac{V}{R}H \\ &= -(4\pi R^2)^{-1} e^{-2\phi} D_r (e^{2\phi} 4\pi R^2 S) - 2\frac{V}{R} H \end{aligned} \tag{31}$$

with

$$\Gamma = D_r R, \qquad D_r = e^{-\Lambda}\partial_r \ . \tag{32}$$

In obtaining (31) we have also made use of Eq. (28).

We note at this point that (32) and (27) imply

$$\Gamma = 4\pi \varrho_0 R^2 R' \quad (R' = \partial R/\partial r). \tag{33}$$

c) Momentum Equation. The 1-component of (2) gives the relation (see Appendix A)

$$(\varrho + P) D_r \phi = -D_r P - D_t S - 2\left(D_t \Lambda + \frac{V}{R}\right) S + 2\frac{\Gamma}{R} H. \tag{34}$$

Instead, we could use the 1-component of (18), which reads

$$[\varrho_0(1+\varepsilon_{\rm m}) + P_{\rm m}] D_r \phi + D_r P_{\rm m} = -q^1. \tag{35}$$

d) Constraint Equations. Next we write down the constraint equations: $G^0_\mu = 8\pi G T^0_\mu$. From [48] we obtain

$$G^0_1 = -2R^{-1} e^{-\phi-\Lambda}(\dot{R}' - \dot{R}\phi' - R'\dot{\Lambda}) = 8\pi G T^0_1 = -8\pi G S$$

or, with (29) and (32),

$$D_t \Lambda = \frac{V'}{R'} - 4\pi G \frac{R}{\Gamma} S. \tag{36}$$

With (28) this can also be written as

$$D_t \ln(\varrho_0 R^2) + \frac{V'}{R'} = 4\pi G \frac{R}{\Gamma} S. \tag{37}$$

Similarly, $G^{00} = 8\pi G T^{00} = 8\pi G \varrho$ gives

$$8\pi G \varrho R^2 = 1 - e^{-2\Lambda}(R'^2 + 2R''R) + V^2 + 2RVD_t\Lambda - R'R(e^{-2\Lambda})'. \tag{38}$$

We can write this in a more convenient form by using (36) and noting that $\Gamma = R'e^{-\Lambda}$ implies

$$\frac{1}{R'}(V^2R - \Gamma^2 R)' = 2\frac{V'}{R'}VR + V^2 - \Gamma^2 - 2e^{-2\Lambda}R(R'' - R'\Lambda').$$

This gives for (38)

$$8\pi G \varrho R^2 = 1 + \frac{1}{R'}(V^2R - \Gamma^2R)' - 8\pi G R^2 \frac{V}{\Gamma} S \tag{39}$$

which can be integrated to

$$2G \int_0^r 4\pi R^2 e^{\Lambda}[\Gamma\varrho + VS]dr = R + V^2R - \Gamma^2 R.$$

Here we have used the boundary condition: $\Gamma = 1$ for $r = 0$. Introducing the function

$$M(r,t) = \int_0^r 4\pi R^2 e^{\Lambda}(\Gamma\varrho + VS)dr \tag{40}$$

we can write the last equation as

$$\Gamma^2 = 1 + V^2 - \frac{2GM}{R}. \tag{41}$$

Clearly we have

$$D_r M = 4\pi R^2(\Gamma\varrho + VS) \tag{42}$$

or with (27)

$$M' = \frac{1}{\varrho_0}(\Gamma\varrho + VS). \tag{43}$$

e) Dynamical Equations. Finally, we work out the dynamical equation

$$G^{00} - G^{11} = 8\pi G(T^{00} - T^{11}) = 8\pi G(\varrho - P).$$

From [48] we find

$$\begin{aligned} 4\pi G(\varrho - P) = & \frac{1}{R^2}[1 - R'^2 e^{-2\Lambda} + \dot{R}^2 e^{-2\phi}] \\ & + \frac{1}{R}e^{-2\phi}\dot{R}\dot{\Lambda} + \frac{1}{R}e^{-2\Lambda}(R'\Lambda' - R'') \\ & + \frac{1}{R}e^{-2\phi}(\ddot{R} - \dot{R}\dot{\phi}) - \frac{1}{R}e^{-2\Lambda}R'\phi'. \end{aligned} \tag{44}$$

This can be rewritten in the form

$$4\pi G R^2(\varrho - P) = 1 + D_t(RV) + RVD_t\Lambda - D_r(R\Gamma) - R\Gamma D_r\phi. \tag{45}$$

Here we use Eq. (36) and compute $D_r\Gamma$ with the help of (41) and (42):

$$D_r\Gamma = \frac{V'}{R'}V + \frac{GM}{R^2} - 4\pi GR\varrho - 4\pi GR\frac{V}{\Gamma}S. \tag{46}$$

After some simplifications we then find the basic equation

$$D_tV = \Gamma D_r\phi - G\frac{M + 4\pi R^3 P}{R^2}. \tag{47}$$

Applying D_t on Eq. (32) und using (36) as well as (41) gives

$$D_t\Gamma = VD_r\phi + 4\pi GRS. \tag{48}$$

Together with (47), we obtain as a further consequence of the previous equations

$$D_tM = -4\pi R^2 PV - 4\pi R^2\Gamma S. \tag{49}$$

f) Liouville Operator. Finally, we have to work out the Liouville operator (8). We use the parametrization

$$p = \left(E,\ E\mu,\ E\sqrt{1-\mu^2}\cos\bar{\varphi},\ E\sqrt{1-\mu^2}\sin\bar{\varphi}\right) \tag{50}$$

and consider f as a function of the independent variables (E, μ, r, t).

The first term on the rhs of (8) is

$$p^\mu e_\mu(f) = ED_tf + E\mu D_rf. \tag{51}$$

With the help of the connection forms [48] we obtain for the last term in (8)

$$\begin{aligned}\omega^i_\alpha(p)p^\alpha\frac{\partial f}{\partial p^i} =& D_r\phi(p^0)^2\frac{\partial f}{\partial p^1} - \frac{\Gamma}{R}\left[(p^2)^2 + (p^3)^2\right]\frac{\partial f}{\partial p^1} \\ &+ \frac{\Gamma}{R}(p^1p^2\frac{\partial f}{\partial p^2} + p^1p^3\frac{\partial f}{\partial p^3}) - \frac{1}{R}\mathrm{ctg}\vartheta\left[(p^3)^2\frac{\partial f}{\partial p^2} - p^2p^3\frac{\partial f}{\partial p^3}\right] \\ &+ D_t\Lambda p^1p^0\frac{\partial f}{\partial p^1} + \frac{V}{R}\left(p^2p^0\frac{\partial f}{\partial p^2} + p^3p^0\frac{\partial f}{\partial p^3}\right).\end{aligned} \tag{52}$$

From

$$0 = \frac{\partial f}{\partial\bar{\varphi}} = -\frac{\partial f}{\partial p^2}p^3 + \frac{\partial f}{\partial p^3}p^2$$

we see that the term proportional to $\mathrm{ctg}\vartheta$ in (52) vanishes. We use also

$$E(1-\mu^2)\frac{\partial f}{\partial\mu} = \left[(p^2)^2 + (p^3)^2\right]\frac{\partial f}{\partial p^1} - p^1p^2\frac{\partial f}{\partial p^2} - p^1p^3\frac{\partial f}{\partial p^3}$$

and

$$E\frac{\partial f}{\partial E}=p^1\frac{\partial f}{\partial p^1}+p^2\frac{\partial f}{\partial p^2}+p^3\frac{\partial f}{\partial p^3}$$

in order to rewrite (52) as follows

$$\begin{aligned}\omega^i_\alpha(p)p^\alpha\frac{\partial f}{\partial p^i}=&\frac{\partial f}{\partial E}\left[\mu E^2D_r\phi+E^2\frac{V}{R}+\left(D_t\Lambda-\frac{V}{R}\right)\mu^2E^2\right]\\&+\frac{\partial f}{\partial\mu}\left[E(1-\mu^2)D_r\phi-E(1-\mu^2)\frac{\Gamma}{R}+\mu E(1-\mu^2)(D_t\Lambda-\frac{V}{R})\right].\end{aligned}$$

Thus, we find for the Liouville operator:

$$\begin{aligned}\frac{1}{E}L_{X_g}f=&D_tf+\mu D_rf-E\left[\mu D_r\phi+\mu^2D_t\Lambda+(1-\mu^2)\frac{V}{R}\right]\frac{\partial f}{\partial E}\\&+(1-\mu^2)\left[-D_r\phi+\frac{\Gamma}{R}+\mu\left(\frac{V}{R}-D_t\Lambda\right)\right]\frac{\partial f}{\partial\mu}.\end{aligned}\tag{53}$$

This agrees with the expresson given by Lindquist [49].

Since $d\Omega(p)=2\pi E\;d\mu dE$ we obtain from (9) for the quantities ϱ_ν, P_ν, etc, in (22):

$$\varrho_\nu=\sum_i\int 2\pi E^3 f_{(\nu_i)}d\mu\,dE,\tag{54}$$

$$P_\nu=\sum_i\int 2\pi E^3\mu^2 f_{(\nu_i)}d\mu\,dE,\tag{55}$$

$$S=\sum_i\int 2\pi E^3\mu f_{(\nu_i)}d\mu\,dE,\tag{56}$$

while H is determined by (23).

The components q^0 and q^1 in (30) and (35) can be obtained from Eqs. (17) and (13):

$$q^0=\sum_i\int\left[-\kappa_{(\nu_i)}\varrho_0(f_{(\nu_i)}-f^{(0)}_{(\nu_i)})\;+\;\text{scattering terms}\right]2\pi E^2d\mu\,dE,\tag{57}$$

$$q^1=\sum_i\int\left[-\kappa_{(\nu_i)}\varrho_0(f_{(\nu_i)}-f^{(0)}_{(\nu_i)})\;+\;\text{scattering terms}\right]2\pi E^2\mu\,d\mu\,dE.\tag{58}$$

(The sums always extend over all neutrino types.) Alternatively, they could be expressed in terms of ϱ_ν, P_ν, S, H by computing the divergence of (22) (see Appendix A).

g) Lepton Number Conservation. Let $\bar{Z}$ be defined by the equation

$$n_{e^-}-n_{e^+}=\frac{1}{m_B}\varrho_0\bar{Z}.\tag{59}$$

Using baryon number conservation (3b) we obtain

$$\nabla \cdot [(n_{e^-} - n_{e^+})U] = \frac{1}{m_B} \varrho_0 D_t \bar{Z}$$

and thus from (4) and (14)

$$\frac{1}{m_B} D_t \bar{Z} = \int \left[\kappa_{(\nu_e)}(f_{(\nu_e)} - f^{(0)}_{(\nu_e)}) - \kappa_{(\bar{\nu}_e)}(f_{(\bar{\nu}_e)} - f^{(0)}_{(\bar{\nu}_e)}) + \cdots \right] 2\pi E d\mu dE. \quad (60)$$

h) Charge Conservation. Defining Z such that

$$n_p = \frac{\varrho_0 Z}{m_B} \quad (61)$$

we can write (5) with the help of (26) in the form

$$D_t \bar{Z} = D_t Z. \quad (62)$$

C. Summary of the Supernova Equations

Metric:

$$g = e^{2\phi} dt^2 - [e^{2\Lambda} dr^2 + R^2 d\Omega^2] \quad (63)$$

Baryon Conservation:

$$e^{-\Lambda} = 4\pi \varrho_0 R^2 \quad (64)$$

Constraint Equations:

$$D_t \ln(\varrho_0 R^2) + \frac{V'}{R'} = 4\pi G \frac{R}{\Gamma} S \quad (65a)$$

with

$$V = D_t R, \quad \Gamma = D_r R \quad (D_t = e^{-\phi}\partial_t, \quad D_r = e^{-\Lambda}\partial_r) \quad (65b)$$

and

$$\Gamma = [1 + V^2 - 2GM/R]^{1/2}, \qquad M = \int_0^r 4\pi R^2 e^{\Lambda}(\Gamma \varrho + VS) dr \quad (65c)$$

Dynamical Equations:

$$D_t V = \Gamma D_r \phi - G \frac{M + 4\pi R^3 P}{R^2} \quad (66)$$

Momentum Conservation:

$$D_r \phi = -(\varrho + P)^{-1} \left[D_r P + D_t S + 2\left(D_t \Lambda + \frac{V}{R} \right) S - 2\frac{\Gamma}{R} H \right] \quad (67)$$

Energy Conservation:

$$D_t \varepsilon_m + P_m D_t(1/\varrho_0) = \sum_i \int \left[\kappa_{(\nu_i)}(f_{(\nu_i)} - f^{(0)}_{(\nu_i)}) + \cdots \right] 2\pi E^2 d\mu \, dE \quad (68)$$

L_e-Conservation:

$$\frac{\varrho_0}{m_B} D_t \bar{Z} = \int \left[\kappa_{(\nu_e)}(f_{(\nu_e)} - f^{(0)}_{(\nu_e)}) - \kappa_{(\bar{\nu}_e)}(f_{(\bar{\nu}_e)} - f^{(0)}_{(\bar{\nu}_e)}) + \cdots \right] 2\pi E \, d\mu \, dE \quad (69)$$

Electric Charge Conservation:

$$D_t \bar{Z} = D_t Z \quad (70)$$

Boltzmann-Equations:

$$L_{X_g} f_{(\nu_i)} = -\kappa_{(\nu_i)} \varrho_0 [f_{(\nu_i)} - f^{(0)}_{(\nu_i)}] \; + \; \text{scattering terms,} \quad (71a)$$

with

$$\begin{aligned} \frac{1}{E} L_{X_g} f = D_t f + \mu D_r f - E \left[\mu D_r \phi + \mu^2 D_t \Lambda + (1-\mu^2) \frac{V}{R} \right] \frac{\partial f}{\partial E} \\ + (1-\mu^2) \left[-D_r \phi + \frac{\Gamma}{R} + \mu \left(\frac{V}{R} - D_t \Lambda \right) \right] \frac{\partial f}{\partial \mu}. \end{aligned} \quad (71b)$$

Up to minor differences, these equations can be found in [26] and have been used as a basis of numerical simulations. Many workers have used only their Newtonian limits, which can immediately been obtained from the equations above.

D. Basic Equations for the Kelvin-Helmholtz Phase

The basic equations that describe the physical state and quasi-static evolution of a nascent neutron star can easily be obtained from the previous results.

Let us first introduce the following changes of notation. The previous radial coordinate r is now denoted by m_0 (enclosed rest mass) and for R we use little r. We also set with the new notation

$$e^{\Lambda} dm_0 = e^{\bar{\Lambda}} dr \quad (72)$$

and thus the metric (63) takes the form

$$g = e^{2\phi} dt^2 - [e^{2\bar{\Lambda}} dr^2 + r^2 (d\vartheta^2 + \sin^2\vartheta d\varphi^2)]. \quad (73)$$

By definition, we have

$$\frac{dm_0}{dr} = 4\pi r^2 \varrho_0 e^{\bar{\Lambda}}. \quad (74)$$

Eqs. (65–67) imply the following well-known stellar structure equations (neglecting terms involving S and H):

$$\frac{dM}{dr} = 4\pi r^2 \varrho, \quad (75)$$

$$\frac{dP}{dr} = -\frac{G(\varrho + P)[M + 4\pi r^3 P]}{r^2(1 - 2GM/r)}, \tag{76}$$

$$e^{-2\bar{\Lambda}} = 1 - 2GM/r, \qquad \frac{d\phi}{dr} = \frac{G(M + 4\pi r^3 P)}{r^2(1 - 2GM/r)}. \tag{77}$$

Next we write the L_e-conservation (69) in the form

$$\frac{1}{m_B} D_t \bar{Z} = -\mathcal{S}, \tag{78}$$

where the source term $-\mathcal{S}$ for the lepton number L_e is, on the one hand, equal to ϱ_0^{-1} times the rhs in (69). On the other hand, it is also given by ϱ_0^{-1} times the divergence in (4), which we write explicitly as follows. Let

$$N_{(\nu_e)} - N_{(\bar{\nu}_e)} = (n_{\nu_e} - n_{\bar{\nu}_e})e_0 + F e_1 ,$$

where F is the lepton number flux. Then $\mathcal{S}$ times the volume form, *vol*, is determined by

$$\begin{aligned}\varrho_0\, \mathcal{S}\, vol &= d\left[(n_{\nu_e} - n_{\bar{\nu}_e})\, \theta^1 \wedge \theta^2 \wedge \theta^3\right] - d(F\theta^0 \wedge \theta^2 \wedge \theta^3) \\ &= \left\{ D_t(n_{\nu_e} - n_{\bar{\nu}_e}) + \partial_r(F e^{\phi} r^2)\frac{1}{e^{\bar{\Lambda}} r^2} e^{-\phi} \right\} vol ,\end{aligned}$$

which gives

$$\mathcal{S} = \varrho_0^{-1} D_t(n_{\nu_e} - n_{\bar{\nu}_e}) + e^{-\phi} \frac{1}{4\pi r^2 \varrho_0 e^{\bar{\Lambda}}} \frac{\partial(e^{\phi} 4\pi r^2 F)}{\partial r}$$

or, with (74),

$$\mathcal{S} = D_t\left(\frac{n_{\nu_e} - n_{\bar{\nu}_e}}{\varrho_0}\right) + e^{-\phi}\frac{\partial(e^{\phi} L_n)}{\partial m_0}, \qquad L_n = 4\pi r^2 F. \tag{79}$$

By definition, L_n is the lepton number luminosity. For the energy equation we use the form (A.11) in Appendix A, giving (with (27)):

$$D_t(\varrho/\varrho_0) + P D_t(1/\varrho_0) = -e^{-2\phi}\frac{\partial(e^{2\phi} L)}{\partial m_0}. \tag{80}$$

The Liouville operator (71b) in the Boltzmann equations (71a) simplifies to

$$\begin{aligned}\frac{1}{E} L_{X_g} f =& D_t f + e^{-\bar{\Lambda}}\left[\mu \frac{\partial f}{\partial r} + \frac{1-\mu^2}{r}\frac{\partial f}{\partial \mu}\right] \\ &- e^{-\bar{\Lambda}}\frac{d\phi}{dr}\left[(1-\mu^2)\frac{\partial f}{\partial \mu} + \mu E \frac{\partial f}{\partial E}\right].\end{aligned} \tag{81}$$

Like S in Eq. (56) the lepton number flux F can also be expressed as a moment of the distribution functions

$$F = \sum_i \int 2\pi E^2 \mu f_{(\nu_i)} d\mu\, dE . \tag{82}$$

Expressions for F and S in the diffusion approximation are derived in [49] (see also Appendix C).

VII. Prompt and Delayed Explosions

A. Damping of the Shock

As the hot shock moves outward, it is strongly damped by dissociation of nuclei into nucleons and alpha particles. Note that this costs about 8 MeV / nucleon, or 8×10^{51} erg per $0.5 M_{\odot}$. Furthermore, we have seen that the deleptonization can easily take away $(1-3) \times 10^{51}$ erg within a couple of milliseconds. On the other hand, we saw that the initial shock energy is typically around 5×10^{51} erg. Thus, both a strong bounce and a thin mantle are required for a successful prompt explosion. Otherwise, the shock stalls and settles into an accretion shock. It is not unlikely that this will indeed happen for relatively massive stars, in which case one would expect the formation of a black hole.

Many different supernova models have been proposed over the years to explain how the gravitational energy released during collapse can partly ($\sim 1\%$) be transmitted to the outer layers of the star and expel them. Unfortunately, the quest for this coupling mechanism has proved elusive and its solution remains controversal, in spite of the enormous efforts of incorporating increasingly realistic microphysics input data into the hydrodynamic computer codes. The general properties of the solutions have been discussed in numerous papers and reviews (see Ref. [38, 50–62] as a guide to the literature). Many workers are now concluding that nuclear dissociation and neutrino losses stall most (all?) bounce shocks into accretion shocks within milliseconds of bounce. In particular, there are at the moment no computer simulations which lead to an explosion for the mass range of progenitor stars which has been inferred from SN 1987 A.

B. Delayed Explosion by Neutrino Heating

Colgate and White suggested already in 1966 in a classical paper [63] a neutrino "energy deposition" model in which neutrino-antineutrino pairs generated in the hot matter behind the shock deposit their energy in the overlying regions of lower density. The idea was that the resulting heating provides sufficient thermal pressure to blow off the outer parts of the star. Later, more detailed hydrodynamical calculations came to the conclusion that this mechanism was inefficient. Wilson [59] has recently revived this neutrino transfer model. Further analysis by him and his collaborators [60] and by others [56] showed that nucleons outside the neutrino sphere may perhaps (!) absorb enough energy from ν_e and $\bar{\nu}_e$ that the shock is revitalized.

Although this delayed explosion model has not been verified convincingly by the community, it remains an interesting possibility. Let us estimate the net

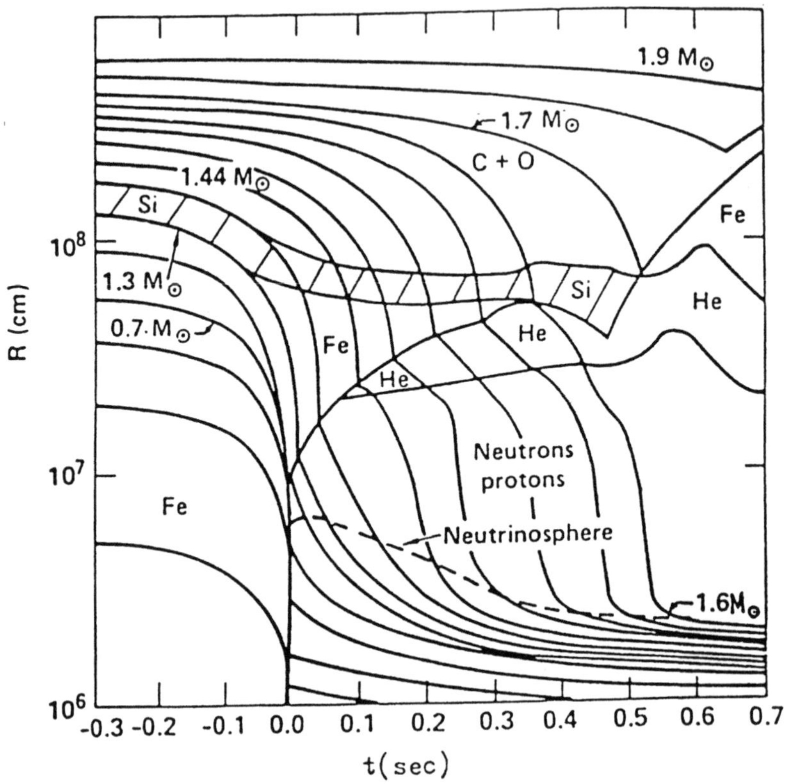

Fig. VII.1. Collapse and bounce of a $15M_\odot$ star. Note the shrinking of the neutrino sphere. (From Ref. [26].)

neutrino heating to see whether it might reenergize the shock. The calculation is similar to the one carried out in Sect. V.4 C.

For orientation see Fig. VII.1 which shows the late time behavior after collapse and bounce of a $15M_\odot$ star.

The region outside the neutrino sphere is composed of neutrons and protons (dissociation by the shock). A piece of matter at perhaps $R = 100 - 300\,\mathrm{km}$ looses energy by the reactions $\mathrm{e}^- + \mathrm{p} \rightarrow \mathrm{n} + \nu_\mathrm{e}$, $\mathrm{e}^+ + \mathrm{n} \rightarrow \mathrm{p} + \bar{\nu}_\mathrm{e}$.

Let us first consider the loss rate per baryon due to $\mathrm{e}^- + \mathrm{p} \rightarrow \mathrm{n} + \nu_\mathrm{e}$. If we neglect the Pauli blocking for the neutrinos, we obtain as in Sect. V.4 C for this quantity

$$\begin{aligned} Q^-(\mathrm{e}^- + \mathrm{p} \rightarrow \mathrm{n} + \nu_\mathrm{e}) &= Y_\mathrm{p}\frac{\sigma_0}{8\pi^2}[1 + 3(g_\mathrm{A}/g_\mathrm{V})^2]\frac{1}{m_\mathrm{e}^2} \cdot \int E_\nu^2 E_\mathrm{e}^2 f_\mathrm{e} dE_\mathrm{e} \\ &= Y_\mathrm{p}\frac{\sigma_0}{8\pi^2}[1 + 3(g_\mathrm{A}/g_\mathrm{V})^2] m_\mathrm{e}^4 \left(\frac{T}{m_\mathrm{e}}\right)^6 F_5(\eta_\mathrm{e}). \end{aligned}$$

For $\mathrm{e}^+ + \mathrm{n} \rightarrow \mathrm{p} + \bar{\nu}_\mathrm{e}$ we must make the replacements $Y_\mathrm{p} \rightarrow Y_\mathrm{n}$, $F_5(\eta_\mathrm{e}) \rightarrow F_5(-\eta_\mathrm{e})$. Since the energy density of electrons is substantially greater than that of the positrons, we write for the total cooling rate per baryon

$$Q^- \simeq Y_p \frac{\sigma_0}{8\pi^2}[1+3(g_A/g_V)^2]m_e^4 \left(\frac{T}{m_e}\right)^6 [F_5(\eta_e)+F_5(-\eta_e)]. \tag{1}$$

From (V.64) we obtain by inegration

$$F_5(\eta)+F_5(-\eta) = \frac{31}{126}\pi^6 + \frac{7}{6}\pi^4\eta^2 + \frac{5}{6}\pi^2\eta^4 + \frac{1}{6}\eta^6. \tag{2}$$

For simplicity, we set $\eta = 0$, changing the result by less than a factor 2 [64].

For the heating rate per baryon due to $\nu_e + n \rightarrow e^- + p$, $\bar{\nu}_e + p \rightarrow n + e^+$ we obtain almost the same expression, but T must be replaced by the temperature T_ν of the neutrino sphere and we have an additional factor $(R_\nu/2R)^2$, because of a different solid angle. The net heating rate per baryon is, therefore,

$$\begin{aligned} Q &\simeq 8.1\sigma_0 c\left(\frac{m_e c}{\hbar}\right)^3 m_e c^2 \left(\frac{kT_\nu}{m_e c^2}\right)^6 \left[\left(\frac{R_\nu}{2R}\right)^2 - \left(\frac{T}{T_\nu}\right)^6\right] \\ &= 3.3\times 10^{-6}(\text{erg/baryon sec})\left[\left(\frac{R_\nu}{2R}\right)^2 T_\nu^6[\text{MeV}] - T^6[\text{MeV}]\right]. \end{aligned} \tag{3}$$

It is natural to consider the heating time scale, defined by $t_H = E_{\text{grav}}/Q$. For typical values $T_\nu = 5\,\text{MeV}$, $T = 1\,\text{MeV}$, $R_\nu = 3\times 10^6\,\text{cm}$, $R = 3\times 10^7\,\text{cm}$, we obtain $t_H = 0.03\,\text{sec}$. This is sufficiently fast to be potentially interesting for shock revival. Note, however, the strong dependence of the effect on T_ν and the matter temperature further out. This requires a very accurate treatment of the neutrino transport. [3]

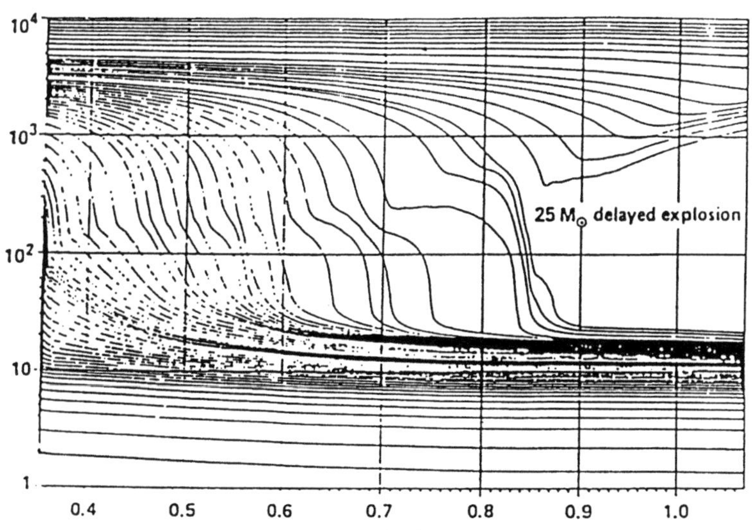

Fig. VII.2. Radius as a function of time in a successful $25M_\odot$ delayed explosion. (From Ref. [60].)

[3] Practial (sufficiently fast) neutrino transport methods can be checked and calibrated by Monte Carlo methods [67].

Wilson and coworkers [60] have indeed obtained delayed explosions for a variety of stellar masses (Fig. VII.2).

The validity of Wilson's delayed mechanism has, however, not yet been verified by other theoreticians [65].

VIII. Conclusions and Supplementary Remarks

What have we learned? For me the most notable result is the good agreement of the theoretically expected properties of the neutrino pulse with what has been observed by Kamiokande II and IMB. We have seen that the total integrated number of events as a function of time, the average neutrino energy and the total neutrino luminosity are all explained reasonably well by the standard model of core collapse and neutron star formation.

The formation of a black hole in SN 1987 A seems to be unlikely. Otherwise, the integrated number of events versus time would be difficult to understand. We are looking forward to see, within a year or so, a neutron star (millisecond pular ?).

No exotic physics beyond the standard model is required. In particular, the data do not imply any nonvanishing neutrino mass. The inferred upper limits for m_{ν_e} are comparable to the laboratory limits from tritium decay and bounds for neutrinoless double β-decay. (We have not discussed the Mont Blanc experiment, because the reported neutrino pulse is still the subject of debate and may be due to background.)

In many aspects, SN 1987 A has behaved differently from other supernovae. In particular, the light curve is most peculiar. Presumably, an unusually large fraction of its energy went into expansion of its atmosphere because the precursor was a blue supergiant at the time of core collapse. This was also a great surprise, but may be due to mass loss and different metallicity. (Convective overshoot may also be relevant.) The suggestion by A. Maeder and other authors that the progenitor of SN 1987 A has lost a substantial fraction of its mass in a previous red supergiant stage, has been indirectly confirmed: UV spectra of nitrogen lines with low velocity dispersion (< 30 km/sec) give evidence for circumstellar matter which may be due to such mass losses. More detailed stellar evolutionary calculations, together with better abundance determinations in LMC, may soon settle these issues.

Spectroscopic and photometric observations for days 135 to 260 after the neutrino burst showed that during this period the bolometric flux has undergone an exponential decline [68]. Furthermore, the e-folding time of decline is very close to the mean life of ^{56}Co (see Fig. VIII.1).

This provides strong evidence that radioactive decay of ^{56}Co is the main source of energy during this period.

The decay energy for ^{56}Co is 3.58 MeV per nucleus and this gives an energy generation rate

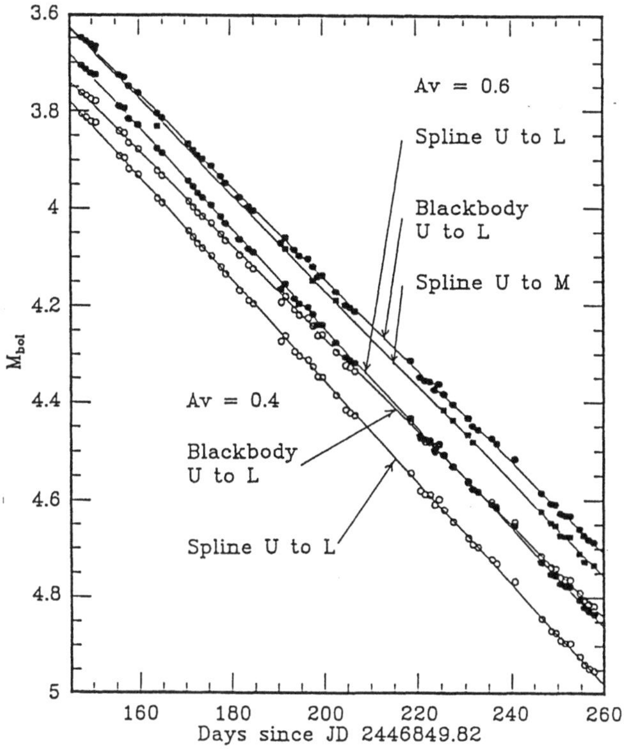

Fig. VIII.1. Bolometric magnitude determined by two fitting procedures as a function of time for two absorption corrections A_V in the V band. (From Ref. [68].)

$$\dot{S} = 7.0 \times 10^9 \; e^{-t/111.3} \text{erg (g of original Ni)}^{-1} \sec^{-1},$$

where the time t is measured in days. (The mean lifes of ^{56}Ni and ^{56}Co are 8.8 and 111.3 days.)

Comparison with the observation shows that about $0.08 M_\odot$ of ^{56}Ni was produced. The uncertainty of this number is at least 15%.

A sad aspect of supernova theory is that at the moment no computer simulations lead to an explosion for the mass range of progenitor stars which has been inferred from SN 1987 A. This state of affair might change with more realistic models which include rotation and/or magnetic fields, etc.

Let me end these lectures with the following amusing remark. The observation of a neutrino burst within 3 hours of the associated optical burst from SN 1987 A provides a new test of the weak equivalence principle [70]. Indeed, it is easy to estimate the general relativistic Shapiro time delay which turns out to be a few months. (The precise number depends on the gravitational potential of the galaxy at large distances.)

In the parametrized post-Newtonian (PPN) formalism, the Shapiro time delay is given by

$$\Delta t = -\frac{1+\gamma}{2}\, 2 \int \phi(\boldsymbol{x}(t))dt,$$

where ϕ is the Newtonian potential and the integral extends over the light path which can be replaced by a straight line.

If the weak equivalence principle is violated, $1+\gamma$ is different for photons and neutrinos. However, the upper limit of 10^4sec on the difference in arrival time implies that

$$\gamma_\gamma - \gamma_\nu \leq 5 \times 10^{-3}.$$

Thus the validity of the weak equivalence principle for photons and neutrinos is verified to better than 0.5% accuracy.

Appendix A. Energy-Momentum Conservation

In this Appendix we derive in detail some results given in Chapter VI.

1. Energy Conservation

We consider an energy-momentum tensor of the following general form (see Appendix B of Ref. [24]):

$$T^{\mu\nu} = (\varrho + P)U^\mu U^\nu - P g^{\mu\nu} + U^\mu Q^\nu + U^\nu Q^\mu + \Delta T^{\mu\nu}, \tag{A.1}$$

where Q^μ is the heat flux which is orthogonal to the four-velocity U^μ and $\Delta T^{\mu\nu}$ is a non-ideal contribution orthogonal to U^μ. Thus

$$U \cdot Q = 0, \quad U_\mu \Delta T^{\mu\nu} = \Delta T^{\mu\nu} U_\nu = 0. \tag{A.2}$$

From $T^{\mu\nu}{}_{;\nu} = 0$ and $N^\mu{}_{;\mu} = 0$ for the particle current $N^\mu = nU^\mu$ one deduces for the energy conservation $U_\mu T^{\mu\nu}{}_{;\nu} = 0$ (see Eq. (B.37) of [24]):

$$n[\nabla_U(\varrho/n) + P\nabla_U(1/n)] = -\nabla \cdot Q + Q \cdot \nabla_U U + \theta_{\alpha\beta}\Delta T^{\alpha\beta}, \tag{A.3}$$

where $\theta_{\alpha\beta}$ is the expansion tensor:

$$\theta_{\alpha\beta} = h_\alpha{}^\mu h_\beta{}^\nu U_{(\mu;\nu)}, \qquad h_\alpha{}^\mu = \delta_\alpha{}^\mu - U_\alpha U^\mu. \tag{A.4}$$

This can be easily worked out for a spherically symmetric collapse in the gauge (VI.19), for which

$$Q = S e_1. \tag{A.5}$$

Clearly,

$$\nabla \cdot Q = (4\pi R^2 e^\Lambda)^{-1} \partial_r (e^\phi L), \tag{A.6}$$

where

$$L = 4\pi R^2 S \tag{A.7}$$

is the luminosity. Furthermore

$$Q \cdot \nabla_U U = S(\nabla_U U)_1 = -S\omega_1^0(e_0) = -S\phi' e^{-\Lambda}$$

and thus

$$\nabla \cdot Q - Q \cdot \nabla_U U = (4\pi R^2 e^{2\phi})^{-1} D_r(L e^{2\phi}). \tag{A.8}$$

Relative to the tetrad (VI.21) we have

$$\theta_{\alpha\beta} \Delta T^{\alpha\beta} = \theta_{ij} \Delta T^{ij} \tag{A.9}$$

with

$$\theta_{ij} = \frac{1}{2}[(e_i, \nabla_{e_j} e_0) + (e_j, \nabla_{e_i} e_0)] = -\frac{1}{2}[\omega^i_0(e_j) + \omega^j_0(e_i)],$$

i.e.,

$$\theta_{11} = -D_t \Lambda, \quad \theta_{22} = \theta_{33} = -\frac{V}{R}, \quad \theta_{ij} = 0 \quad \text{for } i \neq j. \tag{A.10}$$

Inserting these results in (A.3) gives the following expression for the energy conservation

$$\begin{aligned} n[D_t(\varrho/n) + P D_t(1/n)] = &- (4\pi R^2 e^{2\phi})^{-1} D_r(L e^{2\phi}) \\ &- D_t \Lambda \Delta T^{11} - \frac{V}{R}(\Delta T^{22} + \Delta T^{33}). \end{aligned} \tag{A.11}$$

For the energy-momentum tensor (VI.22) we have $\Delta T^{11} = 0$, $\Delta T^{22} = \Delta T^{33} = H$ in which case (A.11) reduces to (VI.31).

2. Momentum Conservation

Introducing the vector-valued one-forms $T^\mu = T^\mu{}_\nu \theta^\nu$, we can write the energy-momentum conservation $T^{\mu\nu}{}_{;\nu} = 0$ as (see Chap. 2 of [24]):

$$d\,{}^*T^\alpha + \omega^\alpha{}_\beta \wedge {}^*T^\beta = 0\,. \tag{A.12}$$

With the help of the connection forms [48], this can immediately be worked out for the energy-momentum tensor (VI.22). For $\alpha = 1$ one obtains the result given in Eq. (VI.34). [For $\alpha = 0$ one finds, on the other hand, again (VI.31).] This calculation provides also the components q^0 and q^1 expressed in terms of ϱ_ν, P_ν, S (recall that q^μ is the divergence of (VI.22)):

$$q^0 = D_t \varrho_\nu + D_r S + \left(D_t \Lambda + 3\frac{V}{R}\right) \varrho_\nu + 2\left(D_r \phi + \frac{\Gamma}{R}\right) S + \left(D_t \Lambda - \frac{V}{R}\right) P_\nu, \tag{A.13}$$

$$q^1 = D_t S + D_r P_\nu + \left(D_r \phi - \frac{\Gamma}{R}\right) \varrho_\nu + 2\left(D_t \Lambda + \frac{V}{R}\right) S + \left(D_r \phi + 3\frac{\Gamma}{R}\right) P_\nu. \tag{A.14}$$

(The quantity H has been eliminated with the help of (VI.23).)

Appendix B. Core Instabilities

In this Appendix we explain the core instabilities, which have been briefly discussed in Sect. III.5, in some more detail.

a) Dissociation of Iron-Peak Elements

It has been pointed out by Hoyle and Fowler long ago that at high temperatures the most tightly bound nuclei of the iron group will be photodisintegrated. (Note that the *free* energy has to have a minimum under equilibrium conditions.) The equilibrium abundances can be calculated with standard statistical mechanics. Consider, for instance, a model for which matter consists entirely of ^{56}Fe, α particles, and neutrons under equilibrium conditions

$$\gamma + {}^{56}_{26}\mathrm{Fe} \rightleftharpoons 13\alpha + 4\mathrm{n}\ .$$

The corresponding Saha equation gives the following temperature density relation for 50 percent dissociation of Fe into He:

$$\log \varrho(\mathrm{g/cm^3}) = 11.6 + 1.5 \log T_9 - \frac{39.2}{T_9}\ . \tag{B.1}$$

(For the partition sum ω_{Fe} of ^{56}Fe the ground state and the first excited levels were taken into account, with the result $\omega_{\mathrm{Fe}} \simeq 1.4$). A few representative values satisfying (B.1) are:

$\varrho(\mathrm{g/cm^3})$	1	10^3	10^6	10^9
T_9	3.43	4.1	5.8	9.6

(Compare this with Fig. III.1.)

We show now with a rough calculation that the dissociation of iron reduces the adiabatic index Γ_1 below 4/3.

For an adiabatic change of the specific internal energy $\mathcal{E}(P, v)$, v = specific volume, we have (for a one component fluid)

$$d\mathcal{E} = \mathcal{E}_{,P}\, dP + \mathcal{E}_{,v}\, dv = -P dv$$

and thus

$$\Gamma_1 = \left(\frac{\partial \ln P}{\partial \ln \varrho}\right)_{\mathrm{ad}} = -\frac{v}{P}\left(\frac{\partial P}{\partial v}\right)_{\mathrm{ad}} = \frac{Pv + v\mathcal{E}_{,v}}{P\mathcal{E}_{,P}}. \tag{B.2}$$

Let us assume that $\mathcal{E}$ is only a function of Pv, as is for instance the case for an ideal gas. Then we obtain

$$\Gamma_1 = 1 + \frac{1}{\mathcal{E}'}\ . \tag{B.3}$$

For the iron-helium transition the changes of $\mathcal{E}$ and Pv are

$$\Delta\mathcal{E} = 2 \times 10^{18} \mathrm{erg/g}$$

$$\Delta(Pv) = \frac{kT}{m_{\mathrm{H}}}\left(\frac{1}{\mu_{\mathrm{He}}} - \frac{1}{\mu_{\mathrm{Fe}}}\right) = \frac{kT}{m_{\mathrm{H}}}\left(\frac{3}{4} - \frac{27}{56}\right)$$
$$= \frac{1}{4}\frac{kT}{m_{\mathrm{H}}} = 2.2 \times 10^{17} \mathrm{erg/g}, \quad \text{for } T \simeq 10^{10}\mathrm{K}.$$

Hence, (B.3) gives

$$\Gamma_1 \simeq 1 + \frac{\Delta(Pv)}{\Delta\mathcal{E}} \simeq 1.1 < 4/3.$$

(In this estimate we have ignored the electrons, whose adiabatic index is very close to 4/3.)

b) Inverse β-Decay

Electron capture at high densities brings Γ_1 also below 4/3. This is seen as follows. Consider, for simplicity, $T = 0$. By definition

$$\Gamma_1 = \frac{d\ln P}{d\ln n_e}\,\frac{d\ln n_e}{d\ln \varrho} = \Gamma_1^{(0)}(1-\delta),$$

where n_e is the electron number density, and $\Gamma_1^{(0)}$ is the value of Γ_1 without electron capture. For relativistic electrons one finds easily

$$\Gamma_1^{(0)} = \frac{4}{3} + \frac{2}{3(P_F/m_e)^2} + \cdots ,$$

which is very close to 4/3 after Si burning, say. With inverse β-decay ($\delta > 0$), Γ_1 drops below 4/3 in the region shown in Fig. III.1.

Appendix C. Diffusion Approximation, Flux-Limited Diffusion

As a starting point for the diffusion approximation and an approximation called flux limited diffusion, we consider the first two moments of (V.43).

Integrating Eq. (V.43) over $d\Omega$ gives

$$\begin{aligned}\partial_t u + \nabla\cdot \boldsymbol{S} = &-\kappa^a(\omega)[4\pi B_\omega - u] \\ &-\int d\omega' d\Omega' d\Omega \kappa^s(\omega, \boldsymbol{n} \to \omega', \boldsymbol{n}')\,\{\cdots\}\,,\end{aligned} \tag{C.1}$$

where the curly bracket is the same as in (V.43). (Note that only inelastic scattering contributes to the integral). Next, we multiply Eq. (V.43) with $\boldsymbol{n}$ and integrate over $d\Omega$. This gives

$$\begin{aligned}\partial_t \boldsymbol{S} + \int \boldsymbol{n}(\boldsymbol{n}\cdot\nabla I)d\Omega = &-[\kappa^a(\omega)+\kappa^s(\omega)]\boldsymbol{S} \\ &+\int \boldsymbol{n}\kappa^s(\omega,\boldsymbol{n}\to\omega',\boldsymbol{n}')\left(\frac{\omega}{\omega'}\right)^3 I(\omega',\boldsymbol{n}')d\omega' d\Omega' d\Omega \\ &+\text{induced scattering terms.}\end{aligned} \tag{C.2}$$

($\kappa^{\mathrm{s}}(\omega)$ is defined in (V.49)). In order to work out the integral on the right hand side we set

$$\kappa^{\mathrm{s}}(\omega, \boldsymbol{n} \to \omega', \boldsymbol{n}') = \kappa^{\mathrm{s}}(\omega \to \omega') p(\boldsymbol{n}, \boldsymbol{n}')$$

with

$$\int p(\boldsymbol{n}, \boldsymbol{n}') d\Omega' = 1.$$

Clearly,

$$\int \kappa^{\mathrm{s}}(\omega \to \omega') d\omega' = \kappa^{\mathrm{s}}(\omega)$$

and

$$\int \boldsymbol{n}\, p(\boldsymbol{n}, \boldsymbol{n}') d\Omega = \langle \cos\theta \rangle \boldsymbol{n}',$$

where $\langle \cos\theta \rangle$ is the average of the cosine of the scattering angle

$$\langle \cos\theta \rangle = \int (\boldsymbol{n} \cdot \boldsymbol{n}')\, p(\boldsymbol{n}, \boldsymbol{n}') d\Omega.$$

Thus

$$\begin{aligned} &\int \boldsymbol{n} \kappa^{\mathrm{s}}(\omega, \boldsymbol{n} \to \omega', \boldsymbol{n}') \left(\frac{\omega}{\omega'}\right)^3 I(\omega', \boldsymbol{n}') d\omega' d\Omega' d\Omega \\ &= \langle \cos\theta \rangle \int d\omega' \left(\frac{\omega}{\omega'}\right)^3 \kappa^{\mathrm{s}}(\omega \to \omega') \boldsymbol{S}(\omega') \\ &\simeq \langle \cos\theta \rangle \kappa^{\mathrm{s}}(\omega) \boldsymbol{S}(\omega). \end{aligned}$$

In the last expression we have neglected off-diagonal ($\omega' \neq \omega$) parts. In this approximation the induced scattering terms vanish also, and we obtain

$$\partial_t \boldsymbol{S}(\omega) + \int \boldsymbol{n}(\boldsymbol{n} \cdot \nabla I) d\Omega = -[\kappa^{\mathrm{a}}(\omega) + \kappa^{\mathrm{s}}(\omega)\langle 1 - \cos\theta \rangle] \boldsymbol{S}(\omega). \qquad \text{(C.3)}$$

The **diffusion approximation** is obtained as follows. First, the intensity I is assumed to be only weakly anisotropic. Then we have in leading order

$$\int \boldsymbol{n}(\boldsymbol{n} \cdot \nabla I) d\Omega \simeq \frac{1}{3} \nabla u. \qquad \text{(C.4)}$$

Furthermore, it is assumed that $\partial_t \boldsymbol{S}$ can be neglected in (C.3) (quasi-stationary situation). Then we obtain **Fick's law** for the spectral energy flux

$$\boldsymbol{S}(\omega) = -\frac{1}{3\kappa(\omega)} \nabla u(\omega) \qquad \text{(C.5)}$$

with

$$\kappa(\omega) = \kappa^{\mathrm{a}}(\omega) + \kappa^{\mathrm{s}}(\omega)\langle 1 - \cos\theta \rangle. \qquad \text{(C.6)}$$

If we write the intensity as a sum

$$I = I^{(0)} + I^{(1)},$$

with an isotropic part $I^{(0)}$, then the leading anisotropy is according to (C.5) given by

$$I^{(1)}(\omega, \boldsymbol{n}) = -\frac{1}{\kappa(\omega)} \boldsymbol{n} \cdot \nabla I^{(0)}. \tag{C.7}$$

We discuss now the **flux-limited diffusion approximation** in more detail than in Sect. V.4B. For simplicity, we consider only elastic scattering, in which case (V.43) reduces to

$$D_t I = -\kappa^{\mathrm{a}}(I - B_\omega) - \kappa^{\mathrm{s}} \left(I - \int p(\boldsymbol{n}, \boldsymbol{n}') I(\boldsymbol{n}') d\Omega' \right). \tag{C.8}$$

(The variable ω is usually suppressed.) The zeroth moment equation (C.1) simplifies to

$$\partial_t u + \nabla \cdot \boldsymbol{S} = \kappa^{\mathrm{a}}(4\pi B_\omega - u). \tag{C.9}$$

We proceed now as in Ref. [37]. (Our basic equation (C.8) is, however, a bit more general.) Let us define the normalized intensity ψ by

$$I = u\psi. \tag{C.10}$$

Clearly,

$$\int_{S^2} \psi d\Omega = 1.$$

Inserting (C.10) in (C.8) and (C.9) yields

$$\begin{aligned} \partial_t(u\psi) + \boldsymbol{n} \cdot \nabla(u\psi) + (\kappa_{\mathrm{a}} + \kappa_{\mathrm{s}})u\psi &= \kappa_{\mathrm{a}} B \\ &\quad + \kappa_{\mathrm{s}} u \int p(\boldsymbol{n}, \boldsymbol{n}')\psi(\boldsymbol{n}')d\Omega', \\ \partial_t u + \nabla \cdot (u\boldsymbol{f}) &= \kappa_{\mathrm{a}}(4\pi B - u). \end{aligned} \tag{C.11}$$

Here

$$\boldsymbol{f} = \int_{S^2} \psi(\boldsymbol{n})\boldsymbol{n}\, d\Omega \tag{C.12}$$

so that

$$\boldsymbol{S} = u\boldsymbol{f}.$$

We use (C.12) in (C.11) to eliminate $\partial_t u$ and obtain

$$\begin{aligned} (\partial_t\psi + \boldsymbol{n} \cdot \nabla\psi)u + (\boldsymbol{n} \cdot \nabla u - \boldsymbol{f} \cdot \nabla u - u\nabla \cdot \boldsymbol{f} + \kappa_{\mathrm{s}} u + \kappa_{\mathrm{a}} 4\pi B)\psi & \\ = \kappa_{\mathrm{a}} B + \kappa_{\mathrm{s}} u \int p(\boldsymbol{n}, \boldsymbol{n}')\psi(\boldsymbol{n}')d\Omega'. & \end{aligned} \tag{C.13}$$

This is still an exact relation. Now, we assume that ψ is slowly varying and set

$$\partial_t\psi + \boldsymbol{n} \cdot \nabla\psi = 0. \tag{C.14}$$

This implies by integration $\nabla \cdot \boldsymbol{f} = 0$ and so we find from (C.13)

$$(\boldsymbol{n} \cdot \nabla u - \boldsymbol{f} \cdot \nabla u + \kappa\varpi u)\psi = \kappa_{\mathrm{a}} B + \kappa_{\mathrm{s}} u \int p\psi' d\Omega', \tag{C.15}$$

where ϖ is the effective albedo given by

$$\varpi = \frac{\kappa_a 4\pi B + \kappa_s u}{\kappa u}, \qquad \kappa = \kappa_a + \kappa_s. \tag{C.16}$$

For an isotropic p ($p = 1/4\pi$) we have

$$\int p(\boldsymbol{n}, \boldsymbol{n}')\psi(\boldsymbol{n}')d\Omega' = 1/4\pi.$$

We replace now the integral on the right hand side in (C.15) by $1/4\pi$ and obtain

$$(\boldsymbol{n} \cdot \nabla u - \boldsymbol{f} \cdot \nabla u + \kappa\varpi u)\psi = \frac{1}{4\pi}\kappa\varpi u. \tag{C.17}$$

This equation has the same form as Eq. (15) in [37]. After a few steps one finds Fick's law of diffusion

$$\boldsymbol{S} = -\frac{1}{\kappa}D_{\mathrm{F}}\nabla u \tag{C.18}$$

with

$$D_{\mathrm{F}} = \frac{1}{X}[\coth(X/\varpi) - \varpi/X], \tag{C.19}$$

where

$$X = \frac{1}{\kappa}\left|\frac{\nabla u}{u}\right|. \tag{C.20}$$

Appendix D. Plasma-Neutrino Process

Since some of the participants found this process quite intriguing, I present here a detailed discussion [66].

1. Polarisation Tensor

Let me first recall the phenomenological definition of the polarisation tensor.

We start from the phenomenological equations of Maxwell in a medium with $\mu = 1$:

$$\begin{aligned} \nabla \wedge \boldsymbol{E} + \partial_t \boldsymbol{B} &= 0 \\ \nabla \cdot \boldsymbol{B} &= 0 \\ \nabla \wedge \boldsymbol{B} - \partial_t \boldsymbol{D} &= \boldsymbol{J} \\ \nabla \cdot \boldsymbol{D} &= \varrho, \end{aligned} \tag{D.1}$$

$\boldsymbol{J}$: external current density, ϱ: external charge density.
Between $\boldsymbol{D}$ and $\boldsymbol{E}$ we have the usual linear nonlocal relation

$$D_i(\boldsymbol{x}, t) = \int dt' d^3x' \varepsilon_{ij}(\boldsymbol{x} - \boldsymbol{x}', t - t')E_j(\boldsymbol{x}', t'), \tag{D.2}$$

where ε_{ij} is the dielectric tensor. In (D.2) we assumed translational invariance. Let us consider an isotropic medium. Then the dielectric tensor can be split into a transversal and a longitudinal part

$$\varepsilon_{ij}(\boldsymbol{k},\omega)=\varepsilon^{\mathrm{t}}(\boldsymbol{k},\omega)\left(\delta_{ij}-\frac{k_ik_j}{|\,\boldsymbol{k}\,|^2}\right)+\varepsilon^{\mathrm{l}}(\boldsymbol{k},\omega)\frac{k_ik_j}{|\,\boldsymbol{k}\,|^2}. \tag{D.3}$$

Note that (D.3) implies $\varepsilon_{ij}=\varepsilon_{ji}$.

Let us next introduce the electromagnetic potentials

$$\boldsymbol{E}=-\partial_t\boldsymbol{A}-\nabla\varphi,\quad \boldsymbol{B}=\nabla\wedge\boldsymbol{A}.$$

Then the inhomogeneous equations can be written in a compact four dimensional way as (details follow)

$$[k^2g_{\mu\nu}-k_\mu k_\nu+\Pi_{\mu\nu}(k)]A^\nu(k)=-J_\mu(k). \tag{D.4}$$

Here $\Pi_{\mu\nu}(k)$ $(k=(\omega,\boldsymbol{k}))$ is the *polarisation tensor*, which can be expressed in terms of the transversal and longitudinal dielectric constants as follows

$$\Pi_{\mu\nu}=-\omega^2(\varepsilon^{\mathrm{t}}-1)(P_{\mathrm{T}})_{\mu\nu}-k^2(\varepsilon^{\mathrm{l}}-1)(P_{\mathrm{L}})_{\mu\nu}, \tag{D.5}$$

where P_{T} and P_{L} denote the following projection operators

$$(P_{\mathrm{T}})_{\mu\nu}=\sum_{\lambda=1,2}e^{\mathrm{t}}_\mu(k,\lambda)e^{\mathrm{t}}_\nu(k,\lambda) \tag{D.6a}$$

and

$$(P_{\mathrm{L}})_{\mu\nu}=e^{\mathrm{l}}_\mu(k)e^{\mathrm{l}}_\nu(k). \tag{D.6b}$$

Here $e^{\mathrm{t}}_\mu(k,\lambda)$, $\lambda=1,2$, are the two transversal polarisation vectors, and $e^{\mathrm{l}}_\mu(k)$ is the longitudinal polarisation vector, i.e.,

$$e^{\mathrm{l}}_\mu(\boldsymbol{k},\omega)=\frac{1}{\sqrt{k^2}}(|\boldsymbol{k}|,\ -\omega\boldsymbol{k}/|\boldsymbol{k}|)\ . \tag{D.7}$$

More explicitly (D.6) can be written as

$$(P_{\mathrm{T}})_{ij}=\delta_{ij}-\frac{k_ik_j}{|\,\boldsymbol{k}\,|^2}, \tag{D.6'}$$

$(P_{\mathrm{T}})_{\mu\nu}=0$ if μ and/or ν equal 0. Note that

$$P_{\mathrm{T}}^2=P_{\mathrm{T}},\quad P_{\mathrm{L}}^2=P_{\mathrm{L}},\quad P_{\mathrm{T}}P_{\mathrm{L}}=0. \tag{D.8}$$

Details of the Derivation. Let us write symbolically

$$\boldsymbol{D}=\boldsymbol{E}+(\boldsymbol{\varepsilon}-\boldsymbol{1})*\boldsymbol{E}$$

($*$: convolution in x-space and contraction of vector indices). Then we can obviously write the inhomogeneous Maxwell equations as follows

$$\begin{aligned}\nabla\wedge\boldsymbol{B}-\partial_t\boldsymbol{E}&=\boldsymbol{J}+\boldsymbol{J}^{\mathrm{Pol}}\\ \nabla\cdot\boldsymbol{E}&=\varrho+\varrho^{\mathrm{Pol}},\end{aligned} \tag{D.1'}$$

where $\boldsymbol{J}^{\mathrm{Pol}}$ and ϱ^{Pol} denote the polarisation current and the polarisation charge, respectively:

$$\boldsymbol{J}^{\rm Pol} = \partial_t((\boldsymbol{\varepsilon} - \mathbf{1}) * \boldsymbol{E})$$
$$\varrho^{\rm Pol} = -\boldsymbol{\nabla}((\boldsymbol{\varepsilon} - \mathbf{1}) * \boldsymbol{E}).$$

Clearly, Eqs. (D.1′) can be put together into one four dimensional equation

$$(g_{\mu\nu} \Box - \partial_\mu \partial_\nu) A^\nu = J_\mu + J_\mu^{\rm Pol}. \tag{D.9}$$

Here we have not yet specified a special gauge. In momentum space we obtain

$$\begin{aligned} \boldsymbol{J}^{\rm Pol} &= -i\omega(\boldsymbol{\varepsilon} - \mathbf{1}) * \boldsymbol{E} = -i\omega(\boldsymbol{\varepsilon} - \mathbf{1}) * (i\omega \boldsymbol{A} - i\boldsymbol{k} A_0) \\ &= \omega(\boldsymbol{\varepsilon} - \mathbf{1}) * (\omega \boldsymbol{A} - \boldsymbol{k} A_0); \end{aligned}$$

similarly

$$\varrho^{\rm Pol} = i\boldsymbol{k} \cdot ((\boldsymbol{\varepsilon} - \mathbf{1}) * \boldsymbol{E}) = \boldsymbol{k} \cdot (\boldsymbol{\varepsilon} - \mathbf{1}) * (\omega \boldsymbol{A} - \boldsymbol{k} A_0) \ .$$

We collect these two equations in the form

$$J_\mu^{\rm Pol}(k) = \Pi_{\mu\nu}(k) A^\nu(k), \tag{D.10}$$

with

$$\begin{aligned} \Pi_{ij} &= -\omega^2(\varepsilon_{ij} - \delta_{ij}) = \Pi_{ji} \\ \Pi_{i0} &= \omega(\varepsilon_{ij} - \delta_{ij}) k^j = \Pi_{0i} \\ \Pi_{00} &= -k^i k^j (\varepsilon_{ij} - \delta_{ij}) \ . \end{aligned}$$

Now we use (D.3) to obtain

$$\Pi_{ij} = -\omega^2 \left[(\varepsilon^{\rm t} - 1) \left(\delta_{ij} - \frac{k_i k_j}{|\, \boldsymbol{k} \,|^2} \right) + (\varepsilon^{\rm l} - 1) \frac{k_i k_j}{|\, \boldsymbol{k} \,|^2} \right],$$

or

$$\Pi_{ij} = -\omega^2 (\varepsilon^{\rm t} - 1)(P_{\rm T})_{ij} - k^2 (\varepsilon^{\rm l} - 1)(P_{\rm L})_{ij}.$$

Similarly

$$\Pi_{i0} = -\omega(\varepsilon^{\rm l} - 1) k_i = -k^2 (\varepsilon^{\rm l} - 1)(P_{\rm L})_{0i}.$$

Finally

$$\Pi_{00} = -|\, \boldsymbol{k} \,|^2 (\varepsilon^{\rm l} - 1) = -k^2 (\varepsilon^{\rm l} - 1)(P_{\rm L})_{00}.$$

This proves Eq. (D.5).

Note also

$$k^\mu \Pi_{\mu\nu}(k) = 0. \tag{D.11}$$

2. The Retarded Photon Green's-Function and Plasmons

The phenomenological Maxwell equations can be obtained from QED by taking a (grand) canonical average. From

$$(g_{\mu\nu} \Box - \partial_\mu \partial_\nu) A^\nu(x) = j_\mu + J_\mu, \tag{D.12}$$

where

$$j_\mu = e\bar{\psi}\gamma_\mu\psi,$$
ψ : electron–positron field,
A^μ : field operator for the radiation field,

we obtain

$$(g_{\mu\nu}\Box - \partial_\mu\partial_\nu)\langle A^\nu(x)\rangle = J_\mu(x) + J_\mu^{\mathrm{Pol}}(x), \tag{D.13}$$

with

$$J_\mu^{\mathrm{Pol}}(x) = \langle j_\mu(x)\rangle. \tag{D.14}$$

Here $\langle \cdots \rangle$ denotes the grand canonical average. If we define the polarisation tensor $\Pi_{\mu\nu}$ through Eq. (D.10), which reads

$$J_\mu^{\mathrm{Pol}}(k) = \Pi_{\mu\nu}(k)\langle A^\nu(k)\rangle \ , \tag{D.15}$$

then we have reached the connection with the phenomenological equations of the first paragraph. In addition Eqs. (D.13), (D.14) and (D.15) define the polarisation tensor microscopically.

The linear connection (D.2) between $\boldsymbol{E}$ and $\boldsymbol{D}$ (with ε independent of $\boldsymbol{E}$) only holds in a linear response approximation. Hence we have to calculate J_μ^{Pol} in this approximation.

Linear Response (General Discussion). Let the Hamilton operator of a system be

$$H = H_0 + V(t).$$

In the Schrödinger picture, the average of an operator Q is

$$\langle Q\rangle = \mathrm{Tr}(Q\varrho(t)),$$

with

$$i\dot{\varrho} = [H, \varrho] \ .$$

Obviously

$$\langle Q\rangle = \mathrm{Tr}(\mathrm{e}^{iH_0t}Q\mathrm{e}^{-iH_0t}\mathrm{e}^{iH_0t}\varrho(t)\mathrm{e}^{-iH_0t}).$$

Let

$$\varrho_{\mathrm{I}}(t) := \mathrm{e}^{iH_0t}\varrho(t)\mathrm{e}^{-iH_0t}$$
$$Q_{\mathrm{I}}(t) := \mathrm{e}^{iH_0t}Q\mathrm{e}^{-iH_0t}$$

(I: interaction picture).
Then

$$\langle Q\rangle = \mathrm{Tr}(Q_{\mathrm{I}}(t)\varrho_{\mathrm{I}}(t)).$$

Now we have

$$\begin{aligned} i\dot{\varrho}_{\mathrm{I}} &= \mathrm{e}^{iH_0t}\{-[H_0, \varrho(t)] + [H, \varrho(t)]\}\mathrm{e}^{-iH_0t} \\ &= \mathrm{e}^{iH_0t}[V, \varrho(t)]\mathrm{e}^{-iH_0t} \\ &= [V_{\mathrm{I}}(t), \varrho_{\mathrm{I}}(t)] \ , \end{aligned}$$

where

$$V_{\mathrm{I}}(t) = \mathrm{e}^{iH_0t}V(t)\mathrm{e}^{-iH_0t}.$$

For $V = 0$ the density matrix $\varrho_I(t)$ would be time independent. We assume that for $t \to -\infty$ ϱ is the Gibbs state, which we denote by ϱ_0. In the first approximation of $V(t)$ we then obtain

$$\varrho_\mathrm{I}(t) = \varrho_0 + (-i)\int_{-\infty}^{t} dt'[V_\mathrm{I}(t'), \varrho_0].$$

Hence the linear response for Q is given by

$$\langle Q\rangle = \langle Q_\mathrm{I}(t)\rangle_0 + (-i)\int_{-\infty}^{t} dt\,\mathrm{Tr}(Q_\mathrm{I}(t)[V_\mathrm{I}(t'), \varrho_0])\ ,$$

where $\langle\ \cdots\ \rangle_0$ denotes the grand canonical average. With well known properties of the trace we can write

$$\langle Q\rangle = \langle Q_\mathrm{I}(t)\rangle_0 + (-i)\int_{-\infty}^{t} dt'\langle[Q_\mathrm{I}(t), V_\mathrm{I}(t')]\rangle_0. \tag{D.16}$$

Now we apply this equation to our problem. The index I in (D.16) means, that we have to take the time dependence of the field operator for the external current equal to zero (Heisenberg fields of QED without external currents). This will be indicated by a "hat" on the operators. In our case the interaction is

$$V_\mathrm{I}(t) = \int d^3x\, J_\mu(x)\hat{A}^\mu(x).$$

Hence we obtain

$$\langle A(x)\rangle = \langle \hat{A}(x)\rangle_0 + \int d^4x'\, D_\mathrm{R}^{\mu\nu}(x-x')J_\nu(x'), \tag{D.17}$$

where $D_\mathrm{R}^{\mu\nu}$ is the retarded photon Green's function

$$D_\mathrm{R}^{\mu\nu}(x-x') = -i\langle\Theta(x-x')[\hat{A}^\mu(x), \hat{A}^\nu(x')]\rangle_0\ . \tag{D.18}$$

Clearly $\langle\hat{A}_\mu(x)\rangle_0 = 0$, hence (D.17) reads

$$\langle A^\mu\rangle = \int d^4x'\, D_\mathrm{R}^{\mu\nu}(x-x')J_\nu(x'). \tag{D.17$'$}$$

Eq. (D.17′) implies that $D_\mathrm{R}^{\mu\nu}$ is the Green's function for the differential equation (D.4) for $\langle A^\mu\rangle$. In momentum space this means

$$(k^2 g_{\mu\nu} - k_\mu k_\nu + \Pi_{\mu\nu})D_{\mathrm{R}\lambda}{}^\nu(k) = -\left(g_{\mu\lambda} - \frac{k_\mu k_\lambda}{k^2}\right)\ . \tag{D.19}$$

Here we have chosen the gauge such that

$$k^\mu D_{\mathrm{R}\mu}{}^\nu = 0. \tag{D.20}$$

$D^\mathrm{R}_{\mu\nu}(k)$ can be decomposed as

$$D^\mathrm{R}_{\mu\nu}(k) = D^\mathrm{t}(k)(P_\mathrm{T})_{\mu\nu} + D^\mathrm{l}(k)(P_\mathrm{L})_{\mu\nu}.$$

Putting this into (D.19) and using (D.5) gives the following equations

$$k^2 + \omega^2(\varepsilon^{\rm t} - 1)D^{\rm t}(\boldsymbol{k},\omega) = 1$$

$$k^2 + k^2(\varepsilon^{\rm l} - 1)D^{\rm l}(\boldsymbol{k},\omega) = 1.$$

Hence we can express $D_{\rm R}^{\mu\nu}(k)$ in terms of the dielectric constants as follows

$$D_{\mu\nu}^{R}(\boldsymbol{k},\omega) = (P_{\rm T})_{\mu\nu}\frac{1}{\omega^2\varepsilon^{\rm t} - |\,\boldsymbol{k}\,|^2} + (P_{\rm L})_{\mu\nu}\frac{1}{k^2\varepsilon^{\rm l}}. \qquad \text{(D.21)}$$

In Section 6, we show that $D_{\rm R}^{\mu\nu}(\boldsymbol{k},\omega)$ is analytic in ω in the upper half of the complex ω-plane. Poles in the lower half plane correspond to *elementary exitations* (plasmons).

Let us determine these poles and their residues under the assumption that the damping is weak (the damping will be discussed later).

Transversal mode: $\omega = \omega^{\rm t} - i\Gamma^{\rm t},\ \Gamma^{\rm t} \ll \omega^{\rm t}$.
The pole position is given by

$$\omega^2\varepsilon^{\rm t}(\boldsymbol{k},\omega) = |\,\boldsymbol{k}\,|^2,$$

which implies the dispersion law

$$(\omega^{\rm t})^2{\rm Re}\,\varepsilon^{\rm t}(\boldsymbol{k},\omega^{\rm t}) = |\,\boldsymbol{k}\,|^2 \qquad \text{(D.22)}$$

and

$$\Gamma^{\rm t} = \left.\frac{\omega^2{\rm Im}\,\varepsilon^{\rm t}(\boldsymbol{k},\omega)}{(\partial/\partial\omega)(\omega^2{\rm Re}\,\varepsilon^{\rm t}(\boldsymbol{k},\omega))}\right|_{\omega=\omega^{\rm t}}. \qquad \text{(D.22')}$$

Longitudinal mode: $\omega = \omega^{\rm l} - i\Gamma^{\rm l},\ \Gamma^{\rm l} \ll \omega^{\rm l}$;

$${\rm Re}\,\varepsilon^{\rm l}(\boldsymbol{k},\omega^{\rm l}) = 0, \qquad \text{(D.23)}$$

$$\Gamma^{\rm l} = {\rm Im}\,\varepsilon^{\rm l}(\boldsymbol{k},\omega^{\rm l})/\left.\frac{\partial}{\partial\omega}{\rm Re}\,\varepsilon^{\rm l}(\boldsymbol{k},\omega)\right|_{\omega=\omega^{\rm l}}. \qquad \text{(D.23')}$$

The residues of the poles are easily determined. In the neighborhood of these poles one finds

$$\begin{aligned} D_{\mu\nu}^{\rm R}(\boldsymbol{k},\omega) \simeq &(P_{\rm T})_{\mu\nu}\frac{1}{(\partial/\partial\omega)(\omega^2{\rm Re}\varepsilon^{\rm t})}\left[\frac{1}{\omega - \omega^{\rm t} + i\Gamma^{\rm t}} - \frac{1}{\omega + \omega^{\rm t} - i\Gamma^{\rm t}}\right] \\ &+ (P_{\rm L})_{\mu\nu}\frac{1}{k^2(\partial\varepsilon^{\rm l}/\partial\omega)}\left[\frac{1}{\omega - \omega^{\rm l} + i\Gamma^{\rm l}} - \frac{1}{\omega + \omega^{\rm l} - i\Gamma^{\rm l}}\right]. \end{aligned} \qquad \text{(D.24)}$$

In Section 6 we also derive the following important formula (Γ: plasmon):

$$\langle 0 \mid A_\mu(x) \mid \Gamma(\boldsymbol{k},\omega)\rangle = (2\pi)^{-3/2}\,\sqrt{\rm residue}\; e_\mu(\boldsymbol{k},\omega){\rm e}^{-ik\cdot x}, \qquad \text{(D.25)}$$

where $e_\mu(\boldsymbol{k},\omega)$ is the polarisation vector. With these preparations we can now express the plasmon-neutrino decay rates in terms of the dielectric constants $\varepsilon^{\rm t}$ and $\varepsilon^{\rm l}$. A microscopic calculation of these "constants" will be given later.

3. Plasmon-Neutrino Decay Rates

We consider first a V-A interaction

$$L_{\mathrm{V-A}} = -\frac{G}{\sqrt{2}}\hat{\bar{\psi}}_e \gamma^\mu (1-\gamma^5)\hat{\psi}_e \bar{\psi}_\nu \gamma_\mu (1-\gamma_5)\psi_\nu$$

Here $\hat{\psi}_e$ denotes the electron-positron field with radiative corrections. It turns out that only the vector part is important (see Ref. [70]). If we forget about the axial part, the T-matrix element for $\Gamma \to \nu + \bar{\nu}$ is

$$T = \frac{G}{\sqrt{2}e}\langle 0 \mid \hat{j}_\mu(0) \mid \Gamma\rangle \, \langle \nu\bar{\nu} \mid \bar{\psi}_\nu(0)\gamma^\mu(1-\gamma^5)\psi_\nu(0) \mid 0\rangle.$$

Using (D.25) and the Maxwell equations we get

$$\langle 0 \mid \hat{j}_\mu(0) \mid \Gamma\rangle = (2\pi)^{-3/2} k^2 \sqrt{\text{residue}}\; e_\mu(k).$$

Hence

$$T = \frac{G}{\sqrt{2}e}\frac{1}{(2\pi)^{9/2}} k^2 \sqrt{\text{residue}}\; \bar{u}_\nu(p_2)\, \not{e}(1-\gamma^5) v_\nu(p_1).$$

In the longitudinal case we obtain from this

$$\begin{aligned} w^{\mathrm{l}} =& \frac{G^2}{2e^2}\frac{1}{(2\pi)^2}\frac{k^2}{\partial\varepsilon^{\mathrm{l}}/\partial\omega} \\ & \times \int \frac{d^3p_1}{2p_1^0}\frac{d^3p_2}{2p_2^0}\delta(k-p_1-p_2)\underbrace{\mathrm{Tr}[\not{e}(1-\gamma_5)\, \not{p}_1\, \not{e}(1-\gamma_5)\, \not{p}_2]}_{8[2(e\cdot p_1)(e\cdot p_2)+(p_1\cdot p_2)]} \,. \end{aligned}$$

Using

$$\int \frac{d^3p_1}{2p_1^0}\frac{d^3p_2}{2p_2^0}\delta(k-p_1-p_2)p_1^\mu p_2^\nu = \frac{\pi}{24}[2k^\mu k^\nu + g^{\mu\nu}k^2]$$

we finally get

$$w^{\mathrm{l}} = \frac{2}{3}\frac{G^2}{4\pi e^2}\frac{(k^2)^2}{\partial\varepsilon^{\mathrm{l}}/\partial\omega}\,. \tag{D.26}$$

Similarly one finds

$$w^{\mathrm{t}} = \frac{2G^2}{3(4\pi e^2)}\left[\omega\left(2\varepsilon^{\mathrm{t}} + \omega\frac{\partial\varepsilon^{\mathrm{t}}}{\partial\omega}\right)\right]^{-1} k^2\,[\omega^2(\varepsilon^{\mathrm{t}}-1)]^2. \tag{D.27}$$

For the total decay rate into all three neutrino flavors we conclude from (III.14) that the expressions (D.26) and (D.27) have to be multiplied by

$$C_{\mathrm{V}}^2 + 2(C_{\mathrm{V}}-1)^2 \simeq 0.9.$$

4. Calculation of the Dielectric Tensor

For an approximate calculation of ε^{t} and ε^{l} we start from Eqs. (D.14) and (D.15) which we repeat here

$$J_\mu^{\mathrm{Pol}} = \langle j_\mu \rangle \tag{D.14}$$

$$J_\mu^{\mathrm{Pol}}(k) = \Pi_{\mu\nu}(k)\langle A^\nu(k)\rangle . \tag{D.15}$$

Both $\langle j_\mu \rangle$ and $\langle A^\nu \rangle$ are only needed in a linear response approximation of the external field $\mathcal{A}^\nu(x)$ (corresponding to the external current $J_\mu(x)$).

Now we suppose we can express J_μ^{Pol} (in linear response approximation) in the following form

$$J_\mu^{\mathrm{Pol}}(x) = \int K_{\mu\nu}(x-x')\mathcal{A}^\nu(x')d^4x' . \tag{D.28}$$

What is the relation between $\Pi_{\mu\nu}$ and $K_{\mu\nu}$?

From Maxwell's equations we obtain

$$A_\mu(x) = A_\mu^{\mathrm{in}}(x) + \int \overset{\circ}{D}{}^{\mathrm{R}}_{\mu\nu}(x-x')[J^\nu(x') + j^\nu(x')]d^4x' ,$$

where A_μ^{in} is the incoming radiation field and $\overset{\circ}{D}{}^{\mu\nu}_{\mathrm{R}}$ the *free* retarded photon Green's function. This equation can also be written as

$$A_\mu(x) = A_\mu^{\mathrm{in}}(x) + \mathcal{A}_\mu + \int \overset{\circ}{D}{}^{R}_{\mu\nu}(x-x')j^\nu(x')d^4x' . \tag{D.29}$$

Hence

$$\langle A_\mu \rangle = \mathcal{A}_\mu + \int \overset{\circ}{D}{}^{R}_{\mu\nu}(x-x')\langle j^\nu(x')\rangle d^4x' .$$

In momentum space this reads

$$\langle A^\mu(k)\rangle = \mathcal{A}^\mu(k) + \overset{\circ}{D}{}^{\mu\nu}_{\mathrm{R}}(k)J_\nu^{\mathrm{Pol}}(k).$$

Consequently we get (in momentum space)

$$J_\mu^{\mathrm{Pol}} = \Pi_{\mu\nu}\langle A^\nu\rangle = \Pi_{\mu\nu}[\mathcal{A}^\nu + \overset{\circ}{D}{}^{\nu\lambda}_{\mathrm{R}} J_\lambda^{\mathrm{Pol}}],$$

or in matrix notation

$$(1 - \Pi \overset{\circ}{D}_{\mathrm{R}})J^{\mathrm{Pol}} = \Pi \mathcal{A}.$$

Using (D.28) this gives

$$(1 - \Pi \overset{\circ}{D}_{\mathrm{R}})K = \Pi . \tag{D.30}$$

Note that current conservation implies $\partial^\nu K_{\mu\nu} = 0$. We calculate Π only in the lowest order in the fine structure constant. In the literature this is often called random phase approximation (see e.g. Ref. [71]).

Note that Π and K are equal if both kernels are taken in lowest order. This remark will allow us to compute Π in lowest order.

For doing this we start from the Dirac equation

$$(i\,\not{\partial}_+ - m)\psi = e\,\not{A}\psi, \tag{D.31}$$

where $\not{\partial}_+ = (\partial_0 + \mu)\gamma^0 - \boldsymbol{\gamma}\cdot\boldsymbol{\partial}$; μ: chemical potential.
Equation (D.31) can be transformed into the following integral equation

$$\psi = \psi^{\text{in}} + e\int d^4x'\, S_{\text{R}}(x - x')\,\not{A}(x')\psi(x'), \tag{D.32}$$

where ψ^{in} is the incoming electron-positron field. The Green's function $S_{\text{R}}(x)$ will be calculated further below. It will turn out that the retarted and the advanced Green's function are related as follows

$$\gamma^0 S_{\text{R}}^*(x)\gamma^0 = S_{\text{A}}(-x).$$

Hence we obtain from (D.32)

$$\bar{\psi}(x) = \bar{\psi}^{\text{in}}(x) + e\int \bar{\psi}(x')\,\not{A}(x')S_{\text{A}}(x' - x)d^4x' \ . \tag{D.33}$$

Let us expand ψ in powers of e:

$$\psi \;=\; \psi^{\text{in}} \;+\; \psi^{(1)} \;+\; \dots \ .$$

Then we obtain from (D.32) and (D.33)

$$\begin{aligned}
j^\mu &= \frac{e}{2}[\bar{\psi}, \gamma^\mu\psi] \\
&= \frac{e}{2}[\bar{\psi}^{\text{in}}, \gamma^\mu\psi^{\text{in}}] + \frac{e}{2}[\bar{\psi}^{\text{in}}, \gamma^\mu\psi^{(1)}] + \frac{e}{2}[\bar{\psi}^{(1)}, \gamma^\mu\psi^{\text{in}}] + \ \dots \\
&= \frac{e}{2}[\bar{\psi}^{\text{in}}, \gamma^\mu\psi^{\text{in}}] + \frac{e^2}{2}\int d^4x'\ \{[\bar{\psi}^{\text{in}}(x), \gamma^\mu S_{\text{R}}(x - x')\gamma^\nu\psi^{\text{in}}(x')] \\
&\qquad + [\bar{\psi}^{\text{in}}(x')\gamma^\nu S_{\text{A}}(x' - x), \gamma^\mu\psi^{\text{in}}(x)]\}\ A_\nu(x') \ .
\end{aligned}$$

If we use (D.29) in lowest order, together with $\langle A_\mu^{\text{in}}\rangle = 0$, we obtain J_μ^{Pol} in the form (D.28) with K in lowest order. In the random phase approximation, Π is equal to this first approximation of K, but itself is given by (D.30). (This approximation is sometimes called the "improved perturbation theory".)

Let $S^{(1)}$ denote the following Green's function

$$\langle[\bar{\psi}_\alpha^{\text{in}}(x), \psi_\beta^{\text{in}}(x')]\rangle = S_{\beta\alpha}^{(1)}(x' - x) \ ; \tag{D.34}$$

then we obtain for Π in lowest order

$$\begin{aligned}
\Pi^{\mu\nu}(x - x') =& \frac{e^2}{2}\Big\{\text{Tr}[\gamma^\mu S_{\text{R}}(x - x')\gamma^\nu S^{(1)}(x' - x)] \\
&+ \text{Tr}\left[\gamma^\nu S_{\text{A}}(x' - x)\gamma^\mu S^{(1)}(x - x')\right]\Big\} \ .
\end{aligned}$$

In momentum space

$$\begin{aligned}\Pi^{\mu\nu}(k) =& \frac{e^2}{2}(2\pi)^{-4}\int d^4p\,\{\,\mathrm{Tr}[\gamma^\mu S_\mathrm{R}(p+k)\gamma^\nu S^{(1)}(p)] \\ &+ \mathrm{Tr}[\gamma^\mu S^{(1)}(p+k)\gamma^\nu S_\mathrm{A}(p)]\,\}\,.\end{aligned} \tag{D.35}$$

We still have to determine the distributions S_R, S_A and $S^{(1)}$. $S_\mathrm{R,A}$ are defined by the equation

$$(i\,\partial\!\!\!/_{+} - m)S_\mathrm{R,A} = \delta^4(x)\,.$$

In momentum space this gives

$$(p\!\!\!/_{+} - m)\tilde{S}_\mathrm{R,A}(p) = 1, \quad p_+ = (p^0 + \mu, \boldsymbol{p})\,.$$

Multiplying with $(p\!\!\!/_{+} + m)$ leads to

$$\tilde{S}_\mathrm{R,A}(p) = \frac{(p\!\!\!/_{+} + m)}{(p_+^2 - m^2)_\mathrm{R,A}}\,. \tag{D.36}$$

The index (R,A) on the right hand side of (D.36) indicates how one has to treat the poles in p^0. The correct boundary conditions are obtained by putting

$$\frac{1}{(p_+^2 - m^2)_\mathrm{R,A}} = \frac{1}{(p^0 + \mu \pm i\varepsilon)^2 - \boldsymbol{p}^2 - m^2}\,. \tag{D.37}$$

Eq. (D.36) can also be written as

$$\tilde{S}_\mathrm{R,A}(p) = \frac{1}{2E_p}\left[\frac{m + \gamma^0 E_p - \boldsymbol{\gamma}\cdot\boldsymbol{p}}{p^0 - E_p + \mu \pm i\varepsilon} + \frac{-m + \gamma^0 E_p + \boldsymbol{\gamma}\cdot\boldsymbol{p}}{p^0 + E_p + \mu \pm i\varepsilon}\right]\,. \tag{D.36'}$$

A simple calculation shows that

$$\begin{aligned}S^{(1)} =& (2\pi)^{-3}\int \frac{d^3p}{2E_p}\,[(p\!\!\!/ + m)(2n_-(\boldsymbol{p}) - 1)\mathrm{e}^{-ip_-\cdot x} \\ &-(p\!\!\!/ - m)(2n_+(\boldsymbol{p}) - 1)\mathrm{e}^{ip_+\cdot x}]\,.\end{aligned}$$

Here $n_\mp(\boldsymbol{p})$ are the electron and positron occupation probabilities[4]

$$\begin{aligned}n_-(\boldsymbol{p}) &= \langle a^*(\boldsymbol{p})a(\boldsymbol{p})\rangle = \left[\mathrm{e}^{\beta(E_p-\mu)} + 1\right]^{-1} \\ n_+(\boldsymbol{p}) &= \langle b^*(\boldsymbol{p})b(\boldsymbol{p})\rangle = \left[\mathrm{e}^{\beta(E_p+\mu)} + 1\right]^{-1}.\end{aligned}$$

[4] Note that $H_0 - \mu N = \int \frac{d^3p}{2E_p}[(E_p - \mu)a^*(\boldsymbol{p})a(\boldsymbol{p}) + (E_p + \mu)b^*(\boldsymbol{p})b(\boldsymbol{p})]$

The 4-dimensional Fouriertransform of $S^{(1)}$ is

$$\begin{aligned} S^{(1)}(p) =& \frac{2\pi}{2E_p}[(m+\gamma^0 E_p - \boldsymbol{\gamma}\cdot\boldsymbol{p})(2n_-(\boldsymbol{p})-1)\delta(p^0-E_p+\mu) \\ & -(-m+\gamma^0 E_p+\boldsymbol{\gamma}\cdot\boldsymbol{p})(2n_+(\boldsymbol{p})-1)\delta(p^0+E_p+\mu)] \,. \end{aligned} \tag{D.38}$$

Now we insert (D.36′) and (D.38) into (D.35). The p^0-integration can trivially be carried out because of the δ-functions in (D.38). It is sufficient to calculate Π_{ij}.

Each of the eight terms which one obtains in carrying out the p^0-integration, is proportional to the trace

$$\begin{aligned} T_{ij}^{(\sigma\sigma')} &= \frac{1}{4}\mathrm{Tr}\,\{\gamma_i[-\sigma m+\gamma^0 E_{p+k}+\sigma\boldsymbol{\gamma}\cdot(\boldsymbol{p}+\boldsymbol{k})] \\ &\qquad \times\gamma_j[-\sigma' m+\gamma^0 E_p+\sigma'\boldsymbol{\gamma}\cdot\boldsymbol{p}]\} \\ &= \sigma\sigma'[\delta_{ij}(\sigma\sigma' E_{p+k}E_p-\boldsymbol{p}\cdot(\boldsymbol{p}+\boldsymbol{k})-m^2) \\ &\qquad +2p_i p_j+k_i p_j+k_j p_i] \,, \end{aligned} \tag{D.39}$$

where $\sigma, \sigma' = \pm 1$.

From (D.5), (D.6), (D.6′) and (D.7) we have

$$\Pi_{ij} = -\omega^2(\varepsilon^{\mathrm{t}}-1)\left(\delta_{ij}-\frac{k_i k_j}{|\boldsymbol{k}|^2}\right)-\omega^2(\varepsilon^{\mathrm{l}}-1)\frac{k_i k_j}{|\boldsymbol{k}|^2} \,.$$

Hence

$$\begin{aligned} -\omega^2(\varepsilon^{\mathrm{t}}-1) &= \frac{1}{2}\sum_{i,j}\Pi_{ij}\left(\delta_{ij}-\frac{k_i k_j}{|\boldsymbol{k}|^2}\right) \\ -\omega^2(\varepsilon^{\mathrm{l}}-1) &= \sum_{i,j}\Pi_{ij}\frac{k_i k_j}{|\boldsymbol{k}|^2} \,. \end{aligned} \tag{D.40}$$

Now

$$\frac{1}{2}\sum_{i,j}T_{ij}^{(\sigma,\sigma')}\left(\delta_{ij}-\frac{k_i k_j}{|\boldsymbol{k}|^2}\right) = E_p E_{p+k}\Delta^{\sigma\sigma'}(\boldsymbol{p},\boldsymbol{k}), \tag{D.41}$$

where

$$\Delta^{\sigma\sigma'}(\boldsymbol{p},\boldsymbol{k}) = 1-\sigma\sigma'\,\frac{\boldsymbol{p}\cdot\boldsymbol{k}+(\boldsymbol{p}\cdot\boldsymbol{k})^2/m^2+m^2}{E_p E_{p+k}} \,.$$

Using (D.40) and (D.41) we obtain

$$-\omega^2(\varepsilon^{\mathrm{t}}-1) = \frac{1}{2}(2\pi)^{-3}\int d^3p\{\,\dots\,\} \,, \tag{D.42}$$

$$\begin{aligned}\{\ \dots\ \} = &(2n_-(p)-1)\left[\frac{\Delta^+(p,k)}{E_p - E_{p+k} + \omega + i\varepsilon} + \frac{\Delta^-(p,k)}{E_p + E_{p+k} + \omega + i\varepsilon}\right] \\ &-(2n_+(p)-1)\left[\frac{\Delta^-(p,k)}{\omega - E_p - E_{p+k} + i\varepsilon} + \frac{\Delta^+(p,k)}{\omega + E_{p+k} - E_p + i\varepsilon}\right] \\ &+(2n_-(p+k)-1)\left[\frac{\Delta^+(p,k)}{E_{p+k} - E_p - \omega - i\varepsilon} + \frac{\Delta^-(p,k)}{E_{p+k} + E_p - \omega - i\varepsilon}\right] \\ &-(2n_+(p+k)-1)\left[\frac{\Delta^-(p,k)}{-E_{p+k} - E_p - \omega - i\varepsilon} + \frac{\Delta^+(p,k)}{-E_{p+k} + E_p - \omega - i\varepsilon}\right] . \end{aligned} \tag{D.43}$$

Now we substitute in the last two terms $\boldsymbol{p}+\boldsymbol{k} \to \boldsymbol{p}$ and in the first two terms $\boldsymbol{p} \to -\boldsymbol{p}$. Furthermore, note the identity

$$\begin{aligned}\Delta^\pm(p-k,k) &= \Delta^\pm(-p,k) \\ &= 1 \mp \frac{m^2 - \boldsymbol{p}\cdot\boldsymbol{k} + (\boldsymbol{p}\cdot\boldsymbol{k})^2/|\boldsymbol{k}|^2}{E_p E_{p-k}} \equiv \Lambda^{\mathrm{t}}_{\mp} .\end{aligned} \tag{D.44}$$

Then (D.43) can be simplified

$$\begin{aligned}\varepsilon^{\mathrm{t}}(\boldsymbol{k},\omega) = &1 - \frac{e^2}{\omega^2}\int d^3p\, f(E_p)\left\{\frac{E_p - E_{p-k}}{(E_p - E_{p-k})^2 - \omega^2}\Lambda^{\mathrm{t}}_- + \right. \\ &\left. + \frac{E_p + E_{p-k}}{(E_p + E_{p-k})^2 - \omega^2}\Lambda^{\mathrm{t}}_+\right\} + \text{vacuum polarisation} ,\end{aligned} \tag{D.45}$$

where

$$f(E_p) = 2(2\pi)^{-3}[n_-(p) + n_+(p)] . \tag{D.46}$$

The vacuum polarisation term in (D.45) represents those contributions which survive for $n_- = n_+ = 0$. They have to be renormalized, of course. Their contribution is, however, negligible compared to the effect of the polarisation of the dense plasma.

For the longitudinal dielectric constant one obtains

$$\begin{aligned}&\varepsilon^{\mathrm{l}}(\boldsymbol{k},\omega) = 1 - \frac{e^2}{\omega^2}\int d^3p\, f(E_p) \\ &\left\{\frac{E_p - E_{p-k}}{(E_p - E_{p-k})^2 - \omega^2}\Lambda^{\mathrm{l}}_- + \frac{E_p + E_{p-k}}{(E_p + E_{p-k})^2 - \omega^2}\Lambda^{\mathrm{l}}_+\right\} ,\end{aligned} \tag{D.47}$$

where

$$\Lambda^{\mathrm{l}}_{\mp} = 1 \mp \frac{E_p^2 + \boldsymbol{p}\cdot\boldsymbol{k} - 2(\boldsymbol{p}\cdot\boldsymbol{k})^2/|\boldsymbol{k}|^2}{E_p E_{p-k}} . \tag{D.48}$$

In the stellar plasmas the conditions $\omega \ll 2m$ and $k \ll m$ are well satisfied. In this case, both dielectric constants reduce to the form

$$\varepsilon(\omega) = 1 - \omega_p^2/\omega^2 , \tag{D.49}$$

where

$$\omega_p^2 = e^2 \int d^3p \, \frac{f(E_p)}{E_p} \left(1 - \frac{p^2}{3E_p^2}\right) . \tag{D.50}$$

Exercise. Derive the nonrelativistic limit of Re $\varepsilon^{\mathrm{l}}(\boldsymbol{k}, \omega)$:

$$\mathrm{Re}\, \varepsilon^{\mathrm{l}}(\boldsymbol{k}, \omega) \simeq 1 - \frac{2e^2}{|\boldsymbol{k}|^2} P\!\int d^3p \, f(E_p) \frac{\varepsilon_p - \varepsilon_{p-k}}{(\varepsilon_p - \varepsilon_{p-k})^2 - \omega^2}$$

where $\varepsilon_p = \boldsymbol{p}^2/2m$, and compute the screening action of the electron gas. Compare the result with that calculated in the Fermi-Thomas approximation.

---

5. Damping of the Plasmons

Finally we discuss the validity of treating the plasmons essentially as normal modes (weak damping). For this we have to consider the absorption processes which contribute to the imaginary part of the dielectric constants $\varepsilon^{\mathrm{t,l}}$. In the lowest order of perturbation theory the imaginary part of $\varepsilon^{\mathrm{t,l}}$ represents transverse photon and longitudinal plasmon absorption by the inverse Čerenkov process (Landau-damping) and plasmon decay into real electron positron pairs.

(inverse Čerenkov absorption) (pair production)

Energy momentum conservation implies for both transverse and longitudinal plasmons that real electron positron pair creation can occur only if $\omega_0 > 2m$, which is never the case in the density and temperature regime of interest in our study.

Čerenkov absorption occurs only if $k > \omega$ (because of energy momentum conservation); this never obtains for transverse plasmons. For longitudinal plasmons this mechanism is relevant, but neutrino pair emission occurs only for $k < \omega$, and Landau damping will not alter the momentum distribution in the region of interest.

6. General Properties of the Retarded Green's Functions

In this Section we derive the spectral representation of the Green's function $D^{\mathrm{R}}_{\mu\nu}$ and discuss its analytic properties, together with some related topics.

The matrix elements of $A_\mu(x)$ between energy momentum eigenstates $|n\rangle$ are given by

$$\langle m|A_\mu(x)|n\rangle = \langle m|A_\mu(0)|n\rangle \mathrm{e}^{-i(p_n-p_m)\cdot x} .$$

Hence

$$\begin{aligned} D^{\mathrm{R}}_{\mu\nu}(x-y) =& - i\Theta(x^0-y^0)\langle[A_\mu(x), A_\nu(y)]\rangle = \\ =& - i\Theta(x^0-y^0)\sum\nolimits_{n,m} \mathrm{e}^{-\beta(\omega_n-F)} \\ &\times [\langle n|A_\mu(0)|m\rangle \cdot \langle m|A_\nu(0)|n\rangle \mathrm{e}^{i(p_n-p_m)\cdot(x-y)} \\ &- \langle n|A_\nu(0)|m\rangle \cdot \langle m|A_\mu(0)|n\rangle \mathrm{e}^{i(p_m-p_n)\cdot(x-y)}] ; \end{aligned}$$

F: free energy.

Next we use the Fourier representation of the Θ-function:

$$\Theta(\tau) = -\frac{1}{2\pi i}\int \frac{\mathrm{e}^{-i\omega\tau}}{\omega+i\varepsilon}\, d\omega$$

and interchange in the second term the summation indices

$$\begin{aligned} D^{\mathrm{R}}_{\mu\nu}(x-y) =& \frac{1}{2\pi}\int \frac{d\omega}{\omega+i\varepsilon}\sum\nolimits_{n,m} \mathrm{e}^{-\beta(\omega_n-F)} \\ &\times \langle n|A_\mu(0)|m\rangle \cdot \langle m|A_\nu(0)|n\rangle \\ &\times \mathrm{e}^{i(p_n-p_m)\cdot(x-y)}\mathrm{e}^{-i\omega(x^0-y^0)}\left(1-\mathrm{e}^{\beta(\omega_n-\omega_m)}\right) . \end{aligned}$$

Now we calculate the Fourier transform

$$\begin{aligned} D^{\mathrm{R}}_{\mu\nu}(k) =& \int d^4x \mathrm{e}^{ik\cdot x} D^{\mathrm{R}}_{\mu\nu}(x) \\ =& (2\pi)^3 \int \frac{d\omega'}{\omega'+i\varepsilon}\sum\nolimits_{n,m} \mathrm{e}^{-\beta(\omega_n-F)} \\ &\times \langle n|A_\mu(0)|m\rangle \cdot \langle m|A_\nu(0)|n\rangle \\ &\times \delta^3(\boldsymbol{k}+\boldsymbol{p}_n-\boldsymbol{p}_m)\delta(\omega-\omega'+\omega_n-\omega_m)\left(1-\mathrm{e}^{\beta(\omega_n-\omega_m)}\right) . \end{aligned}$$

Substituting $\omega-\omega' \to \omega'$ gives

$$\begin{aligned} D^{\mathrm{R}}_{\mu\nu}(k) =& \int \frac{d\omega'}{\omega-\omega'+i\varepsilon}(2\pi)^3\sum\nolimits_{n,m} \mathrm{e}^{-\beta(\omega_n-F)} \\ &\times \langle n|A_\mu(0)|m\rangle \cdot \langle m|A_\nu(0)|n\rangle \\ &\times \delta^3(\boldsymbol{k}+\boldsymbol{p}_n-\boldsymbol{p}_m)\delta(\omega'+\omega_n-\omega_m)\left(1-\mathrm{e}^{-\beta\omega'}\right) . \end{aligned}$$

We put

$$\begin{aligned} \mathcal{G}_{\mu\nu}(\boldsymbol{k},\omega) =& -(2\pi)^3\sum\nolimits_{n,m} \mathrm{e}^{-\beta(\omega_n-F)}\langle n|A_\mu(0)|m\rangle \\ &\times \langle m|A_\nu(0)|n\rangle \delta^3(\boldsymbol{k}+\boldsymbol{p}_n-\boldsymbol{p}_m)\delta(\omega+\omega_n-\omega_m) \end{aligned} \qquad \text{(D.51)}$$

and obtain the following spectral representation

$$D^{\mathrm{R}}_{\mu\nu}(\boldsymbol{k},\omega) = \int \frac{d\omega'}{\omega' - \omega - i\varepsilon} \left(1 - \mathrm{e}^{-\beta\omega'}\right) \mathcal{G}_{\mu\nu}(\boldsymbol{k},\omega') \, . \tag{D.52}$$

From (D.52) one sees that $D^{\mathrm{R}}_{\mu\nu}(\boldsymbol{k},\omega)$ is analytic in ω in the upper half plane. This is a direct consequence of causality: $D_{\mathrm{R}}(\boldsymbol{x},x^0) = 0$ for $x^0 < 0$. One can show that the poles in the lower half plane correspond to elementary exitations (quasi particles). This is discussed for instance in Ref. [72], page 57.

Remark. One can derive an analogous spectral representation for the time-ordered Green's function. In doing this one finds

$$\begin{aligned} \operatorname{Re} D^{\mathrm{F}}_{\mu\nu}(\boldsymbol{k},\omega) &= \operatorname{Re} D^{\mathrm{R}}_{\mu\nu}(\boldsymbol{k},\omega) \\ \operatorname{Im} D^{\mathrm{F}}_{\mu\nu}(\boldsymbol{k},\omega) &= \coth\left(\frac{\beta\omega}{2}\right) \operatorname{Im} D^{\mathrm{R}}_{\mu\nu}(\boldsymbol{k},\omega) \, . \end{aligned} \tag{D.53}$$

Finally we calculate the contribution of the quasi particle Γ (plasmon) to $D^{\mathrm{R}}_{\mu\nu}(\boldsymbol{k},\omega)$. Let us thereby neglect the damping. If $(E_p, \boldsymbol{p})$ denote energy and momentum of the plasmon we obtain for $T = 0$

$$\mathcal{G}^{\Gamma}_{\mu\nu}(\boldsymbol{k},\omega) = -\sum_{\mathrm{Pol}} \int d^3p f^2(\boldsymbol{p}) e_\mu(\boldsymbol{p}) e_\nu(\boldsymbol{p}) \delta^3(\boldsymbol{k}-\boldsymbol{p}) \delta(\omega - E_p),$$

where e_μ = polarisation vector of Γ and

$$\langle 0|A_\mu(0)|\Gamma\rangle = (2\pi)^{-3/2} f(\boldsymbol{p}) e_\mu(\boldsymbol{p}) \, .$$

If Γ is for instance a transversal plasmon, we obtain

$$\mathcal{G}^{\Gamma}_{\mu\nu}(\boldsymbol{k},\omega) = (P_{\mathrm{T}})_{\mu\nu} f^2(\boldsymbol{p}) \delta(\omega - E_p)$$

and hence

$$D^{\Gamma}_{\mathrm{t}}(\boldsymbol{k},\omega) = \frac{1}{E_p - \omega - i\varepsilon} f^2(\boldsymbol{p}) \, .$$

This shows that $f^2(\boldsymbol{p})$ is the residue of the pole in $D^{\Gamma}_{\mathrm{t}}(\boldsymbol{k},\omega)$ and hence we obtain the important equation

$$\langle 0|A_\mu(0)|\Gamma\rangle = (2\pi)^{-3/2} \sqrt{\text{residue}}\; e_\mu \, .$$

This relation was used in the calculation of the decay rate for $\Gamma \to \nu\bar{\nu}$.

References

1. Q.B. Li, in Proceedings of the "2nd Workshop on High Energy Astropysics", Max-Planck-Academia Sinica, July 1987, to be published by Springer-Verlag.
2. D.H. Clark, F.R. Stephenson, "The Historical Supernovae", Pergamon Press (1977).
3. W. Baade, F. Zwicky, Proc.Acad.Sci., 20, 254 (1934).
4. F.D. Seward, "Supernova Remnants Containing Neutron Stars", Comments Astrophys. 11, 15 (1985).
5. R. Barbon, F. Ciatti and L. Rosino, Astron. Astropys. 25, 241 (1973).
6. J. Craig Wheeler and R.P. Harkness, "Helium-rich Supernovas", Scientific American, Nov. 1987.
7. M. Oda, Physics Today, Dec. 1987.
8. J. Craig Wheeler, Rep.Prog.Phys. 44, 85 (1981).
9. S.L. Shapiro and S.A. Teucholski, "Black Holes, White Dwarfs, and Neutron Stars", John Wiley & Sons (1983).
10. K. Nomoto, in IAU Symp. No. 93: "Fundamental Problems in the Theory of Stellar Evolution", (ed.: D. Sugimoto, D.Q. Lamb and D.N. Schramm), Reidel (1982).
11. For a review see: C. Chiosi and A. Maeder, Ann.Rev. Astron. Astrophys. 24, 329 (1986).
12. K. Nomoto and M. Hashimoto, to be published in Physics Reports (1988).
13. S.E. Woosley and T.A. Weaver, Ann.Rev.Astr.Ap. 24, 205 (1986).
14. K. Nomoto, T. Shigeyama and M. Hashimoto, in Proceedings of the ESO Workshop on SN 1987 A, (ed.: I.J. Danziger), ESO (1987).
15. G. Beaudet et al., Ap. J. 174, 79 (1972).
16. W.M. Brunish and J.W. Truran, Ap. J. 256, 247 (1982); Ap. J. Suppl. 49, 447 (1982).
17. A. Maeder, in Proceedings of the ESO Workshop on SN 1987 A, loc.cit., p. 251.
18. H.A. Bethe et al., Nucl.Phys. A 324, 487 (1979).
19. G.M. Fuller, Ap. J. 252, 741 (1982).
20. G.E. Brown et al., Nucl.Phys. A 375, 481 (1982).
21. R.I. Epstein and C.J. Pethick, Ap. J. 243, 1003 (1982).
22. D.Q. Lamb et al., Phys.Rev.Lett. 41, 1623 (1978).
23. A. Yahil, Ap. J. 265, 1047 (1983).
24. N. Straumann, "General Relativity and Relativistic Astrophysics", Springer (1984).
25. P. Goldreich and S.V. Weber, Ap. J. 238, 991 (1980).
26. R. Mayle, Ph.D. dissertation, UC Berkley (1985).
27. J.M. Lattimer, A. Burrows and A. Yahil, Ap. J. 288, 644 (1985).
28. Kamiokande II Collab., K. Hirata et al., Phys.Rev.Lett. 58,1490 (1987); IMB Collab., R.M. Bionta et al., Phys.Rev.Lett. 58, 1494 (1987).
29. M.T. Koshiba, Physics Today, Dec. 1987, p. 38.
30. A. Burrows and J.M. Lattimer, Ap. J., L 63 (1987).
31. K. Sato and H. Suzuki, Phys.Rev.Lett. 58, 2722 (1987).
32. Proceedings of the ESO Workshop on the SN 1987 A, (ed.: I.J. Danziger), ESO (1987).
33. L.F. Abott, A. De Rujula and T.P. Walker, CERN preprint TH 4799/87.
34. D.N. Spergel and J.N. Bahcall, Phys.Lett. B 200, 366 (1988).
35. SIN Collab., M. Fritschi et al., Phys.Lett. B 173, 485 (1986).
36. S.A. Bludman and K.A. Van Riper, Ap. J. 224, 631 (1978).
37. C.D. Levermore and G.C. Pomraning, Ap. J. 248, 321 (1981).
38. R.L. Bowers and J.R. Wilson, Ap. J. Suppl. 50, 115 (1982); Ap. J. 263,366 (1982).
39. A. Burrows and J.M. Lattimer, Ap. J. 307, 178 (1986).
40. A. Burrows, Arizona Theoretical Astrophysics Preprint 87 – 34.
41. S.W. Bruenn, Phys.Rev.Lett. 59,938 (1987).
42. R. Mayle and J.R. Wilson, Preprint (1987).
43. W.D. Arnett and R.L. Bowers, Ap. J. Suppl. 33, 415 (1977).
44. E. Müller and W. Hillebrandt, Astr.Astrophys. 80, 147 (1981).
45. R. Abraham and J.E. Marsden, "Foundations of Mechanics", 2nd ed., Benjamin (1978).
46. J. Ehlers, "General Relativity and Kinetic Theory", in K. Sachs, "General Relativity and Cosmology", Academic Press (1971); A. Aeppli, Diploma work, University of Zürich (1986).

47. R. Durrer and N. Straumann, to be published in Helv.Phys.Acta.
48. See Appendix to Chapter 3 in Ref. [24].
49. R.W. Lindquist, Ann. of Phys. 37, 487 (1966).
50. G.E. Brown et al., Nucl.Phys. A 375,481 (1982).
51. W. Hillebrandt, Astron. Ap.Lett. 110, L 3 (1982); Ann. N.Y.Acad. Sci. 422,197 (1984).
52. W.D. Arnett, Ann. N.Y.Acad. Sci. 336, 366 (1980); Ap. J. Lett. 263, L 55 (1983).
53. T.J. Mazurek, J. Cooperstein and S. Kahana, "Supernovae: A Survey of Current Research", (ed.: M.J. Rees and R.J. Stoneham), Cambridge University Press (1982), p. 71.
54. S.W. Bruenn, Ap. J. Suppl. 58,771 (1985).
55. I. Lichtenstadt and S.A. Bludman, Ap. J. 276, 246 (1984).
56. J.M. Lattimer and A. Burrows, "Problems of Collapse and Numerical Relativity", (ed.: D. Bancel and M. Signore), D. Reidel, Dordrecht (1984), p. 147.
57. A. Burrows and J.M. Lattimer, Ap. J. Lett. 299, L19 (1985).
58. E. Baron, J. Cooperstein and S. Kahana, Nucl.Phys. A 440, 744 (1985).
59. J.R. Wilson, "Numerical Astrophysics", (ed.: J.M. Centrella, J.M. LeBlanc and R.L. Bowers), Jones and Bartlett, Boston (1985), p. 422;
R.L. Bowers, loc.cit., p. 403.
60. J.R. Wilson et al., Proc. of the 12th Texas Symp. on Rel. Ap., Ann. N.Y.Acad. Sci. 470, 267 (1986).
61. W. Hillebrandt, K. Nomoto and R. Wolff, Astron.Ap. 133, 175 (1984).
62. E.S. Myra et al., Ap. J., in press.
63. S.A. Colgate and R.H. White, Ap. J. 143, 626 (1966).
64. H.A. Bethe and J.R. Wilson, Ap. J. 295, 14 (1985).
65. W. Hillebrandt, in "Cosmogonical Processes", Proceedings of A.G.W. Cameron's 60th Birthday Symposium, Boulder, Colorado (1985).
66. N. Straumann, "Neutrino Emission and Stellar Evolution", Proceedings of the Spring School on Weak Interactions and Nuclear Structure, SIN (1972).
67. H.T. Janka, MPI preprint, München (1987).
68. R.M. Catchpole et al., South African Astronomical Observatory, preprint, Nov. 1987.
69. M.J. Longo, Phys.Rev.Lett. 60, 173 (1988);
L.M. Krauss and S. Tremaine, Phys.Rev.Lett. 60,176 (1988).
70. Adams et al., Phys.Rev. 129, 1383 (1963).
71. D. Pines, "Elementary Exitations in Solids", Benjamin (1963).
72. Abrikosov, Gorkov and Dzyaloskinsky, "Quantum Field Theoretical Methods in Statistical Physics", Pergamon Press (1965), p. 57.

Evidence for Black Holes

N. Straumann

Institut für Theor. Physik, Universität Zürich, Schönberggasse 9, CH–8001 Zürich, Switzerland

I. Introduction

The evidence for the existence of black holes is, unfortunately, still indirect and not conclusive. During the past few years, the amount of circumstantial evidence for the presence of black holes of stellar mass in our Galaxy and in neighboring galaxies has, however, increased with the addition of new candidates. The organizers of this school have asked me to present the current situation in the black hole search. An unambiguous proof that black holes are as real as neutron stars, or solar-type stars, would definitely be one of the most important events in the history of astronomy.

The evolution of our ideas since the age of Laplace concerning the dark objects that populate the universe has recently been beautifully described by W. Israel [1]. Not all of you may know that even before P.S. Laplace the Reverend John Michell (1724 – 1793), a teacher at Cambridge (1749–1763) and afterwards Rector at Thornhill, Yorkshire, proposed already in 1783 the existence of black holes [2]. His paper was communicated to the Royal Society by his friend Cavendish on November 27, 1783, and caused an enormous excitement in London circles. Michell used not only the same arguments as Laplace later on in 1796, which were based on Newton's universal gravitation and the corpuscular theory of light, but suggested also how stars for which the velocity of escape exceeds the speed of light may be detected:

"If there should really exist in nature any bodies whose density is not less than that of the sun, and whose diameters are more than 500 times the diameter of the sun, since their light could not arrive at us ... we could have no information from sight; yet, if any other luminous bodies should happen to revolve about them we might still perhaps from the motions of these revolving bodies infer the existence of the central ones with some degree of probability, as this might afford a clue to some of the apparent irregularities of the revolving bodies ...: but as the consequences of such a supposition are very obvious, and the consideration of them somewhat beside my present purpose, I shall not prosecute them any further."

In 1964, long before the X-ray window was really opened by the famous satellite UHURU, Zel'dovich first suggested the possibility that black holes may manifest themselves as compact X-ray sources in close binary systems. Cygnus X-1, for a long time the solitary example of a black-hole candidate with firm supporting evidence, was located by UHURU within a few minutes of arc. By 1971 Cyg X-1 was linked with a variable radio source which was then more accurately pin-pointed by radio astronomers in Green Bank and Holland. It was finally identified with the one-line optical binary HDE 226868 by L. Webster and P. Murdin and also by C.T. Bolton and a mass estimate of $\sim 6M_{\odot}$ was made. (For reviews of this system, see Refs. [3–6].)

In December 1982, evidence for a massive invisible component in the extragalactic X-ray binary LMC X-3 was found [7]. J.E. McClintock and R.A. Remillard [8] identified a third black-hole candidate, A0620-00, in 1985. The properties of these three X-ray binaries (Cyg X-1, LMC X-3, and A0620-00) will be discussed in some detail in Sect. IV.

I conclude this introduction with a few remarks on the formation of black holes. In my previous lectures [9] on core collapse and neutron star formation, we have seen that the iron cores of the more massive stars are already large at collapse and thus one expects that the prompt mechanism of exploding the outer part of these cores fails, because the shock fizzles out. With increasing stellar mass, the cores will therefore fatten appreciably.

There are now two possibilities. For some mass range, neutron star residues with increasing masses of $1.4 - 2.0M_{\odot}$ say, will be left behind a prompt or delayed explosion. For sufficiently massive stars, the core will, however, most likely accrete too much mass to be stable and will then collapse very quickly to a black hole.

In this picture we do not expect a collapse directly to a black hole. A protoneutron star is formed first, which accretes sufficient mass through the stalled shock until it becomes unstable and undergoes a general relativistic collapse. It is difficult to say for which mass range of stars this is going to happen, but we do expect the formation of black holes for some very massive stars. There may be a million of such stellar mass black holes in our Galaxy.

I shall not talk about the evidence that active galactic nuclei are powered by supermassive black holes, a possibility that was already proposed in 1964 by Zel'dovich and Novikov, shortly after the discovery of quasars [11]. This scenario has become very suggestive but there is certainly no convincing evidence for the existence of supermassive black holes and the case remains weak. Alternative models are still viable.

II. Mechanics of Binary Systems

So far all the evidence that black holes exist comes from dynamical studies of X-ray binaries. Therefore, it is necessary to make a few preliminary remarks about the mechanics of such systems.

Suppose for simplicity that the two components of the binary system move about each other in a circular orbit. The orbital period P and the angular frequency Ω are given by Kepler's third law:

$$\Omega^2 = \left(\frac{2\pi}{P}\right)^2 = G\,\frac{M_1 + M_2}{a^3}\ , \tag{1}$$

where M_1 and M_2 are the masses of the two stars and a is the distance between their centers of mass. We now consider a single gas particle in the field of the two masses (restricted three body problem). Its motion is most conveniently described in a co-rotating system:

$$\dot{\boldsymbol{v}} = -\nabla\psi - 2\boldsymbol{\Omega} \wedge \boldsymbol{v}. \tag{2}$$

Here $\boldsymbol{v}$ is the velocity and ψ is the sum of Newtonian and gravitational potentials:

$$\psi(\boldsymbol{x}) = -\frac{GM_1}{r_1} - \frac{GM_2}{r_2} - \frac{1}{2}(\boldsymbol{\Omega} \wedge \boldsymbol{x})^2. \tag{3}$$

The equilibrium positions in the co-rotating system are thus precisely the critical points of ψ. There are five of these, of which three are collinear with M_1 and M_2 (Euler 1767) and two are equilateral (Lagrange 1772).

Moreover, the three collinear equilibrium points are unstable (Plummer 1901). The equilateral equilibrium points can be stable if the mass ratio is sufficiently small ("troyan asteroids"). (See, e.g., Ref. [11])

After taking the inner product of (2) with $\boldsymbol{v}$, one finds the Jacobi integral

$$\frac{1}{2}\,\boldsymbol{v}^2 + \psi(\boldsymbol{x}) = \text{const}\ . \tag{4}$$

Obviously ψ cannot increase above the value of this integral as the particle moves. For this reason, the structure of the equipotential surfaces of ψ is extremely important. These are shown in Fig. 1, along with the five equilibrium positions. The equipotential surface passing through the "inner Lagrange point" L_1 is particularly important. Inside this critical Roche surface the equipotential surfaces, which enclose the two centers of mass, are disjoint; outside, this is not the case. The Roche limit thus determines the maximum volume (Roche volume) of a star. If this is exceeded, a portion of the outer layer of the star will flow over to the other star.

For a gas, the Navier-Stokes equation in the co-rotating system reads

$$D_t\boldsymbol{v} = -\nabla\psi - 2\boldsymbol{\Omega} \wedge \boldsymbol{v} - \nabla P + \text{friction terms}\ , \tag{5}$$

where $D_t\boldsymbol{v}$ denotes the hydrodynamic derivative of the velocity field. For a stationary situation ($\boldsymbol{v} = 0$ in the co-rotating system), one obtains, as expected,

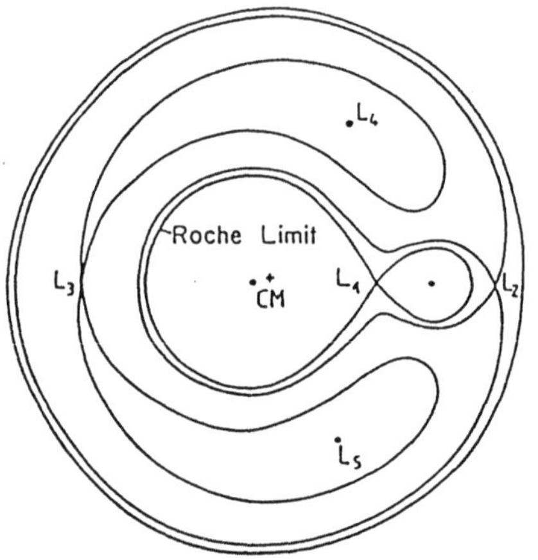

Fig. 1. Equipotential lines in the orbital plane of a binary star system, equilibrium points and Roche limit for the restricted three body problem. L_1, L_2 and L_3 are unstable equilibrium positions.

$$\nabla P = -\varrho \nabla \psi. \tag{6}$$

Hence the surfaces of constant pressure, and in particular, the surface of the star, are equipotential surfaces of ψ.

We now imagine a situation in which the more massive of the two stars enters an evolutionary phase (red giant phase) during which it fills up the Roche limit. If this happens, matter flows through the inner Lagrange saddle point into the Roche volume of its partner. Part of it will escape and the rest will be accreted by the secondary. This mechanism will play an important role in the following.

Eq. (2) is the Euler-Lagrange equation of the Lagrange function

$$L = \frac{1}{2}\,\dot{\boldsymbol{x}}^2 + (\boldsymbol{\Omega} \wedge \boldsymbol{x}) \cdot \boldsymbol{x} - \psi(\boldsymbol{x}).$$

The Hamiltonian of the corresponding autonomous Hamiltonian system is just the Jacobi integral.

The **mass function** of a binary system plays also some important role. This is defined by

$$F(M_1, M_2, i) = M_1^3 \sin^3 i/(M_1 + M_2)^2, \tag{7}$$

where i is the inclination angle (see Fig. 2). From Kepler's third law, one obtains immediately

$$F(M_1, M_2, i) = \frac{4\pi^2}{GP^2}\,(a_2 \sin i)^3. \tag{8}$$

Here a_1 and a_2 denote the semimajor axes of the two orbits. If one now calculates the velocity of M_2 in the direction of observation, one finds for the arithmetic mean (semiamplitude) V_2, of the maximum and minimum values

$$V_2 = \frac{2\pi}{P}\frac{a_2 \sin i}{\sqrt{1-e^2}}, \tag{9}$$

where e is the eccentricity. Inserting this in (8) gives

$$F(M_1, M_2, i) = \frac{P}{2\pi G}\,(1-e^2)^{3/2}\,V_2^3. \tag{10}$$

Except for the eccentricity, the right-hand side contains only directly measurable quantities. In order to determine e one needs to know the precise form of the velocity curve. The component in the direction of observation is

$$(v_2)_{\rm obs} = \frac{M_1}{[(M_1+M_2)a(1-e^2)]^{1/2}}\,\sin i\,[e\cos\omega + \cos(\phi+\omega)], \tag{11}$$

where ϕ is the true anomaly and ω is the argument of the periastron (angular distance from the node; see Fig. 2).

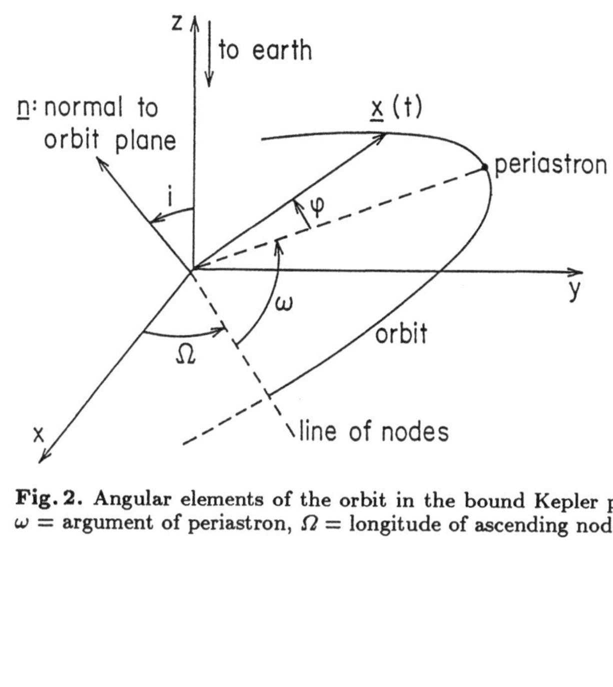

Fig. 2. Angular elements of the orbit in the bound Kepler problem (i = inclination angle, ω = argument of periastron, Ω = longitude of ascending node, ϕ = true anomaly).

III. Upper Bound for the Mass of a Neutron Star

The value of the largest possible mass, $M_{\rm max}$, of a neutron star plays a decisive role in the observational identification of black holes. If one can find a very compact object which can be shown convincingly to have a mass which is definitly larger than $M_{\rm max}$, then one has a serious candidate for a black hole. Precisely such arguments have been used to identify the X-ray sources Cygnus X-1, LMC X-3, and A0620-00 with black holes, as will be discussed in the next Section.

In view of the large uncertainties of the equation of state for high densities, it is important to find reliable limits for $M_{\rm max}$. This can be achieved by assuming that for densities less than some given density ϱ_0 (which may depend on our current knowledge), the equation of state is reasonably well known, but for densities higher than ϱ_0, the equation of state will be required to satisfy only quite general conditions.

We consider only "slowly" rotating neutron stars and make the following minimal assumptions about the matter:

(1) The energy-momentum tensor is described in terms of the mass-energy density and an isotropic pressure p, which satisfies an equation of state $p(\varrho)$. Stresses which may have arisen as a result of the slowing down of the star's rotation are neglected.
(2) The density is positive: $\varrho \geq 0$.
(3) Matter is microscopically stable: $dp/d\varrho \geq 0$. Since the pressure is certainly positive at low density, we then have $p \geq 0$.
(4) The equation of state is known for $\varrho \leq \varrho_0$. We may choose for ϱ_0 a value which is not significantly below nuclear densities.

Some authors also assume $dp/d\varrho \leq 1$, since $(dp/d\varrho)^{1/2}$ is the phase velocity of sound waves. However, sound waves propagating through matter in neutron stars are subject to dispersion and absorption, and hence $(dp/d\varrho)^{1/2}$ is not the velocity at which a signal propagates. Thus causality does not require $dp/d\varrho \leq 1$. On the other hand, most of the "realistic" equations of state do satisfy this inequality. Although we shall not require this "causality assumption", we shall mention the implications of this additional assumption.

In addition, we have the well-known general relativistic equations of stellar structure [12].

One can show easily that $d\varrho/dr \leq 0$ and this allows us to divide the star into two regions.

The **envelope** is the part having $\varrho < \varrho_0 \quad (r > r_0)$, for which the equation of state is known. The interior region with $\varrho > \varrho_0$ and $r < r_0$ is called the **core**. In this region we assume only the general requirements (1–4).

The mass M_0 of the core is given by

$$M_0 = \int_0^{r_0} \varrho \, 4\pi r^2 dr \; . \tag{12}$$

If M_0 is given, the structure equations can be integrated outward from r_0. The corresponding boundary conditions are $p(r_0) = p(\varrho_0) =: p_0, \; m(r_0) = M_0$. The total mass M of the star is

$$M = M_0 + M_{\text{env}}(r_0, M_0). \tag{13}$$

We may regard the mass of the envelope as a known function of r_0 and M_0, since the equation of state is known there. We may now obtain bounds for M by following the strategy:

(i) Determine the range of values (M_0, r_0) which is allowed by the assumptions (1)–(4). This range of possible core values will be known as the allowed region in the (r_0, M_0) plane.

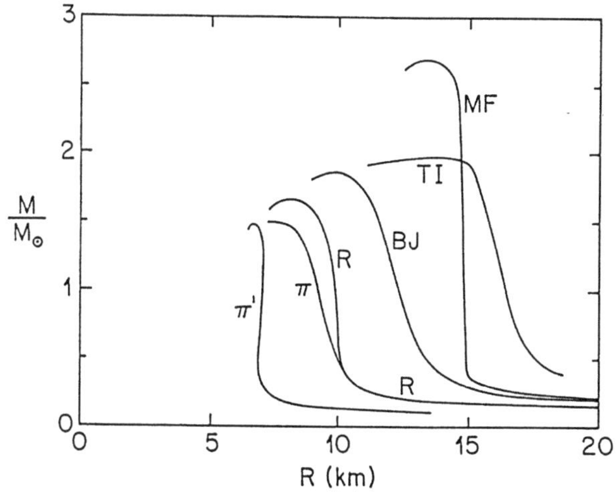

Fig. 3. Gravitational mass versus radius for various models of high density matter. (Adapted from Ref. [22].)

(ii) Look for the maximum of the function M defined by (13), in the allowed region of the variables r_0 and M_0.

One finds the largest core mass for the stiffest equation of state, namely, for incompressible matter with constant energy-mass density ϱ. Even then the mass of the core inside the radius r_0 is limited; otherwise the pressure becomes infinite at the center and equilibrium is lost.

The allowed core region (step (i)) can be exactly determined (see, e.g., Sect. 6.6 of Ref. [12], or [13]). As long as ϱ_0 is not significantly larger than nuclear density, then $p(\varrho_0) \ll \varrho_0$, and we have to high accuracy the bounds

$$M_0 < \frac{4}{9}\left(\frac{1}{3\pi\varrho_0}\right)^{1/2}, \tag{14}$$

$$r_0 < \left(\frac{1}{3\pi\varrho_0}\right)^{1/2}. \tag{15}$$

Calculations show [13] that the contribution of the envelope to the limiting mass is very tiny ($\leq 1\%$). This is true as long as $p(\varrho_0) \ll \varrho_0$.

Thus the bound (14) is, to a good approximation, also a bound for the total gravitational mass M. Numerically this bound reads

$$M \leq 6.8\left(\frac{\varrho_{\text{nucl}}}{\varrho_0}\right)^{1/2} M_\odot. \tag{16}$$

The additional "causality assumption" $dp/d\varrho \leq 1$ would have the effect of replacing the factor 6.8 by 4.0.

These considerations show that a neutron star cannot be more massive than $5M_\odot$. "Realistic" equations of state usually satisfy $dp/d\varrho \leq 1$ and thus give maximal masses below $3.2M_\odot$. A sample of such model calculations is shown in Fig. 3.

IV. Three Black Hole Candidates

We have not yet uncovered any unique characteristics of a black hole which would allow us to unmistakably recognize the presence of such an object. One reason for this is that stellar black holes are difficult to distinguish from weakly magnetized neutron stars. For instance, the radius and gravitational potential well, as well as the period of the last stable orbit of an extreme $10M_\odot$ Kerr black hole are nearly the same as for a $1.4M_\odot$ neutron star [14]. So far all the evidence for black holes rests on mass determinations of compact objects in X-ray binaries. I discuss now the situation for the best three candidates.

1. Cygnus X-1

First of all, let us describe the observed properties of the system [3–6]. As previously mentioned, we are dealing with a single-line spectroscopic binary system. The optical partner is a blue supergiant of ninth magnitude ($V = 8.87$, $B - V = 0.81$, $U - B = -0.30$), spectral type 0 9.7 ab and is catalogued as HDE-226868. The orbital period is 5.6 days and the X-ray luminosity L_X is about $10^4 L_\odot$. The X-ray source fluctuates irregularly over all time intervals between 20 sec and the shortest resolved time of about 0.01 sec. No periodicity is observed and also no clear signal for X-ray eclipses has been detected. It is also worth noting that the X-ray spectrum is unusually hard, corresponding to a temperature of 3×10^8K, with roughly equal amounts of energy in the bands $1\ \mathrm{keV} < E < 10\ \mathrm{keV}$ and $10\ \mathrm{keV} < E < 100\ \mathrm{keV}$.

The only reasonable explanation of these observations is that the X-radiation is emitted by gas falling onto a highly compact object, which can only be a neutron star or a black hole. The first possibility would be excluded if we can show convincingly that the compact object is too massive to be a neutron star. We give now the arguments which show that this is most probably the case. From the velocity curve one can conclude with the help of (11) that $e \leq 0.02$ [5]. The observed value of the semiamplitude is $V_2 = 76 \pm 1$ km/sec and thus one finds for the mass function ($F_X = F(M_X, M_{\mathrm{opt}}, i)$

$$F_X = (0.25 \pm 0.01) M_\odot. \tag{17}$$

This value for the mass function would allow us to establish a lower bound for M_X from

$$M_X = F_X (1+q)^2 / \sin^3 i, \qquad q := \frac{M_{\mathrm{opt}}}{M_X}, \tag{18}$$

if M_{opt} were known. However, it is dangerous to infer the mass from the spectral type in the case of binary stars. We show now that it is possible to infer a (conservative) lower limit from the fact that prominent X-ray eclipses are not observed, without making any assumptions about the mass of the optical star. The following argument is due to Paczynski [15]. From geometric considerations, the radius R_2 of the visible star must satisfy the inequality $R_2 \leq a \cos i$, where $a = a_1 + a_2$. Using (9), we can write this inequality in the form

$$x \leq (1+q)\,\frac{\cos i}{\sin i}\,, \tag{19}$$

where x is defined by

$$x = \frac{2\pi R_2}{PV_2}.$$

We next recall that if D is the distance from the source, the Stefan-Boltzmann law implies

$$4\pi D^2 L_{\rm obs} = 4\pi\sigma R_2^2\, T_e^4.$$

The observed values give from the apparent magnitude with bolometric and absorption corrections

$$R_2/R_\odot \simeq 9.5\left(\frac{D}{1\rm kpc}\right) \tag{20}$$

and thus

$$x = 1.14\left(\frac{D}{1\rm kpc}\right). \tag{21}$$

Inequality (19) is equivalent to

$$\sin^2 i \leq \frac{(1+q)^2}{x^2+(1+q)^2}.$$

If we now insert this into (18), we obtain

$$M_{\rm X} = (1+q)^2 F_{\rm X}/\sin^3 i \geq \frac{F_{\rm X}}{1+q}[x^2+(1+q)^2]^{3/2}. \tag{22}$$

The right-hand side has a **minimum** with respect to q for $(1+q)^2 = \frac{1}{2}x^2$ and hence, together with (17) and (21)

$$M_{\rm X} \geq \frac{3\sqrt{3}}{2} F_{\rm X}^2 x^2 = 3.4 M_\odot \left(\frac{D}{2\rm kpc}\right)^2. \tag{23}$$

From studies [16] of the absorption with distance for a large sample of stars in the direction of HDE-226868, the distance of Cyg X-1 is known to be approximately 2.5 kpc and probably larger than 2 kpc. In this way, we arrive at the lower limit $M_{\rm X} \geq 3.4 M_\odot$.

Unfortunately, there are significant uncertainties associated with all reddening-distance measurements, due to patchiness in the distribution of dust or the possibility that the source is shrouded in dust and intrinsically reddened. Moreover, the reddening studies [16] have serious shortcomings. Neither of them included more than three stars beyond 2 kpc and both studies relied heavily an ancient HD classifications.

It should be possible to improve the case for lower bounds on the distance, for instance, by measuring the annual parallax with the VLBI network [16], or from 21-cm absorption studies.

One finds a tighter bound by assuming in addition that the optical star is confined to its Roche lobe:

$$R_2 \leq R_{cr} \,. \tag{24}$$

In the mass range of interest, one can use the formula (Paczynski)

$$R_{cr} = a \left[0.38 + 0.2 \log\left(\frac{M_1}{M_2}\right)\right] , \tag{25}$$

which is valid for $0.3 < M_1/M_2 < 20$.

From the definition of x and (9) (for $e \simeq 0$) the inequality (24) can be written as

$$x \leq (1+q)\frac{0.38 + 0.2\log q}{\sin i}. \tag{26}$$

The problem is now to find the minimum of the expression (18) for M_X with respect to q and $\sin i$, subject to the constraints (19) and (26). The minimum occurs when equality holds in (19) and (26). One finds the numerical inequalities

$$M_X \geq \begin{cases} 6.5 M_\odot & \text{for } D = 2.0 \text{ kpc} \\ 9.5 M_\odot & \text{for } D = 2.5 \text{ kpc}. \end{cases} \tag{27}$$

More precise determinations make use of the observed optical light variations. This is a rather complicated matter and the interested reader is referred to the original literature. Even so, the determination of the mass involves rather large uncertainties. As an example, we mention that the optical velocity curve, which is inferred from the absorption lines, is subject to all sorts of disturbances, including X-ray heating of the optical companion's photosphere, tidal deformations of its surface, emission and absorption from gas streaming between the two stars. In addition, erratic variations of unknown origin occur.

The following result is given in [17]:

$$M_X = (9 - 15) M_\odot .$$

If this is correct, then, according to the discussion of Sect. III the compact object in Cyg X-1 cannot be a neutron star.

In conclusion, we can say that the mass M_X is probably too large for a neutron star, but that not all doubts have been removed. (It is somewhat disappointing that the situation has not improved since a number of years.)

2. LMC X-3

This relatively new candidate [7] is definitely in the Large Magellanic Cloud – as can be inferred from its systematic velocity – and, therefore, the distance of this X-ray source is known ($D \simeq 50$ kpc). The system is also superior to Cyg X-1 because the mass function is relatively high. From $P = 1.70$ days, $V_2 = (235 \pm 11)$km/sec, and $e \simeq 0$ one finds [7]

$$F_X = (2.3 \pm 0.3) M_\odot .$$

Thus, if the compact object should be a neutron star, it would certainly be quite massive.

Paczynski [18] has repeated his argument, which we have discussed for Cyg X-1. For LMC X-3 the most uncertain parameter is the radius R_2 of the B type companion. From all the data one finds [18]

$$R_2/R_\odot = 6 \pm 2. \tag{28}$$

On the other hand, from the absence of eclipses and the requirement that the B type star should not be larger than its Roche lobe, Paczynski finds numerically [18]:

$$M_X/M_\odot \geq 2.3 + 0.32(R_2/R_\odot)^{1.8}. \tag{29}$$

For the estimated value (28) this gives

$$M_X/M_\odot \geq 10 \pm 4. \tag{30}$$

This looks almost like a conclusive proof that LMC X-3 does comprise a black hole.

Unfortunately, Paczynki's argument rests on the assumption that the companion star is the only source of optical radiation. In contrast to Cyg X-1, this is not true for LMC X-3. Observed secular light variations (with a maximum excursion of 0.5 magnitudes) are probably due to an accretion disk [19]. There is also the possibility that the radial velocity data have large systematic uncertainties, for reasons that we have already mentioned earlier. Therefore, a heavy neutron star model for LMC X-3 cannot be completely excluded [20].

More detailed observations (high-resolution spectra) could improve the situation. In particular, the emission-line contamination could be determined. This is for the observers one of the most perfidious effects that can affect the absorption-line velocities.

3. A0620-00

This system, which has been discovered by McClintock and Remillard [8], provides at present the best evidence for the existence of black holes. The salient points have recently been summarized by McClintock [21].

A0620-20 is an X-ray nova. In the fall of 1975, it was for two months the brightest celestical X-ray source and was then identified with a blue star of 12th magnitude. After a bit more than a year, the star returned to its pre-outburst brightness ($V \simeq 18.3$). Since then no X-ray emission could be registered in numerous observations of the A0620-00 region. The quiescent optical spectrum consists of two nearly equal components: A K4V to K7V **stellar** spectrum and an emission-line component which is probably coming from an accretion disk.

The period is very short

$$P = 7.75 \text{ hours}$$

and the semiamplitude of the radial velocity is very large

$$V_2 = 457 \pm 8 \text{ km/s}.$$

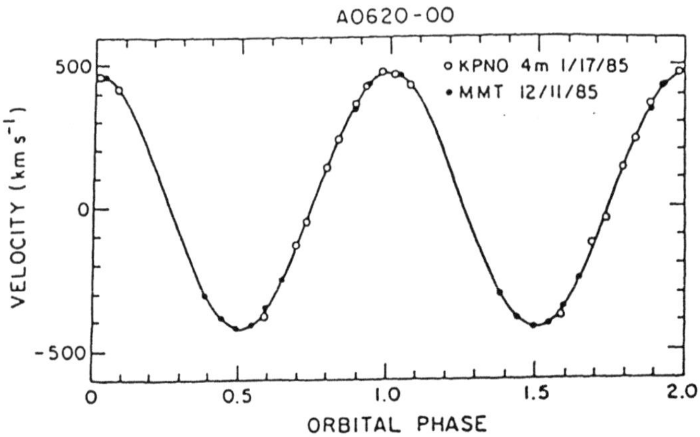

Fig. 4. Radial velocity curve of the optical companion of A0620-00. The smooth curve is a fit to a circular orbit. (From Ref. [21])

The velocity curve is very well fitted by a circular orbit (Fig. 4).

These observations imply a remarkably high mass function

$$F_X = (3.18 \pm 0.16) M_\odot \tag{31}$$

and thus, without any further assumptions,

$$M_X \geq 2.70 M_\odot (3\sigma). \tag{32}$$

This is a beautiful discovery.

One is still on firm grounds by assuming that $M_{\rm opt} \geq 0.25 M_\odot$ and $i \leq 85^o$, because of the absence of X-ray eclipses. This leads to

$$M_X \geq 3.2 M_\odot (3\sigma). \tag{33}$$

With additional assumptions, one can establish higher mass limits which are more realistic, but not really hundred percent safe [8].

It appears that potential systematic errors that may affect the velocity amplitude are not important for this system. For example, the center of an absorption line cannot shift by more than a fraction of its $\simeq 1.5$ Åwidth and this is very small compared to the observed Dopper modulation (16 Åor 900 km/sec).

The surprising discovery of A0620-00 has raised the hope that one day an X-ray binary with a mass function $F_X \geq 5 M_\odot$ may be found. Only then would we be really sure that a black hole has been located. This will, however, not be easy, since several requirements must be fulfilled [21].

References

1. W. Israel, in "Three Hundred Years of Gravitation", ed. S.W. Hawking, W. Israel, Cambridge Univ. Press (1987).
2. J. Michell, Phil. Trans. R. Soc. 74, 35 (1784). Reprinted in: S. Detweiler, "Black Holes": Selected Reprints. Am. Assoc. of Physics Teachers, Stony Brook (1982).
3. C.T. Bolton, Ap.J. 200, 269 (1975).
4. M. Oda, Space Sci. Rev. 20, 757 (1977).
5. D.R. Gies and C.T. Bolton, Ap.J. 260, 240 (1982).
6. D.R. Gies and C.T. Bolton, Ap.J. 304, 371 (1986).
7. A.P. Cowley and et al., Ap.J. 272, 118 (1983).
8. J.E. McClintock and R.A. Remillard, Ap.J. 308, 110 (1986).
9. N. Straumann, "Physics of Type II Supernova Explosions", this volume.
10. Ya.B. Zel'dovich and I.D. Novikov, Sov. Phys. Dokl. 9, 834 (1964).
11. N. Straumann, "Klassische Mechanik", Springer Lecture Notes, Vol. 289, Sect. 6.4, (1987).
12. N. Straumann, "General Relativity and Relativistic Astrophysics", Springer (1984).
13. J.B. Hartle, Phys. Rep. 46, 201 (1978).
14. K.S. Thorne, Ap.J. 191, 507 (1974).
15. B. Paczynski, Astron. Astrophys. 34, 161 (1974).
16. N. Bartel et al., Nature 319, 733 (1986).
17. J.N. Bahcall, Ann. Rev. Astron. Ap. 16, 241 (1979).
18. B. Paczynski, Ap.J. (Lett.) 273, L 81 (1983).
19. Van der Klies et al., Astron. Astrophys. 126, 265 (1983).
20. T. Mazeh et al., Astron. Astrophys. 157, 113 (1986).
21. J.E. McClintock, "The Physics of Accretion onto Compact Objects", Lecture Notes in Physics, Vol. 266, Springer Verlag (1986), p. 211.
22. W.D. Arnett and R.L. Bowers, Ap.J. Suppl. 33, 415 (1977).

Principles of Black Hole Energetics

Petr Hájíček

Inst. for Theoretical Physics, University of Berne, Sidlerstraße 5,
CH–3012 Berne, Switzerland

Abstract: These lectures introduce first the important notions of the theory of gravitational collapse and black hole physics like that of trapped surface, of apparent and event horizons. Then, the concepts of energy density and of the total energy in general relativity are explained, Nester's expression for the total 4-momentum is written down and the positive energy theorems are reviewed. Finally, a report on the Hawking effect, on the calculations of the expected value of the stress-energy tensor, and on the back reaction problem is given. The lectures resort nowhere to a complicated mathematics, as all important logical relations can be clarified with the help of simple, spherically symmetrical models. Arguments are presented that the event horizon need not play such a crucial role in black hole physics as it is usually assumed.

1. Introduction

Ideally, black hole energetics ought to be presented in two parts. The volume 1 would contain general principles: definition of energy, questions of its balance and positivity, theory of black hole evaporation. The volume 2 would be devoted to applications, describing how black holes could power highly active systems like quasars, centres of galaxies, etc. It can sound like a paradox that *black holes* would give energy to the *brightest* objects in the sky. However, black holes are sitting at the bottom of a potential well that, in a sense, is the deepest one possible in nature. Indeed, letting a weight of mass m fall slowly down towards a Schwarzschild black hole, and letting the gravitational attraction do work, one would, in principle, extract the total rest energy m of the weight at the moment as it reaches the horizon. Thus, there is one hundred percent efficiency, at least theoretically possible! In realistic processes, the efficiency is, as a rule, much smaller, but still considerably larger that one percent (i.e., the efficiency of nuclear burning).

This lecture will contain some material of the "volume 1" only; volume 2 had better to be written by some astrophysicist (in fact, such books are available, for example [1]). Let us briefly describe the central problem; we can do it using the Newton theory of gravitation.

In the Newton theory, one can in principle extract an unlimited amount of energy from an isolated system (see, e.g. [2]). Consider, for example, a system consisting of two small masses in orbit around each other. Since the system is gravitationally bound, the total energy is negative. Since the Newtonian potential is unbounded from below, we can make the total energy as large and negative as we wish by making the orbit smaller and smaller.

The existence of isolated systems with arbitrarily large and negative energy would, however, lead to paradoxes in theoretical physics. (One has to keep in mind that it is not allowed to subtract anything from the energy. At the moment one switches on the gravitational interaction, not only differences in energy but the value of energy is measurable.) For example, if we start to include the effects of special relativity so that we naively add the rest mass energy of the two small masses, it seems that the total relativistic energy of such a system could be negative. Special relativity tells us that energy is equivalent to inertial mass. However, such an isolated system would have a total energy seen by a distant observer of

$$E = E_0\sqrt{1-v^2}$$

with $E_0 < 0$, where v is the system's velocity relative to the observer. Thus, accelerating such a system, one can extract energy, and accelerating to the light velocity, one obtains an infinite amount of it. This is, still, only a perpetuum mobile of the second kind: the internal state of the machine changes as it works. A perpetuum mobile of the first kind can be constructed, if we assume that the equivalence principle is valid.

The weak equivalence principle states that the acceleration $\boldsymbol{a}$ of a test particle in a gravitational field is independent of the test particle itself, and is completely determined by the gravitational potential ϕ of the field:

$$\boldsymbol{a} = -\nabla\phi.$$

The gravitational potential of any mass m is given by

$$\phi = -\frac{Gm}{r}.$$

Hence, positive masses exert a universal attraction, whereas the negative ones repel everything. Consider the corresponding movement of a mass dipole. The positive mass will be accelerated away from the negative and the negative one towards the positive one. Thus, the whole system will be accelerated in the direction showing from the negative to the positive mass (the force will be comparable to that measured by Cavendish with help of a torsion balance!). The system will do work without changing its internal configuration (distance of the two masses).

The conclusion is that negative masses in gravity theory are much more "exotic" than e.g., negative charges in electrodynamics. Can we prevent their formation in Newton's theory? Adding quantum effects, such as the zero point energy or Pauli principle, will not help very much [3].

Thus we have a fundamental problem: how can the laws of physics prevent extracting too much energy from gravitational attraction?

General relativity is supposed to cure the inconsistency between Newton's theory and special relativity. Indeed, here the situation looks more hopeful since it seems that if we make our two-body system sufficiently small, we will create a black hole which should act so as to limit the amount of energy which can be extracted, and hence ensure that negative energy systems do not exist.

However, the horizons of black holes seem to be unstable against Hawking evaporation. What is the end result of this process? Of course, the end of the process comes only after a very large lapse of time. A black hole of mass M evaporates as a black body with the temperature

$$T_H = 10^{-7} \frac{M_\odot}{M} \text{ K} . \tag{1}$$

If the hole is in *vacuum*, then the process lasts for about

$$t = 10^{66} \left(\frac{M}{M_\odot} \right)^3 \text{ years} \tag{2}$$

(see, e.g. [4]). If the temperature T of the surroundings is larger than T_H (e.g., 3 K), then the hole will grow. Only after T becomes smaller, the evaporation can begin. The problem is, however, only delayed, not solved. In this lectures, we briefly describe what is known and what can be said today about all these things.

We shall use Planck units ($G = c = \hbar = 1$), and the metric will have the signature $(+ , - , - , -)$.

2. Model of Gravitational Collapse

In this section, we will explain the most important notions of the modern theory of gravitational collapse. The strategy we adopt will be to work with simple models so that we can avoid the complicated mathematics necessary for general definitions and still convey some qualitative understanding of the main concepts and their properties. The Oppenheimer-Snyder [5] model we are going to describe here is, in fact, a paradigm: the ideas about collapse and black holes have been shaped according to it. Contemporary numerical simulations (see, e.g., [6]) only confirm the model. More recent calculations than that of [5] can be found, e.g. in [7], p. 851.

The model consists of a perfectly spherical star made of homogeneous isotropic dust (no pressure) collapsing to a point from a momentarily static state. Outside the star, there will be a vacuum (no or negligible radiation, asymptotically flat spacetime (isolated system)).

A) Inside the Star

Let us first describe the metric inside the star. It must be spatially homogeneous and isotropic, and for some value of time coordinate it must be momentarily static. Thus, it is a closed cosmos Friedmann metric (the momentarily static state will be the "state of maximum expansion"), one of the following family

$$\begin{aligned} ds^2 &= d\tau^2 - a^2(\tau)\,[d\chi^2 + \sin^2\chi\,(d\theta^2 + \sin^2\theta\,d\phi^2)], \\ a &= \frac{1}{2}a_m(1+\cos\eta), \\ \tau &= \frac{1}{2}a_m(\eta+\sin\eta), \end{aligned} \tag{3}$$

where a_m is the parameter of the family with the meaning of the maximum expansion radius. The maximum expansion corresponds to $\eta = 0$, and the final singularity to $\eta = \pi$. We cut out a sphere, $\chi \leq \chi_0$ with boundary $\chi = \chi_0$. The surface area F of the boundary is given by

$$F = 4\pi\,a^2(\eta)\,\sin^2\chi_0,$$

the centre of the sphere is at $\chi = 0$. The angular variables θ and ϕ run over the surface of a unit sphere. By that, the range of all coordinates and the metric inside the star is settled.

It is important to remember that it is the geometry of null hypersurfaces, which plays the most important role in the theory of gravitational collapse and black holes. For spherically symmetric spacetimes, it is sufficient to study the behaviour of the *spherically symmetric null hypersurfaces* (abbreviation: NS). Such a study will, therefore, fill in a great part of these lectures.

Consider the NS's inside the matter. Each NS must have its vertex at the centre of the star, $\chi = 0$, and at a particular value of the time parameter $\eta = \eta_v$. The two radial null directions starting at this point are given by the equations

$$\begin{aligned} \chi &= \pm(\eta - \eta_v), \\ \tau &= \frac{1}{2}a_m(\eta + \sin\eta). \end{aligned}$$

Here, "+" corresponds to outgoing ($\eta > \eta_v$) and "−" to ingoing NS ($\eta < \eta_v$). We will say that an NS is *expanding* (*contracting*), if its surface area *increases* (*decreases*) towards future. Let us denote the surface area, at the time η, of an NS with vertex at η_v and which is out- (+) or ingoing (−), by $F_\pm(\eta, \eta_v)$. A simple calculation leads to the following formula

$$F_\pm(\eta,\eta_v) = 4\pi\,[\pm a_m \sin(\eta - \eta_v)\,(1+\cos\eta)]^2.$$

The sign of the derivative

$$\frac{\partial F}{\partial \eta}$$

will determine whether we have expansion or contraction. In this way, we find that the ingoing NS's are always contracting, whereas there are two cases of the outgoing NS's:

$\eta_v < \pi - 3\chi_0$: These NS's expand till they reach the surface of the star.

$\eta_v \geq \pi - 3\chi_0$: These NS's expand till they reach the hypersurface with

$$\frac{\partial F}{\partial \eta} = 0,$$

which is timelike and given by the equations

$$\eta = \frac{1}{3}(\pi + 2\eta_v), \quad \chi = \frac{1}{3}(\pi - \eta_v).$$

This hypersurface will be denoted by $\mathcal{A}$ (see Fig. 1). Afterwards, these NS's contract (with the exception of the first one).

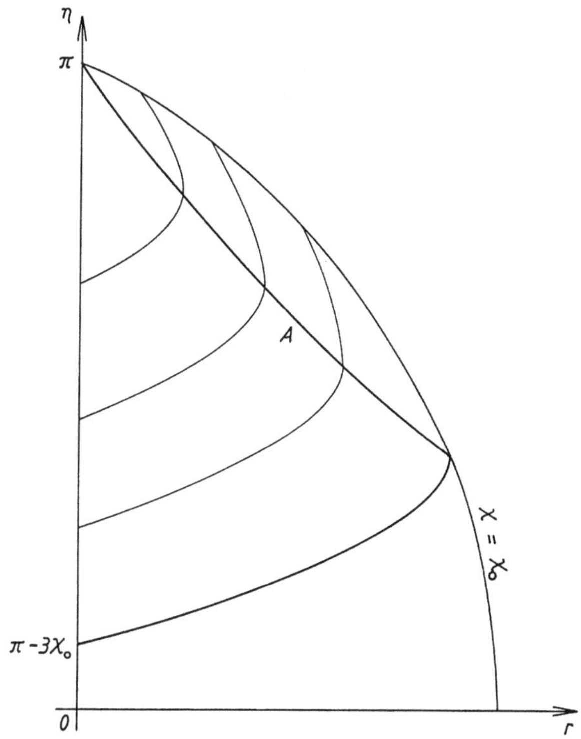

Fig. 1. Inside the matter: outgoing null hypersurfaces starting at the centre $r = 0$. If the time parameter η_v of its start satisfies $\eta_v < \pi - 3\chi_0$, then the hypersurface expands forever, if $\eta_v = \pi - 3\chi_0$, then it ceases to expand just at the surface of the star (boldface curve), and if $\eta_v > \pi - 3\chi_0$, then it reaches a maximum surface apparent horizon at $\mathcal{A}$ and converges afterwards. The coordinate r is defined by $r = a \sin \chi$.

B) Outside the Star

Outside the matter, the metric has to be:

a) spherically symmetric,

b) asymptotically flat,

c) vacuum solution of Einstein equations.

It follows from the so-called Birkhoff theorem (see, e.g. [8]) that it must be the Schwarzschild solution:

$$ds^2 = \left(1 - \frac{2m}{r}\right) dv^2 - 2\, dv\, dr - r^2 \left(d\theta^2 + \sin^2\theta\, d\phi^2\right). \tag{4}$$

Here, v, r, θ and ϕ are the so-called Eddington-Finkelstein coordinates. They have the following meaning:

$v =$ const is the equation of the ingoing NS; v increases towards future;
$v =$ const and $r =$ const is the 2-dimensional sphere with surface area $4\pi r^2$,
θ and ϕ are the spherical coordinates on the spheres.

Finally, m is a parameter with the meaning of total mass. (Schwarzschild's solution is a one parameter family of solutions.)

The metric (3) inside and (4) outside the star must be glued smoothly together at the boundary of the star. This is only possible, if the mass parameter of the Schwarzschild metric satisfies the condition

$$m = \frac{1}{2} a_m \sin^3\chi_0$$

(see, e.g. [7]). The explicit form [7] of the boundary in the Eddington-Finkelstein coordinates will not be needed.

The metric being specified, we can pass to the study of the NS. The ingoing NS are given by the coordinate v. The outgoing NS will be described by a function $r(v)$, satisfying the following differential equation (which is easily obtained from the metric (4)):

$$\frac{dr}{dv} = \frac{r - 2m}{2r}.$$

This equation implies immediately that the outgoing NS will contract, if $r < 2m$, expand, if $r > 2m$, and will have a constant surface area for $r = 2m$. Corresponding to these three cases, there are three classes of solutions:

$$\begin{aligned} r < 2m: &\quad v = 2r + 4m \ln\left(1 - \frac{r}{2m}\right) - u\ , \\ r = 2m: &\quad r = 2m\ , \\ r > 2m: &\quad v = 2r + 4m \ln\left(\frac{r}{2m} - 1\right) + u\ , \end{aligned} \tag{5}$$

where u is a constant whose values discriminate between different outgoing NS's. The solutions satisfy the following boundary conditions:

$$
\begin{aligned}
r < 2m: \quad & r \to 2m - 0 : u \to -\infty, \\
& r \to 0 : v \to -u. \\
r > 2m: \quad & r \to \infty : \ u \text{ is the proper time along the} \\
& \qquad\qquad \text{curves } r = \text{const}, \ \theta = \text{const}, \ \phi = \text{const}, \\
& r \to 2m + 0 : \ u \to +\infty
\end{aligned}
$$

A schematic picture of the behaviour of the outgoing NS is given in Fig. 2. The coordinates τ and χ are transformed to Eddington-Finkelstein-like coordinates v and r inside the star so that these are well-defined and continuous in the whole spacetime.

The geometry of the NS's in our collapse spacetime exhibits some very general and important features. These are:

1. singularity,
2. trapped surfaces,
3. apparent horizons,
4. event horizon,
5. red shift.

1. Singularity. In Fig. 2, the singularity is situated at $r = 0$, if $v \geq v_c$. At $v = v_c$, all matter of the star is compressed into a zero volume and its density diverges. Along the whole singularity, the curvature is infinite: for example,

$$R_{\mu\nu\rho\sigma}R^{\mu\nu\rho\sigma} \sim r^{-6}.$$

The singularities are "ugly": general relativity breaks down there, because some of its axioms (like smooth locally Lorentzian spacetime) are not satisfied. The neighbourhood of a singularity is a place with diverging tidal forces, it is the domain of high energy physics (Planck energy and higher). Presumably, some sort of quantum gravity (string theory?) will be able to predict what happens there. At the present time, we are rather ignorant about it.

2. Trapped Surfaces. Every 2-dimensional spacelike surface S determines uniquely two null hypersurfaces which intersect each other at S. For example, a 2-sphere in Minkowski spacetime is the intersection of two light-cones, one outgoing and one ingoing. A *trapped surface* S is a 2-dimensional *compact* spacelike surface such that the two null hypersurfaces intersecting at S *both contract.*

For example, in Fig. 2, any surface S given by

$$r = r_0, \ v = v_0, \ \theta \in [0, \pi], \ \phi \in [0, 2\pi)$$

with $r_0 < 2m$ and $v_0 > v_c$ will be a trapped surface, because the two null hypersurfaces intersecting at it are the ingoing and outgoing NS's, and they both contract. There is no trapped surface in Minkowski spacetime. We can still construct a 2-dimensional spacelike surface S such that the two null hypersurfaces intersecting at it will contract. For example, the past null cone of

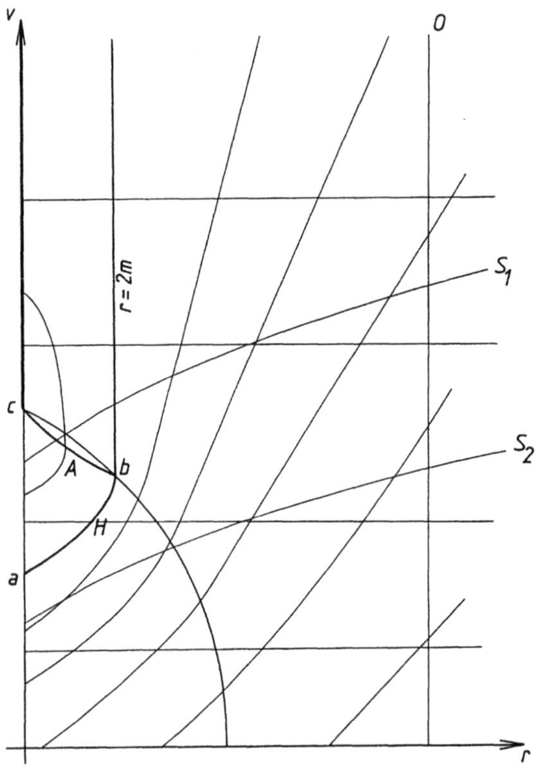

Fig. 2. Geometry of the spherically symmetrical null hypersurfaces in the spacetime $\mathcal{M}$. The apparent horizons lie at the hypersurface A and at the Schwarzschild horizon $r = 2m$. The event horizon $\mathcal{H}$ starts at the point a as an ordinary null hypersurface, reaches the surface of the star at b, and then coincides with the Schwarzschild horizon $r = 2m$. The singularity is the double line. O is the distant observer, S_1 and S_2 are typical spacelike hypersurfaces.

a point p is a contracting null hypersurface. Thus, the intersection of two such null cones, corresponding to two points p and q with spacelike separation, will be the desired surface. However, this surface is not compact!

Trapped surfaces are intimately related to singularities. According to the so-called singularity theorems of Penrose, Hawking, and Geroch (see, e.g. [8]), presence of a trapped surface, together with some "reasonable" further assumptions, always leads to singularities.

One can weaken the conditions on the trapped surface as follows: we assume everything as before, but do not require contraction for the ingoing null hypersurface. Such a compact spacelike surface S lying at a contracting outgoing null hypersurface is called *outer trapped surface.* Trapped surfaces are always outer trapped, but not vice versa. However, in Fig. 2, all spherically symmetrical outer trapped surfaces are trapped, because all ingoing NS's contract.

3. Apparent Horizons. Let S be a spacelike hypersurface in a spacetime $\mathcal{M}$ with metric $g_{\mu\nu}$, given parametrically by

$$x^\mu = f^\mu(y^1, y^2, y^3)\ , \qquad \mu = 0, 1, 2, 3,$$

where y^a, $a = 1,2,3$, are some coordinates on $\mathcal{S}$. The metric $g_{\mu\nu}$ induces a metric γ_{ab} on $\mathcal{S}$:

$$\gamma_{ab}(y^c) = -g_{\mu\nu}(f^\rho(y^c))\frac{\partial f^\mu}{\partial y^a}(y^c)\frac{\partial f^\nu}{\partial y^b}(y^c) .$$

The induced metric is called the *first fundamental form* of the hypersurface $\mathcal{S}$. We will also need the *second fundamental form*, κ_{ab}, of $\mathcal{S}$, which can be described as the rate of change of γ_{ab} in the direction orthogonal to $\mathcal{S}$. κ_{ab} is defined as follows: One can show that there is a neighbourhood of $\mathcal{S}$ (for any spacelike hypersurface $\mathcal{S}$) which can be covered by the so-called Gauss coordinates (t, y^a). These coordinates are adapted in a special way to the spacelike hypersurface so that $\mathcal{S}$ is given by $t = 0$, the metric of the spacetime has the form

$$ds^2 = dt^2 - g_{ab}(t, y^c)dy^a dy^b ,$$

and

$$\gamma_{ab}(y^c) = g_{ab}(0, y^c) .$$

Then, κ_{ab} is given by

$$\kappa_{ab} = \frac{\partial g_{ab}}{\partial t}\bigg|_{t=0} .$$

The pair γ_{ab}, κ_{ab} of symmetric covariant tensors on $\mathcal{S}$ can be used as initial data for Einstein equations (if they satisfy some constraints).

Consider a spacelike hypersurface $\mathcal{S}$ with a first, γ_{ab}, and second, κ_{ab}, fundamental form in a spacetime $\mathcal{M}$, and let S be a 2-dimensional surface in $\mathcal{S}$. Then, working just with γ_{ab} and κ_{ab}, we can decide, whether or not S is a trapped surface. The set $\mathcal{T}$ of all points $p \in \mathcal{S}$ which lie at a trapped surface S in $\mathcal{S}$ is called *trapped set* on $\mathcal{S}$. The boundary A of a trapped set on $\mathcal{S}$ is called an *apparent horizon* at $\mathcal{S}$. The decision whether a point $p \in \mathcal{S}$ lies at a trapped surface or at an apparent horizon cannot be made if one only knows the initial data at p or even in a neighbourhood of p. In this sense, trapped surfaces and apparent horizons are non-local notions.

One can show (see, e.g. [8]) that one of the two null hypersurfaces intersecting at an apparent horizon A has zero contraction. For a spherically symmetrical A this means that

$$\frac{\partial r}{\partial v} = 0 \quad \text{or} \quad \frac{\partial r}{\partial u} = 0$$

at A. In this way, there are outgoing and ingoing apparent horizons. In Fig. 2, the hypersurface $r = 2m$ outside the star together with the hypersurface A inside the star is a locus of apparent horizons. They are all outgoing.

The name "apparent horizon" is not very suitable: an apparent horizon has not much of a horizon on one hand, and it is not just "apparent", but quite a real geometrical structure on the other.

4. Event Horizon. The event horizon $\mathcal{H}$ is defined as the boundary of that part of spacetime that can be seen by asymptotic observers. Thus, spacetime has to be asymptotically flat in order that the event horizon is well-defined. An event horizon is always a null hypersurface. In Fig. 2, the outgoing null hypersurface $\mathcal{H}$ which coincides with $r = 2m$ outside the star is an event horizon. This can be seen as follows. A point $p \in \mathcal{M}$ can be seen by distant observers, if the outgoing NS through p reach $r = \infty$ in the future of p. As we have seen, only those NS's reach the infinity, which pass the region outside the star and with $r > 2m$ (u is the retarded time – that is the time of asymptotic observers). Thus, $\mathcal{H}$ is the event horizon. There are no event horizons in Minkowski spacetimes: any event of it can be seen by distant observers.

The notion of *black hole* is associated with that of event horizon. A black hole on a spacelike hypersurface $\mathcal{S}$ is defined as the inside of the intersection of $\mathcal{S}$ with the event horizon $\mathcal{H}$. Thus, $\mathcal{H}$ is the evolution of a black hole boundary. What happens inside black holes is absolutely inobservable for those which stay outside of it. One can see what happens inside, if one falls into it, but it is impossible to send a preprint out of it. Thus, investigations of geometry and of matter processes inside black holes defined in the above way do not seem very sensible.

In order to understand better the nature of black holes, let us compare them with apparent horizons: an apparent horizon is a *weakly non-local object*, whereas a boundary of black hole is a *strongly non-local object*. This means: if one wants to find an apparent horizon on a given spacelike hypersurface $\mathcal{S}$, then the knowledge of the initial data along $\mathcal{S}$ is sufficient. To find a black hole boundary, however, one must know the spacetime in the *whole future* of $\mathcal{S}$. Fig. 3 illustrates an "experiment": we try to shift an apparent horizon and a boundary of a black hole at a given spacelike hypersurface $\mathcal{S}$ by changing the spacetime in the far future of $\mathcal{S}$. The spacetime $\mathcal{M}_1$ of Fig. 3 differs from the spacetime $\mathcal{M}$ of Fig. 2 by a spherically symmetrical static shell of mass δm at a large distance from the collapsing star. The spacetime geometry of $\mathcal{M}_1$ inside the shell is exactly identical to that of $\mathcal{M}$; outside the shell, it is described by the Schwarzschild metric corresponding to the value of mass parameter $m+\delta m$. There are two different scenarios:

Scenario 1. The shell remains static for all times. Then, the event horizon is the outgoing NS $\mathcal{H}$ and there is no change of event horizon and of black hole surface compared to $\mathcal{M}$. The shell has no influence on the property of outgoing NS's to reach distant observers. The apparent horizon at $\mathcal{S}$ also remains as before, because the geometry at $\mathcal{S}$ did not change inside the shell.

Scenario 2. The shell collapses in a spherically symmetrical way in the far future of $\mathcal{S}$. The collapse need not be predictable from the initial data at $\mathcal{S}$, if one uses some truly random process to decide whether the shell collapses or not. Now, the event horizon changes its position from $\mathcal{H}$ to $\mathcal{H}_1$ and the black hole boundary at $\mathcal{S}$ jumps from B to B_1 , even if there is no change of geometry or matter properties at $\mathcal{S}$. Thus, one cannot send any signal to the observers at $\mathcal{S}$ from the future by that; for them, the position of the black hole surface is absolutely inobservable. For *local observers*, the event horizon of the black hole is a rather fictitious object: any null hypersurface through Schladming can be an event horizon. (But it is less fictitious than,

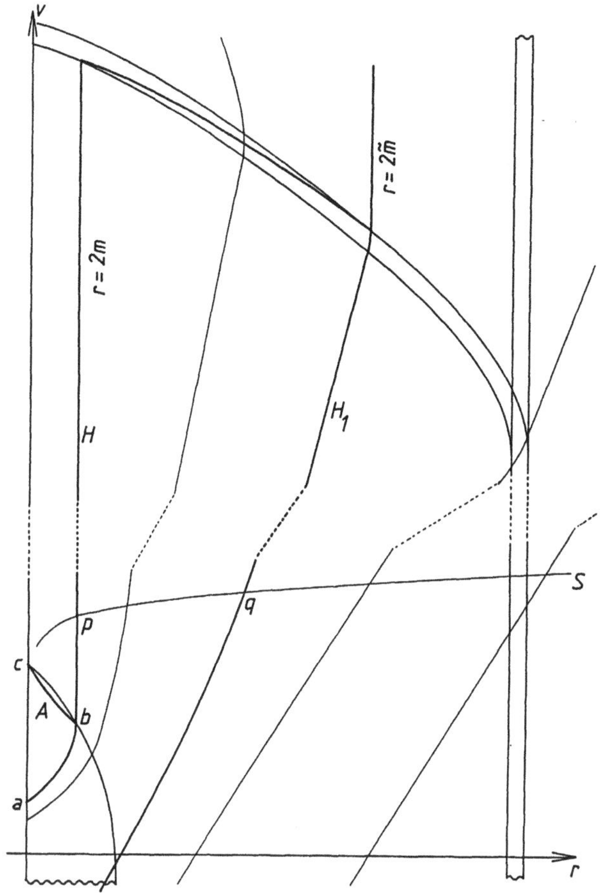

Fig. 3. Spacetime $\mathcal{M}_1$ which differs from $\mathcal{M}$ by the spherically symmetrical and stationary shell of mass δm. If the shell does not collapse, then the event horizon will be the null hypersurface H which coincides with $r = 2m$ in the future of the spacelike hypersurface S. If the shell collapses, then the event horizon will be the null hypersurface H_1 which coincides with $r = 2\tilde{m} = 2(m + \delta m)$ in the far future of S. The points p and q denote the boundaries of the corresponding black holes. The boldface timelike curves inside the star or the shell consist of apparent horizons.

e.g. a choice of gauge: an event horizon is, after all, an invariant geometrical object from the point of view of the spacetime as a whole.) In comparison to it, the apparent horizon at $\mathcal{S}$ does not change and remains exactly as in the first scenario, because it is a function of the local geometry at $\mathcal{S}$ which did not change. In the future of $\mathcal{S}$, the locus of apparent horizons changes only where the local geometry changes.

We observe, however, that the relative positions of the apparent horizon and the black hole boundary in Fig. 3 are always the following: no apparent horizon ever lies outside the event horizon. This is a quite general behaviour because of the following theorem:

Theorem 1. (Hawking) Let $\mathcal{M}$ be an asymptotically flat spacetime and let $\mathcal{S}$ be a spacelike hypersurface in $\mathcal{M}$. Let the stress-energy tensor $T_{\mu\nu}$ satisfy the weak energy condition (WEC) in the future of $\mathcal{S}$. Then, no outer trapped surface is visible from infinity.

For proof, see [8]. The WEC is the following condition on the stress-energy tensor:

$$T_{\mu\nu}t^\mu t^\nu \geq 0$$

for all timelike vectors t^μ. This means that the energy density is non-negative for any observer. Thus, if the premises of the theorem are satisfied, the apparent horizon lies *inside* a black hole. In particular, the existence of an apparent horizon *implies* that of a black hole in this case. However, some assumption about the whole future of $\mathcal{S}$ must again be made.

5. Red Shift. Consider an observer O_1 falling through the event horizon $\mathcal{H}$ in Fig. 2 along a smooth timelike curve $\mathcal{C}$ given parametrically by

$$v = v(s), \quad r = r(s), \quad \theta = \theta(s), \quad \phi = \phi(s),$$

where s is the proper time of O_1. As O_1 approaches the event horizon $\mathcal{H}$, his retarded time u diverges according to the relation (5) or, near the horizon:

$$u(s) \approx -4m \ln\left(\frac{r(s)}{2m} - 1\right) .$$

This means that the interval of the retarded time between any event at the world line of O_1 and his intersection with $\mathcal{H}$ is infinite. If O_1 sends some periodic signal, then this signal will arrive at a distant observer O_2 with longer and longer periods. There is an infinite *time delay* between O_1 and O_2. Because of this delay, O_2 will never see the star crossing the horizon; rather, the star becomes "frozen" at some radius just above the horizon.

An important aspect of this time delay is that the frequency of any signal from O_1 will appear to O_2 more and more red shifted as O_1 approaches the horizon. The red shift will be infinite at the horizon. Thus, not only will every object which is going to cross the horizon be seen from infinity as frozen just above the horizon radius, it will also quickly redden and disappear, because the intensity of its signal will be damped by the red shift. One can calculate the rate of the damping from the above relation with the result:

$$I(u) \approx I_0 \exp\left(-\frac{u}{4m}\right),$$

where $I(u)$ is the intensity measured by O_2 and I_0 is a constant dependent on the source.

3. Positive Energy Theorems

In the late seventies and early eighties, some old conjectures about gravitational energy have been proved. Our understanding of energy and momentum in general relativity theory increased dramatically, and some new and elegant techniques have become available. There is still much activity in the field, but it brings only small improvements. In this section, we will describe these important results. The section is divided into two parts: in the first part, the theorems are introduced and explained, whereas in the second one we will show some technical aspects.

Historically, two completely independent proofs of the theorems have been given. The first one by Schoen and Yau [9] made use of the global differential geometry and was a logical continuation of the development which started with the famous singularity theorems of Penrose and Hawking [8]. It was based on properties of extremal objects (minimal 2-surfaces, maximal 3-surfaces) and it was a typical proof by contradiction. The second one originated in supergravity. Positivity of energy in supergravity was proved by Deser and Teitelboim [10]. This proof suggested that some properties of spinors could be crucial for the positivity of energy in general relativity. Such an idea has been realized by Witten [11], whose proof was based on properties of ordinary spinors (as opposed to Grassmannian ones). It was a constructive proof: the energy was written as an integral over an explicitly non-negative integrand. Since then, many improvements and modifications were published. The present report is based on the review by Horowitz [12].

In general relativity theory, the total energy and energy density have some unusual properties and one must be cautious in using these notions. We have to explain these "curiosities" before we pass to the positivity theorems.

There is no well-defined and general notion of the energy density of the *gravitational field.* This is due to the equivalence principle: at any point, all gravitational forces can be transformed away so that locally, there is *no gravitation* at all. Thus, in the corresponding system of coordinates, the energy density of gravitation at this point must be zero, if such an object exists. In the literature, many different "pseudotensors" or "complexes" are proposed, but they always have too much dependence on the reference system – much more than the energy density of the customary fields has. Indeed, in any relativistic theory, the energy density of matter is the 00-component, in an orthonormal frame, of the so-called stress-energy tensor, $T_{\mu\nu}$. The tensorial transformation law of this quantity expresses important physical properties of the energy density – this is not just mathematics. Although these pseudotensors *are* useful for some purposes, they definitely cannot be considered as local energy and momentum currents of the gravitational field. Concluding, we can say: a) energy and momentum of the gravitational field is not localizable and b) for other fields, the energy and momentum currents are described by a stress-energy tensor, $T_{\mu\nu}$.

In general, no reasonable *total energy* of the system of all fields and gravitation exists, either. Indeed, in any relativistic theory, the total energy is a 0-component of the total 4-momentum, P_μ. We need a frame to interpret the components of a 4-vector. In Minkowski spacetime, a frame has a point-independent, global existence (there are globally defined parallel frame fields). In curved spacetimes, however, a frame is, in general, well-defined only if a point is specified at which the frame vectors are tangent. Frames at different points are not related – existence of a global parallel frame field would imply that the spacetime is flat. Thus, the total energy has to be associated with a particular point of the spacetime, and this contradicts the nature of total energy. Even the notion of total rest mass has not much sense, if the corresponding rest frame is not given. Thus, it is idle to say, e.g. that the total energy of the closed cosmological model is zero, etc. (for more detail, see [7]).

In one special case, however, the notion of total energy is sensible and useful; this is the case of the so-called *isolated systems* in general relativity. Heuristically, an isolated system has a finite radius and is the only source of the gravitational field in the spacetime. The field must, therefore, fall off at a characteristic rate with the distance from the system, and the spacetime is flat at infinity. Such spacetimes are called *asymptotically flat*. Thus, asymptotically flat spacetimes are used as models of isolated systems in general relativity. (An exact definition of asymptotically flat spacetimes is difficult; see, e.g. [8].)

In asymptotically flat spacetimes, one can approach infinity in two different ways: along a spacelike hypersurface $\mathcal{S}$ (this is a sort of limit $r \to \infty$, $t = \text{const.}$) and along a null hypersurface $\mathcal{N}$ (roughly $r \to \infty$, $t \to \infty$). Accordingly, we can define two different kinds of 4-momenta. There is the so-called *Arnovitt-Deser-Misner* (ADM) 4-momentum, $P_\mu^{(ADM)}$, which is associated with the infinity of an asymptotically flat spacelike hypersurface $\mathcal{S}$, and the so-called *Bondi-Sachs* (BS) 4-momentum, $P_\mu^{(BS)}$, associated with the infinity of an asymptotically flat null hypersurface $\mathcal{N}$. At the infinity of $\mathcal{S}$ or $\mathcal{N}$, there are parallel (or, as one says, constant) frame fields, because the spacetime is flat there. These constant asymptotical frames are determined up to a rigid Lorentz transformation, and they play an important role in the definition of the 4-momenta. In particular, because of the existence of global constant frames there, the 4-momenta are not associated with any particular point at infinity.

The physical meaning and relation between $P_\mu^{(ADM)}$ and $P_\mu^{(BS)}$ can be understood as follows (Fig. 4). $P_\mu^{(ADM)}$ represents the totality of 4-momentum of the source + its radiation as contained in the spacelike hypersurface $\mathcal{S}$. $P_\mu^{(ADM)}$ is *constant* (independent of $\mathcal{S}$). $P_\mu^{(BS)}$ is the totality of 4-momentum left in the spacetime at the retarded time u corresponding to the null hypersurface $\mathcal{N}$ after some radiation went away through infinity. $P_\mu^{(BS)}$ is, therefore, not constant, it depends on the retarded time u, and the conservation law for it has the form:

$$\frac{\partial P_\mu^{(BS)}}{\partial u} = -p_\mu,$$

where p_μ is the 4-momentum of the radiation going out at infinity at the corresponding retarded time. After the basic notions have been explained, we can turn to the positive energy theorems.

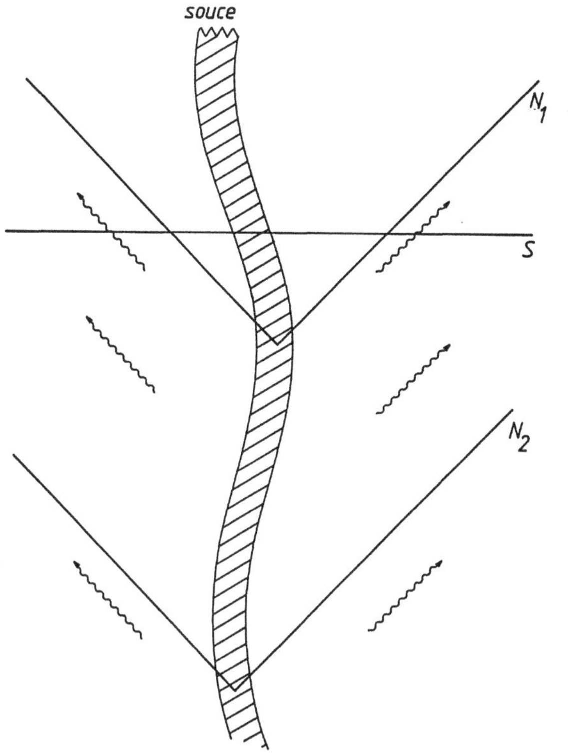

Fig. 4. Difference between the total 4-momentum associated with a spacelike hypersurface S (ADM) and that associated with null hypersurfaces N_1 and N_2 (BS). As some radiation of the source goes away between the two null hypersurfaces the BS energy must decrease. The hypersurface S contains, however, the source together with all of its radiation emitted in the past of S, so the ADM 4-momentum will be constant.

Roughly, the positive energy theorems state that it is impossible to construct an isolated object from "reasonable classical" matter bound together by gravity, whose total energy would be negative, or would become negative at some retarded time after some of its energy has been radiated away. "Reasonable classical" matter satisfies the so-called *dominant energy condition* (DEC).

Definition: The stress-energy tensor $T_{\mu\nu}$ is said to satisfy DEC at a point p, if

$$T_{\mu\nu}t^{\mu}t'^{\nu} \geq 0 \tag{6}$$

for any two future oriented timelike vectors t^{μ} and t'^{ν} at p.

The meaning of DEC is the following. We can write $t^{\mu} = au^{\mu}$, where u^{μ} is a unit timelike vector – 4-velocity of an observer. The vector $E^{\mu} = T^{\mu}{}_{\nu}u^{\nu}$ is the energy current of matter as seen by the observer with the 4-velocity u^{μ}. Thus, (6) requires that E^{μ} is non-spacelike and future oriented. The name "dominant energy condition" stems from the observation that (6) is equivalent to

$$|T_{00}| \geq |T_{\mu\nu}| \qquad \forall\, \mu, \nu,$$

if the components of $T_{\mu\nu}$ are taken in an orthonormal frame: the energy density dominates all other components of the stress-energy tensor. We observe that the DEC is stronger than the WEC: the WEC is obtained as a special case, if $t^\mu = t'^\mu$. Finally, we are prepared to formulate the theorem.

Theorem 2. (Positive energy theorem) Let $\mathcal{S}$ be a nonsingular spacelike surface which either

A) is asymptotically flat, or

B) approaches asymptotically a null cone $\mathcal{N}$ of retarded time u.

If the dominant energy condition is satisfied along $\mathcal{S}$, then

A) $P_\mu^{(ADM)}$ corresponding to $\mathcal{S}$,

B) $P_\mu^{(BS)}$ corresponding to $\mathcal{N}$,

is a *future oriented timelike or zero vector*. Furthermore, $P_\mu = 0$, iff the spacetime is flat in a neighbourhood of $\mathcal{S}$.

The theorem requires that there is no singularity at $\mathcal{S}$. The presence of a singularity can make the total mass negative. A well-known example is the so-called "anti-Schwarzschild solution" with the metric

$$ds^2 = \left(1 + \frac{2m}{r}\right) dv^2 - 2dvdr - r^2(d\theta^2 + \sin^2\theta d\phi^2)$$

for $m > 0$. This metric is regular except for $r = 0$, and has *negative* ADM (or BS) mass: $P_\mu^{(ADM)}$ ($P_\mu^{(BS)}$) $= (-m, 0, 0, 0)$ in a suitable asymptotic frame. The metric is a vacuum ($T_{\mu\nu} = 0$) solution of Einstein's equation, so the DEC is everywhere satisfied. The singularity at $r = 0$ intersects any spacelike hypersurface, so there is no regular $\mathcal{S}$ in the spacetime. One can, however, prove modified positive energy theorems which allow, under certain conditions, for the presence of singularities at $\mathcal{S}$:

Theorem 3. If the assumptions of theorem 2 hold on $\mathcal{S}$ *outside an outer trapped surface*, then $P_\mu^{(ADM)}$ or $P_\mu^{(BS)}$ must be a future directed timelike 4-vector.

This means, roughly speaking, that an outer trapped surface can be caused only by a highly concentrated positive piece of mass, independent of whether there is a singularity inside it or not. It follows from theorem 3 that the mass will also be positive, if the conditions of theorem 2 hold outside and at an apparent horizon. Observe that neither the conditions of theorem 3 imply the existence of a black hole in $\mathcal{S}$ (for this, WEC had to be satisfied in the whole *future* of $\mathcal{S}$, not only at $\mathcal{S}$), nor does the existence of a black hole at $\mathcal{S}$ plus reasonable matter around it imply positivity of the energy. Thus, the "textbook" black hole has nothing to do with positivity of energy. It is tempting to redefine the black hole so that it will be associated, say, with an outermost apparent horizon rather than with an event horizon. We will see later that there are more reasons for that.

The problem set in the introduction is only conditionally solved by the positive energy theorems – the questions about the possible existence of singularities which are not surrounded by trapped surfaces and of negative energy currents remain open.

Let us now go into more detail and show some formulas. We start with a very short introduction to two-component spinors in curved spacetimes. For still more details, see, e.g. [13].

2-Component Spinors

There are four types of 2-component spinors: α^A, $\beta^{A'}$, γ_A, $\delta_{A'}$. They are distinguished from each other by different position and type of the index. The corresponding complex 2-dimensional vector spaces realize representations of the group Sl(2,C): if $\boldsymbol{S} \in \mathrm{Sl}(2,\mathrm{C})$, then the four spinors transform, in turn, by $\boldsymbol{S}$, $\bar{\boldsymbol{S}}$, $(\boldsymbol{S}^{-1})^T$, $(\bar{\boldsymbol{S}}^{-1})^T$, where the bar denotes complex conjugation. The representations can be mapped into each other, e.g. $\overline{(\alpha^A)} = \bar{\alpha}^{A'}$ or $\alpha_A = \epsilon_{AB}\alpha^B$, where

$$\epsilon_{AB} = \epsilon^{AB} = \begin{pmatrix} 0 & -1 \\ 1 & 0 \end{pmatrix}$$

is an invariant spinor. The relation to tensors is based on the existence of invariant spinor-tensors (quantities with mixed spinorial and tensorial indices) $\sigma^\mu_{AA'}$, $\sigma_\mu^{AA'}$ (Pauli matrices). All indices can be raised and lowered by the invariant spinors and tensors ϵ_{AB}, ϵ^{AB}, $\epsilon_{A'B'}$, $\epsilon^{A'B'}$, $\eta^{\mu\nu}$ and $\eta_{\mu\nu}$, where $\eta_{\mu\nu}$ is the Minkowski metric, and the tensorial indices must be transformed by Lorentz matrices $\Lambda^\mu{}_\nu(\boldsymbol{S})$. For example, if α^A and β^A are any two spinors, then the four complex numbers l^μ defined by

$$l^\mu = \sigma^\mu_{AA'}\alpha^A\bar{\beta}^{A'}$$

are components of a null vector:

$$l^\mu \to \Lambda^\mu{}_\nu(\boldsymbol{S})l^\nu, \quad \eta_{\mu\nu}l^\mu l^\nu = 0.$$

In particular, if $\beta^A = \alpha^A$, then the corresponding l^μ is a *real* (σ's are hermitean) and *future oriented* null vector (orientation must be specified before the spinors are defined!).

Let $\mathcal{M}$ be a curved spacetime with the metric $g_{\mu\nu}$, $x \in \mathcal{M}$, and e^μ_α an orthonormal frame at x, which is future oriented and right-handed. If the components of tensors and spinors at x are understood w. r. to e^μ_α then all above relations remain unchanged. The only new concept is that of the *covariant derivative.*

Let $\mathcal{C}$ be a curve in $\mathcal{M}$, parametrically given by

$$x^\mu = x^\mu(\lambda)$$

in some coordinates x^μ. Let $e^\mu_\alpha(\lambda)$ be a parallel frame along $\mathcal{C}$. Then any tensor or spinor field is said to be *parallel* along $\mathcal{C}$, if its components w. r. to $e^\mu_\alpha(\lambda)$ are constant (independent of λ). Let $\alpha^A(x)$ be a spinor field in $\mathcal{M}$ and let $\alpha^A(\lambda)$ be the components of this spinor field along $\mathcal{C}$ taken w. r. to the parallel frame $e^\mu_\alpha(\lambda)$. Then, the spinor, whose components w. r. to $e^\mu_\alpha(\lambda)$ are given by

$$\frac{d\alpha^A}{d\lambda}$$

is called the *covariant derivative* of $\alpha^A(\lambda)$ along $\mathcal{C}$. It is a linear function of the tangent vector,

$$\frac{dx^\mu}{d\lambda},$$

to the curve $\mathcal{C}$ so that we can write

$$\frac{d\alpha^A}{d\lambda} = \frac{dx^\mu}{d\lambda}(\nabla_\mu \alpha)^A ,$$

where $(\nabla_\mu \alpha)^A$ are some coefficients. They form a mixed spinor-tensor quantity which is called the *covariant derivative* of the spinor field $\alpha^A(x)$. In the following, we will omit the brackets. One can show the following formula

$$\nabla_\mu \alpha^A = \partial_\mu \alpha^A + \omega_\mu{}^A{}_B \alpha^B$$

where the matrix field $\omega_\mu(x) \in \mathrm{Sl}(2, \mathrm{C})$ for all μ and x is determined by the frame field $e^\mu_\alpha(x)$ and its derivatives at x and is called a *spin connection.*

Nester Expression

One can use spinors to express the 4-momenta of isolated systems as follows: Let $\mathcal{S}$ ($\mathcal{N}$) be an asymptotically flat spacelike (null) hypersurface and let α^A be an asymptotically parallel spinor field at $\mathcal{S}$ ($\mathcal{N}$). In suitable coordinates and frame fields along $\mathcal{S}$ ($\mathcal{N}$), we will have

$$\alpha^A = \alpha_0^A + \mathcal{O}(r^{-1}),$$

where α_0^A is a constant spinor. The spinor α^A defines a real, future oriented null vector l^μ along $\mathcal{S}$ ($\mathcal{N}$) by

$$l^\mu = \sigma^\mu_{AA'} \alpha^A \bar{\alpha}^{A'} ,$$

as well as a constant null vector l_0^μ by

$$l_0^\mu = \lim_{r\to\infty} l^\mu .$$

Nester's expression for the total 4-momentum of the system is given by

$$l_0^\mu P_\mu = \frac{1}{8\pi} \lim_{r\to\infty} \oint_{S(r)} dS^{\mu\nu} \omega_{\mu\nu},$$

where

$$\omega_{\mu\nu} = i\epsilon_{\mu\nu\kappa\lambda} \sigma^\kappa_{AA'} \alpha^A \nabla^\lambda \bar{\alpha}^{A'}$$

is a covariant 2-form defined by the spinor field α^A, $S(r)$ is a 2-surface $r =$ const at $\mathcal{S}$ ($\mathcal{N}$) and $dS^{\mu\nu}$ is the surface element of $S(r)$. The right hand side of the Nester expression can be shown to be a linear function of the null vector l_0^μ.

The corresponding coefficients, denoted here by P_μ, are the components of the ADM (BS) 4-momentum. We can consider the Nester expression as a definition of these quantities. There are other definitions ([14], [15]) and some of them can be more suggestive, but Nester's one has two advantages: it has the same integrand for ADM and BS 4-momenta and the positivity can be most directly proved for it.

Positivity Proof

We observe easily that the Nester expression is not positive for all asymptotically flat metrics. However, it will be positive, if the metric satisfies the constraints. Let us sketch the proof of the positivity. There are the following steps.

1. Using Stokes theorem, we can write

$$\lim_{r\to\infty} \oint_{S(r)} dS^{\mu\nu}\omega_{\mu\nu} = 2\int_{\mathcal{S}} d\mathcal{S}t^\nu\nabla^\mu\omega_{\mu\nu} \left(+\oint_{S} dS^{\mu\nu}\omega_{\mu\nu}\right), \tag{7}$$

where $\mathcal{S}$ is either the original spacelike hypersurface or the spacelike hypersurface which approaches sufficiently quickly the null hypersurface $\mathcal{N}$ at infinity, $d\mathcal{S}$ is its volume element, t^μ is the future oriented unit vector normal to $\mathcal{S}$, and S is a trapped surface (which need not be there).

2. We transform the integrand of the volume integral in (7) as follows:

$$\begin{aligned} t^\nu\nabla^\mu\omega_{\mu\nu} &= 4\pi T_{\mu\nu}t^\mu t^\nu - t^\nu\sigma_\nu^{AA'}(D_\kappa\alpha_A)(D^\kappa\bar{\alpha}_{A'}) \\ &\quad - 2t^\nu\sigma_\nu^{AA'}\sigma^\kappa_{A'B}(D_\kappa\alpha^B)\sigma^\lambda_{AB'}(D_\lambda\bar{\alpha}^{B'}), \end{aligned} \tag{8}$$

where D_μ is the projection of the 4-dimensional covariant derivative into the hypersurface $\mathcal{S}$ defined by

$$D_\mu = \nabla_\mu - t_\mu t^\nu\nabla_\nu.$$

Calculating the covariant derivative of ω in (8), we obtain two terms. The first contains two covariant derivatives of the spinor α^A which are antisymmetrized by $\epsilon_{\mu\nu\kappa\lambda}$. Thus, the curvature tensor will appear. Using the constraints we obtain the first term on the right-hand side of (8) which is non-negative because of the DEC. The second term contains covariant derivatives of both spinors, α^A and $\bar{\alpha}^{A'}$, and because of the expression $t^\mu\epsilon_{\mu\nu\kappa\lambda}$, only the projected derivatives D_μ remain. This leads to the last two terms on the right-hand side of (8). The first one is non-negative (the metric induced at $\mathcal{S}$ being negative definite). The last term is, however, negative definite.

3. So far, the only condition on the spinor α^A was that it be asymptotically parallel. Thus, we are free to impose a further condition with the limitation that it allows at least four independent solutions. Such a condition is the so-called *Witten's equation*,

$$\sigma^{\kappa}_{A'B}(D_{\kappa}\alpha^{B}) = 0,$$

which makes the last term vanish. It is an elliptical differential equation along $\mathcal{S}$ and it will have enough solutions, if the boundary conditions are not too severe.

4. Some boundary conditions for Witten's equation are already imposed at the infinity of $\mathcal{S}$: the solution must be asymptotically parallel. Moreover, if we want to show the modified positive energy theorem with a trapped surface S, then we have to make the last term in (7) non-negative, and for that some more boundary conditions will be necessary. If we use the fact that S is an outer trapped surface, we can find such a condition (the explicit form of this condition can be found e.g. in [12]).

5. Next, one must show that Witten's equation has a sufficient number of independent solutions with the above boundary conditions. These proofs have been given by Reula [16], by Parker and Taubes [17] and by Reula and Tod [18].

6. Finally, since there are four linearly independent l_0^{μ} , we have the inequality

$$l_0^{\mu} P_{\mu} \geq 0$$

for each of them. Then, however, P_{μ} must be non-spacelike and future oriented, Q.E.D..

4. Hawking Effect

Consider the collapse spacetime $\mathcal{M}$ with metric $g_{\mu\nu}$ as given by Fig. 2. Let us complete this spacetime by a static past: let the star to the past of the momentarily static state be static and let the metric outside of it be the static and spherically symmetrical Schwarzschild one with the corresponding mass parameter m. This spacetime will again be denoted by $\mathcal{M}$ and its metric by $g_{\mu\nu}$.

In the rest of the lectures, we are going to consider the following problem: are there any interesting quantum effects or corrections to the classical picture which would be visible from infinity? At the first sight, it seems that such corrections must be negligible. Indeed, outside the event horizon $\mathcal{H}$, the curvature R satisfies the inequality

$$R \lesssim \left.\frac{m}{r^3}\right|_{\mathcal{H}} \sim \frac{1}{m^2} \sim \left(\frac{M_{\odot}}{M}\right)^2 \times 10^{-10}\mathrm{cm}^{-2},$$

whereas quantum corrections seem to be important only, if

$$R \gtrsim R_{\mathrm{Planck}} \sim 10^{66}\mathrm{cm}^{-2}.$$

However, this estimate, being local, clearly concerns only local effects. It can not say anything about "accumulation", integration of the local quantum corrections over the whole space or over the whole time. That is why the answer to the above question is affirmative after all.

Indeed, there is the so-called *Hawking effect* [19]: the gravitational field outside the horizon creates particle pairs, some of these particles escape towards infinity resulting in a small but steady current of energy, and this energy integrates, over the whole time, to an infinite value. Thus, the corresponding quantum corrections to the spacetime as a *whole* must be *large*.

Let us recall the principal points of Hawking's original calculation. He has made the following assumptions:

1. The studied quantum field is a scalar one, φ, of zero rest mass, conformally coupled to gravity:

$$\frac{1}{\sqrt{|g|}}\partial_\mu\left(\sqrt{|g|}\;g^{\mu\nu}\partial_\nu\varphi\right)-\frac{1}{6}R\varphi=0,$$

   where R is the curvature scalar ($R=0$ for the Schwarzschild solution). This assumption just simplifies calculations.
2. There is no interaction between φ and the matter of the star – this guarantees that we obtain a pure gravitational effect, and it is not unrealistic because of the infinite red shift.
3. There is no self-interaction of φ ("linear field"). This is a simplifying assumption; there will be no divergences.
4. The spacetime $\mathcal{M}$ with the metric $g_{\mu\nu}$ is considered as external potential: it is a classical, fixed background, which is not influenced by φ.
5. At the initial hypersurface, $u=-\infty$, the state of φ was the vacuum $|\mathrm{in},0\rangle$. At infinity of asymptotically flat spacetimes, the vacuum and the particle states of quantum fieds are well-defined.

With these assumptions, no boundary conditions at the horizon $\mathcal{H}$ are necessary: the evolution of the field φ is well-defined from $u=-\infty$ up to $v=\infty$.

Then, Hawking could derive the following results:

1. Let u_1 denote the retarded time of the onset of collapse. Then, for large retarded times,

$$u \gg u_1,$$

   there is a steady current of energy going out through $v=\infty$.
2. This radiation has a slightly distorted thermal spectrum corresponding to the temperature

$$T_H=\frac{1}{8\pi k}\frac{1}{m}\sim 1.6\times 10^{-7}\,\frac{M_\odot}{M}\;{}^\circ\mathrm{K},$$

   where k is the Boltzmann constant. T_H is called *Hawking temperature.*

The calculation was an application of the in-out method; the out-modes of the field were decomposed into the in-modes, and the particles created were obtained from the coefficients of this Bogoliubov transformation. There was no information about what happens near the horizon and the star. However, no regularization or renormalization was necessary.

Hawking's calculation was extremely proper and convincing. It left, however, many questions open.

1. What is the source of Hawking's radiation? Is it the dynamical gravitational field inside the star or the strong static gravitational field just outside the horizon?
2. Correspondingly, what will be the quantum corrected metric? Will the star be converted directly into Hawking's radiation before any horizon can form [20], or will a black hole be produced as in the classical collapse scenario and will it subsequently slowly evaporate [19]?

In order to answer these questions, we have to calculate the so-called *back reaction* of the created particles to the metric. The problem can be roughly illustrated by an electrodynamical example. Suppose that a strong uniform electrostatic field between two plates creates electron-positron pairs with a tiny but steady rate. If the field of the created particles is neglected, there will be no change of the electrostatic field during the process and the total number of created particles over all times will diverge. In reality, however, the positrons will go to the negative plate and the electrons to the positive one so that the potential difference will slowly decrease and the static field will decay. To obtain this effect, we must calculate the charge current of the created particles and couple it back to the field. This is the back reaction.

After Hawking's discovery, workers calculating the corresponding current for gravity, the stress-energy tensor of the quantum field, experienced about six years (1975 - 1980) of stormy development. At the present time, there is quite a consensus reached on the methods and results, so that books are, or are going to be, written ([21], books which are not yet published but one hears about: Parker & Hu & Thom, Gibbons & Perry, and Fulling).

The basis for these calculations is the so-called *semiclassical approximation*. The semiclassical approximation consists, very roughly, of the following steps:

1. One chooses a *background field* $\mathrm{g}_{\mu\nu}$ which satisfies the classical Einstein equations
$$G_{\mu\nu}[\mathrm{g}_{\kappa\lambda}] = 8\pi T_{\mu\nu},$$
where the left-hand side is formed by the Einstein tensor of the metric $g_{\mu\nu}$ and the right-hand side contains the stress-energy tensor of the classical matter (e.g. the collapsing star).
2. Along this fixed background, one calculates the expectation value $\langle T_{\mu\nu}\rangle$,
$$\langle T_{\mu\nu}\rangle = \langle \mathrm{in}, 0|\boldsymbol{T}_{\mu\nu}|\mathrm{in}, 0\rangle,$$
of the stress-energy tensor operator $\boldsymbol{T}_{\mu\nu}$ of the quantum field φ.
3. The correction, $h_{\mu\nu}$, to the classical metric, $g_{\mu\nu}$, can then be calculated from the equation
$$G_{\mu\nu}[g_{\kappa\lambda} + h_{\kappa\lambda}] = 8\pi(T_{\mu\nu} + \langle T_{\mu\nu}\rangle). \tag{9}$$

This method will have problems, whenever the correction $h_{\mu\nu}$ becomes large in comparison to $g_{\mu\nu}$. For our collapse spacetime, we expect that the corrections will be small in the region

$$v < v_m, \quad r > r_{m'},$$

where the maximum advanced time v_m can be quite large, $v_m \gg v_1$, w. r. to the advanced time v_1 of the onset of collapse, and where r_m is some minimal radius, which ought not to be much smaller than the radius of the horizon $2m$.

First, let us review the calculations of the expectation value $\langle T_{\mu\nu} \rangle$ for our collapse spacetime $\mathcal{M}$.

5. Calculations of $\langle T_{\mu\nu} \rangle$

There is a relatively large literature on $\langle T_{\mu\nu} \rangle$ in the *Kruskal spacetime* (this is the maximum analytic extension of the Schwarzschild spacetime) and for the so-called *Hartle-Hawking state* (see, e.g. [22]). The Hartle-Hawking state is a unique [23] state of a quantum field on the Kruskal spacetime which satisfies the conditions:

1. It is a quasifree state (i.e., very roughly, a state which can play the role of a formal ground state in a Fock space).
2. It is a static, spherically symmetrical state.
3. It is a Hadamard state at both the past and future horizons, $r = 2m$. The condition to be "Hadamard" means, roughly, that the corresponding propagator has the "usual" short distance expansion [23].

The Hartle-Hawking state is not empty at infinity; it describes a sort of thermal equilibrium between the hole and the quantum field at Hawking's temperature T_H. Thus, far away from the hole, we have a tiny but non-zero and constant energy density, and this integrates to give an infinite total energy in this quantum state.

There are attempts to determine the quantum corrections to the Kruskal metric corresponding to the Hartle-Hawking state [25] showing that the corrections diverge as r^2 as one approaches infinity; the "corrected" spacetime will differ considerably from the Schwarzschild one at infinity – it cannot be asymptotically flat. However, the flat infinity was crucial for the definition of the event horizon. This is the first suggestion that quantum corrections, although small locally, can lead to large changes as far as strongly non-local notions like that of an event horizon are concerned. In [25], one tries to remove this difficulty by enclosing the hole in a box. This is rather non-realistic, however, because the wall of the box has to reflect perfectly (or at least very effectively) everything including gravitons as well as "dark matter" particles.

The calculation for a collapse spacetime with the state $|\text{in}, 0\rangle$ on it is difficult and could not be accomplished as yet. There are only model calculations for less dimensional spacetimes, namely for a 2-dimensional model of collapse by Davies, Fulling and Unruh [21]. As we have nothing better, we describe the results of these calculations. They are believed to yield a qualitatively true picture of what happens in the 4-dimensional case. Let us first study the 2-dimensional spacetime, denoted by $\mathcal{M}^2$, and its metric, γ_{ab}, in some detail.

$\mathcal{M}^2$ is nothing but the v-r-surface in our collapse spacetime $\mathcal{M}$. Thus, the metric γ_{ab} is given by:

1. inside the star,
$$ds^2 = d\tau^2 - a^2(\tau)d\chi^2,$$
2. outside the star,
$$ds^2 = \frac{r-2m}{r}dv^2 - 2dv\,dr.$$

Let us denote by $\mathcal{O}$ the part of the spacetime outside the star where $r > 2m$. In $\mathcal{O}$, the so-called double null coordinates, u and v, will be useful. u is defined by equation (5). The metric γ_{ab} has, in these coordinates, the form

$$ds^2 = \frac{r-2m}{r}du\,dv, \quad r = r(u,v). \tag{10}$$

The coordinate u is not regular at $r = 2m$ ($u = \infty$). We will need, therefore, a null frame l^a, n^a, which remains regular at $r = 2m$. A null frame must satisfy the following orthonormality conditions:

$$\gamma_{ab}l^a l^b = \gamma_{ab}n^a n^b = 0,$$

$$\gamma_{ab}l^a n^b = 1.$$

W. r. to the double null coordinates u, v a suitable frame has the components:

$$l^a = (0,1),$$

$$n^a = \left(\frac{2r}{r-2m}, 0\right). \tag{11}$$

At $r = 2m$ the u-component of n^a diverges, because n^a is a regular vector, whereas the coordinate u is singular.

The region $\mathcal{O}$ with the metric γ_{ab} is a *static* spacetime. The Killing vector ξ^a that generates the time translations is given by its components:

$$\text{w. r. to } v,r: \qquad \xi^a = (1,0),$$

$$\text{w. r. to } u,v: \qquad \xi^a = (1,1).$$

The existence of the Killing vector ξ^a enables us to define a conserved energy current, E^a, if a conserved stress-energy tensor, T_{ab}, is given:

$$E^a = T^a{}_b\xi^b. \tag{12}$$

From

$$\nabla_a T^a{}_b = 0,$$

and from the Killing equation (which any Killing vector must satisfy)

$$\nabla_a \xi_b + \nabla_b \xi_a = 0,$$

it follows easily

$$\nabla_a E^a = 0.$$

The conserved total energy E of the *matter* in such a static spacetime can be defined as follows:

$$E = \int_{\mathcal{S}} dS\, t_a E^a,$$

where one integrates over a spacelike hypersurface $\mathcal{S}$ with the future directed unit normal t^a.

Let us now turn to the expectation value $\langle T_{ab} \rangle$. The important general properties of $\langle T_{ab} \rangle$ on $\mathcal{M}^2$ are the following (see, e.g. [21]).

1. Conservation:

$$\nabla_a \langle T^a{}_b \rangle = 0 \quad \forall\, b.$$

2. Anomalous trace:

$$T = \langle T_{ab} \rangle \gamma^{ab} = -\frac{1}{24\pi} \frac{m}{r^3}.$$

3. Boundary conditions: vacuum at $u \to -\infty$, or

$$\langle T_{ab} \rangle \to 0 \quad \text{for } u \to -\infty.$$

4. Causality: The value of $\langle T_{ab} \rangle$ at a point $p \in \mathcal{M}^2$ depends only on
   a) the metric γ_{ab} in the *causal past* of p in $\mathcal{M}^2$, and
   b) the state of the quantum field chosen at $u \to -\infty$.

The causal past of a point p in a spacetime $\mathcal{M}$ is the set of all points q which can be joined to p by causal, future directed non-spacelike curves (see, e.g. [8]).

The first three properties determine $\langle T_{ab} \rangle$ uniquely: $\langle T_{ab} \rangle$ has three independent components; the two conservation equations together with the boundary condition determine two of them; the third is given by the trace. Thus, the value of the trace anomaly is the source of all "numerology" in what follows.

The results of the $\langle T_{ab} \rangle$-calculations in the region $\mathcal{O}$ can be summarized as follows:

$$\langle T_{ab} \rangle = T^{(1)}_{ab}(r) + T^{(2)}_{ab}(u). \tag{13}$$

$T^{(1)}_{ab}(r)$ is a static (time independent, zero radial currents), conserved stress-energy tensor, which will be present also around static stars and which is called *vacuum polarization*. It is given by

$$T^{(1)}_{vv} = T^{(1)}_{uu} = -\frac{1}{24\pi}\left[\frac{m^2}{2r^4} + \frac{m}{r^2}\frac{r-2m}{r}\right],$$
$$T^{(1)}_{uv} = -\frac{1}{24\pi}\frac{m}{4r^3}\frac{r-2m}{r}.$$

We will find that $T^{(1)}_{ab}(r)$ is singular at $r = 2m$, if we calculate the scalar $T^{(1)}_{ab}(r)n^a n^b$. Using eq. (11) and the above value of $T^{(1)}_{ab}(r)$, we have

$$T^{(1)}_{ab}(r)n^a n^b = \frac{4r^2}{(r-2m)^2}\, T^{(1)}_{uu} \to -\infty .$$

The asymptotic behaviour of the components of vacuum polarization in a regular frame are as follows:

$$r \to 2m+0: \quad T^{(1)}_{ab}(r) \sim \frac{1}{(r-2m)^2},$$

$$r \to \infty: \quad T^{(1)}_{ab}(r) \sim r^{-3}.$$

The second part of the stress-energy tensor is given by

$$T^{(2)}_{ab}(u) = T^{(2)}(u) l_a l_b,$$

where $T^{(2)}(u)$ is some function of the coordinate u. Thus, $T^{(2)}_{ab}(u)$ represents a flux of null (light-like), pressureless matter along the u = const hypersurfaces – *radiation* with u-dependent intensity. The function $T^{(2)}(u)$ depends in a complicated fashion on the motion of the collapsing star.

What is the behaviour of $T^{(2)}(u)$ near the horizon? If the radiation originated in the star and if $\langle T_{ab} \rangle$ was everywhere regular, then one would expect that $T^{(2)}(u)$ would fall off exponentially as $u \to \infty$ because of the infinite red shift between the star surface near the horizon and infinity. However, we have:

$$\lim_{u\to\infty} T^{(2)}(u) = \frac{1}{768\pi m^2}. \tag{14}$$

Hence, for large u, there is a *steady* (time independent) current of energy through the infinity $v = \infty$. This is the Hawking radiation. Incidentally, 768 is a Bernouli number:

$$\frac{1}{768\pi m^2} = \frac{1}{2\pi}\int_0^\infty \frac{\omega d\omega}{e^{8\pi m\omega} - 1},$$

so that $T^{(2)}_{ab}(u)$ coincides, for late times, with the stress-energy tensor of the current of thermal radiation corresponding to the temperature

$$T_H = \frac{1}{8\pi k}\,\frac{1}{m}.$$

In this way, the 2-dimensional model confirms the results of Hawking's calculations.

The region with radiation, $T^{(2)}_{ab}(u) \neq 0$, lies in the future of the retarded time u_1 of the onset of the collapse – for $u < u_1$, the spacetime is static. For $u \to \infty$, we have

$$T^{(2)}_{ab}(u)n^a n^b \sim \frac{1}{(r-2m)^2} \longrightarrow \infty,$$

so $T^{(2)}_{ab}(u)$ is *singular* at $r = 2m$. The sum (13), however, is finite, and, for stellar values of m, very tiny.

The full stress-energy tensor near $r = 2m$, where we can use the value (14) for $T^{(2)}_{ab}(u)$, is given by

$$\begin{aligned}
\langle T_{uu}\rangle &= \frac{1}{24\pi m^2}\left(\frac{r-2m}{r}\right)^2 \frac{r^2+4mr+12m^2}{32r^2},\\
\langle T_{uv}\rangle &= -\frac{1}{24\pi m^2}\left(\frac{r-2m}{r}\right)\frac{m^3}{r^3},\\
\langle T_{vv}\rangle &= -\frac{1}{24\pi m^2}\,\frac{2m^3r-3m^4}{2r^4}.
\end{aligned}$$

The corresponding energy current E^a, calculated according to (12), can be decomposed into the regular frame l^a, n^a:

$$E^a = E_n l^a + E_l n^a,$$

where

$$E_n = -\frac{1}{24\pi m^2}\left(\frac{m^3}{r^3} - \frac{r-2m}{r}\,\frac{r^2+4mr+12m^2}{16r^2}\right), \tag{15}$$

$$E_l = -\frac{1}{24\pi m^2}\left(\frac{m^4}{2r^4} + \frac{r-2m}{r}\,\frac{3m^3}{2r^3}\right). \tag{16}$$

This energy current is perfectly regular, time-like, but it is *past-directed*. Hence, near the horizon $r = 2m$, not even the WEC is satisfied. This fact has deep roots: it is impossible to create particles from the vacuum so that the corresponding stress-energy tensor satisfies the DEC [26]. One of the important consequences of this result is that one cannot apply the positive energy theorems (which uses DEC) and also Hawking's theorem 1 about the apparent horizon, implying the existence of an event horizon (which uses WEC), becomes inconclusive. The above negative energy currents can be quite large, if the radius of the apparent horizon is small, and this can happen in the far future, when the hole is almost evaporated.

Very important conclusions can be drawn from the value of the energy current $E(r)$ going in through the time-like hypersurface $r =$ const per unit of advanced time. Near the horizon $r = 2m$, we can use the formulas (15) and (16) to obtain:

$$E(r) = -\frac{1}{24\pi m^2}\left[\frac{m^4}{2r^4} + \frac{r-2m}{r}\,\frac{m^3}{r^3} + \left(\frac{r-2m}{r}\right)^2 \frac{r^2+4mr+12m^2}{32r^2}\right].$$

$E(r)$ is negative, so positive energy is going out of $r =$ const. In particular, at the horizon $r = 2m$, we have

$$E(2m) = -\frac{1}{768\pi m^2}.$$

Comparing this to the amount of energy going out through infinity per unit of retarded time (14), we see that there is an exact balance: the energy of Hawking's radiation going out at infinity equals the energy leaking out of the horizon $r = 2m$! This, together with the regularity of the total stress-energy tensor everywhere (except maybe at the singularity), is considered as the main result of the $\langle T_{\mu\nu} \rangle$-calculations. One concludes that the *energy necessary to power the Hawking effect comes out of the Schwarzschild horizon* in a constant (and, for stellar values of the mass m, very tiny) rate. In this way, the calculations seem to clarify where the origin of Hawking's radiation is.

The above conclusion, however, concerns only the very special collapse spacetime $\mathcal{M}$. In particular, $\mathcal{M}$ is special in the exact coincidence of the event horizon, $\mathcal{H}$, with the locus of apparent horizons at the Schwarzschild horizon, $r = 2m$ (which is, at the same time, also the so-called Killing horizon [8]) in the region $\mathcal{O}$. This leads to some confusion in the literature, if one tries to generalize the result to more general collapse spacetimes: as if this were self-evident, one mostly places the negative energy currents at the event horizon. This, however, is *wrong*, because the strongly non-local nature of the event horizon is incompatible with the causality axiom which must be satisfied by the stress-energy tensor. To understand this point, let us return to the two cases of Fig. 3.

In the first case, the shell does not collapse, the event horizon is $\mathcal{H}$, and its intersection with the space-like hypersurface $\mathcal{S}$ coincides with the apparent horizon A. It seems, therefore, that the stress-energy tensor $\langle T_{\mu\nu} \rangle$ at A (and $\mathcal{H}$) will contain the negative energy currents, which will balance the Hawking radiation going through the infinity. At the point q, where $\mathcal{H}_1$ intersects $\mathcal{S}$, $\langle T_{\mu\nu} \rangle$ will consist only of the vacuum polarization part, because the retarded time corresponding to $\mathcal{H}_1$ is smaller than that of the onset of the stellar collapse and the spacetime is static in a neighbourhood (and the past) of $\mathcal{H}_1$.

Let us consider the other case, in which the shell does collapse in the far future. The geometry in a neighbourhood of A and q and in their causal pasts is not changed at all by this collapse. The state of the quantum field chosen at the past null infinity is not changed, either. Thus, according to the property 4 (causality), the form of the stress-energy tensor cannot change: again, there are negative energy currents at A, but no such currents at q. However, the event horizon lies, now, at q. Thus, the source of Hawking's radiation cannot be, in general, the event horizon.

The following hypothesis seems to be plausible [27].

Conjecture: The origin of Hawking's radiation lies near apparent horizons, at least if the locus of apparent horizons is null or nearly null.

Indeed, the event horizon is excluded, and Killing horizons are not generic – they can be destroyed by small perturbations, whereas we would expect that the value of the stress-energy tensor $\langle T_{\mu\nu} \rangle$ is stable against such perturbations. Nobody could calculate as yet what happens at those apparent horizons which lie on a hypersurface which is far from being null (like the hypersurfaces $\mathcal{A}$ and $\mathcal{A}_1$ in Fig. 3).

6. Model of Back Reaction

There is a spherically symmetrical model of back reaction due to York [28], which describes all aspects of black hole evaporation in a small neighbourhood of apparent horizons in an approximative way. The approximation is based on the observation that the "black hole luminosity" L, which is defined by

$$L = \frac{dm}{dv},$$

satisfies the inequality

$$L \ll 1$$

almost until the end of evaporation. Indeed, L is taken in Planck units, and the Planck luminosity, L_P, is enormous:

$$L_P \approx 10^{58} \text{erg/sec}.$$

Thus, everything can be expanded in powers of the small parameter L, and higher powers can be neglected; in fact, York calculates the terms linear in L.

For the purposes of the present section, not all aspects are important, but only the behaviour of the outgoing NS's. However, we will need the behaviour everywhere, not only in a neighbourhood of apparent horizons. Hence, the following results are important for us:

1. The behaviour of the outgoing NS's in a small neighbourhood of apparent horizons is satisfactorily described by the so-called *Vaidya metric*, whose energy current is determined just by the component $\langle T_{vv} \rangle$ of the black hole stress-energy tensor [28].
2. The separatrice of the outgoing null vector field of the Vaidya metric lies in a very small neighbourhood of apparent horizons. Thus, there is some chance that the qualitative behaviour of the outgoing NS's in the whole spacetime will be adequately described by the Vaidya model [29].

The influence of the negative energy currents on the behaviour of the outgoing NS's, and with it the position of apparent and event horizons, in the Vaidya model of back reaction has been studied in [29]. The Vaidya metric is an exact solution of the Einstein equations, which is spherically symmetric, asymptotically flat, but non-static and non-empty. The metric is given by:

$$ds^2 = \frac{r - 2m}{r} dv^2 - 2\, dv\, dr - r^2 d\theta^2 - r^2 \sin^2\theta\, d\phi^2. \tag{17}$$

This looks like the Schwarzschild metric in Eddington-Finkelstein coordinates, but there is a small difference: m is not a constant, but a function of v:

$$m = m(v).$$

If we calculate the Einstein tensor $G_{\mu\nu}$ for the metric (17), we obtain the stress-energy tensor $T_{\mu\nu}$:

$$T_{\mu\nu} = \frac{1}{8\pi} G_{\mu\nu} = \frac{\dot{m}}{4\pi r^2} \partial_\mu v \, \partial_\nu v. \tag{18}$$

Thus, only the component T_{vv} of the stress-energy tensor w. r. to the coordinates v and r is different from zero. The stress-energy tensor (18) describes a conserved flux of a null pressureless fluid (radiation) along the ingoing NS's $v =$ const.. The energy density of the fluid is positive, if $\dot{m} > 0$, and negative, if $\dot{m} < 0$. The stress-energy tensor of the Vaidya metric does not contain any outgoing radiation whatsoever. Thus, Hawking's radiation is not among those aspects of black hole evaporation which would be described properly by the model even in a qualitative way.

Some general properties of the outgoing NS's can be obtained immediately from the metric (17). The differential equation for the function $r(v)$ which describes these surfaces given by

$$\frac{dr}{dv} = \frac{r - 2m(v)}{2r} \tag{19}$$

This equation implies immediately that the points with $r < 2m(v)$ lie at outer trapped surfaces, and the points with $r = 2m(v)$ lie at apparent horizons. It is interesting to calculate the signature of the locus of apparent horizons. Inserting the relations

$$r = 2m(v),$$

and

$$dr = 2\dot{m}dv$$

into the metric (17), we obtain

$$ds^2 = -4\dot{m}dv^2.$$

Thus, the locus of apparent horizons is *time-like*, if $\dot{m} < 0$, *space-like*, if $\dot{m} > 0$, and *null*, if $\dot{m} = 0$.

If we want to go into more detail, we have to choose the function $m(v)$. We suppose that there will be a negative inflow of thermal radiation at the apparent horizon of radius $2m(v)$ with the Hawking temperature T_H, given by

$$T_H = \frac{1}{8\pi k m(v)}. \tag{20}$$

The thermal energy current corresponding to that temperature is proportional to T_H^4, the surface area of the apparent horizon to $m^2(v)$, so that, using (20), we obtain for the total energy influx:

$$\dot{m} = -\alpha m^{-2}(v),$$

where α is some constant. α depends on the number of fields whose particles are created by the gravitational field. In fact, α will grow with temperature, when thresholds for creation of more and more massive particles will be exceeded [4]. At the end of the evaporation, the form of α will strongly depend on the high energy physics of elementary particles. We assume, however, that this is not important for the qualitative understanding of black hole evaporation, and so we take α to be constant. Then, a solution of the above equation reads:

$$m(v) = (3\alpha)^{1/3}(-v)^{1/3}.$$

The integration constant can be given the meaning of the origin of the advanced time v, and our solution corresponds to the choice $v = 0$ at the moment when $m = 0$. As the value of α has no consequences for the qualitative picture, we set $(3\alpha)^{1/3} = 1/2$, or

$$2m(v) = (-v)^{1/3}. \tag{21}$$

The Vaidya metric (17) with the value (21) of the function $m(v)$ will be valid in the following part of the spacetime:

1. outside the star, and
2. for the values

$$v_0 < v < v_s$$

of the advanced time v. Here, v_0 is the advanced time, at which the first apparent horizon formed, and v_s that of the end of the evaporation.

For $v < v_0$, the spacetime is just our old collapse spacetime $\mathcal{M}$. The inside of the star for $v > v_0$ will be practically unchanged, as the Vaidya energy current is very weak at the moment of formation of the first apparent horizon and for the time $\sim 2m$ thereafter. For $v > v_s$, we assume the Vaidya metric with a constant mass function, $m(v) = m_0$. This is a Schwarzschild metric with the mass parameter m_0.

There are no compelling reasons why v_s ought to be zero, that is, why the evaporation is to cease at the moment, when the energy of the star is exhausted. We can distinguish the following cases.

1. The evaporation ceases earlier; then $m_0 > 0$ and the final issue of the evaporation will be a Schwarzschild black hole with the mass m_0.
2. The evaporation ceases later; then $m_0 < 0$ and the final issue will be the anti-Schwarzschild metric. The spacetime is singular at $r = 0$; the singularity has negative mass and does not lie inside an event horizon; it is not surrounded by an apparent horizon or an outer trapped surface, either.
3. $v_s = 0$ and so $m_0 = 0$. The final issue of the evaporation will be the flat spacetime with the metric

$$ds^2 = dv^2 - dr^2 - r^2 d\theta - r^2 \sin^2\theta\, d\phi^2,$$

which holds for all $v > 0$ and all $r \geq 0$.

Let us consider the case 3. The outgoing NS's are given by the formula (19) with $m(v)$ given by (21). They are schematically depicted by Fig. 5. There are six different kinds of these hypersurfaces $\mathcal{N}$ in the past of $v = 0$.

1. $\mathcal{N}$ lies in the past of $\mathcal{H}_2$. It starts at $r = 0$ and expands until $v = 0$.
2. The hypersurface $\mathcal{H}_2$ starts at $r = 0$, expands until the point a, which is an apparent horizon, and then expands again until $v = 0$.

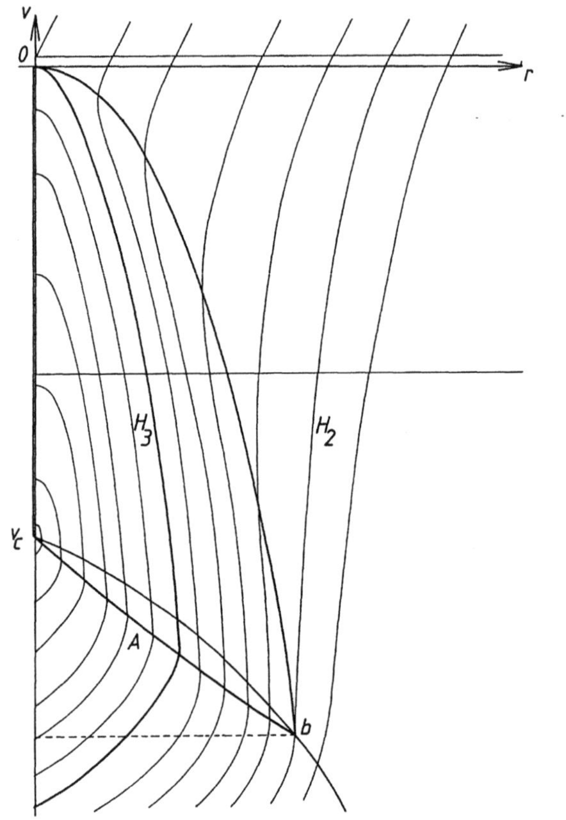

Fig. 5. Geometry of spherically symmetrical null hypersurfaces in the Vaidya model of back reaction. The apparent horizons lie at the timelike hypersurfaces A inside the star, and $r = 2m(v)$ (boldface curve joining b and O) outside of the star. H_2 is the first outgoing null hypersurface which contains an apparent horizon (point b), H_3 is the separatrice of the outgoing null vector field. The double line is the singularity.

3. $\mathcal{N}$ lies between $\mathcal{H}_2$ and $\mathcal{H}_3$. It starts at $r = 0$, expands, reaches a maximum area apparent horizon at $\mathcal{A}$, contracts, reaches a minimum area apparent horizon at $r = 2m(v)$, and expands again until $v = 0$.
4. The hypersurface $\mathcal{H}_3$ starts at $r = 0$, expands, reaches a maximum area apparent horizon at $\mathcal{A}$, and contracts until $r = 0$ and $v = 0$.
5. $\mathcal{N}$ lies in the future of $\mathcal{H}_3$. Its behaviour is qualitatively similar to that of $\mathcal{H}_3$, but its future end point is at $r = 0$ and $v < 0$.
6. The point v_c is a limit point of the family of outgoing NS's, and may be considered as a singular point-like null hypersurface.

All points with the coordinates $r = 0$, $v_c \leq v \leq 0$, are singular. In the future of $v = 0$, we obtain from equation (19) and $m(v) = 0$ for the outgoing NS's:

$$\frac{dr}{dv} = \frac{1}{2}.$$

Thus, they will expand forever and, in Fig. 5, they will appear like straight lines. Fig. 5 shows immediately where the outer trapped surfaces and apparent horizons will be. One can see that the curve $\mathcal{H}_3$ is a separatrice of the vector field given by (19). One can estimate the distance of this separatrice and the locus of apparent horizons at $r = m(v)$ [29]: it will be negligible for large negative v's. One can also estimate the duration $|v_c|$ of the evaporation [29]: in order of magnitude, it coincides with the evaporation time (2). (The duration also depends, of course, on which of the three cases is chosen.)

The next important point is to determine the position of the event horizon. To begin with, we have to follow back the outgoing NS's which intersect the future null infinity, $v = \infty$. Only the points which lie at these hypersurfaces are visible from infinity. Consider an arbitrary outgoing NS $\mathcal{N}$ within $v > 0$; $\mathcal{N}$ is represented by a straight line with the tangent $\frac{1}{2}$ in the Fig. 5, so there are the following three possibilities:

1. $\mathcal{N}$ intersects the curve $r = 0$ at $v > 0$,
2. $\mathcal{N}$ intersects the hypersurface $v = 0$ at $r > 0$,
3. $\mathcal{N}$ approaches the point p with the coordinates $r = 0$ and $v = 0$.

In case 1, $\mathcal{N}$ bounces and becomes an ingoing NS, $v =$ const.. In case 2, $\mathcal{N}$ continues as one of the outgoing NS's in the past of $\mathcal{H}_3$. In case 3, there is no uniquely determined continuation, because p is singular. If we approach p from $r = 0,\ v > 0$, the limit is the ingoing NS $v = 0$, but if we approach it from $v = 0,\ r > 0$, the limit is $\mathcal{H}_3$. There are no reasons why the continuation could not be any of the outgoing or ingoing NS which intersect the singularity at $v < 0$! Hence, we have the result: *the position of the event horizon depends on the behaviour of the singularity and its neighbourhood.* If, for example, we assume that

1. the neighbourhood of the singularity is properly described by the Vaidya metric, and
2. the singularity is perfectly "opaque" for any causal signal (i.e., all causal signals along outgoing and ingoing NS's are lost at the moment they arrive at the singularity),

then the event horizon will coincide with the hypersurface $\mathcal{H}_3$. In order to make this point more transparent, one can change the metric in a small neighbourhood of the singularity and calculate the corresponding changes in the position of the event horizon. We have the following theorem [27]:

Theorem 4. For any $\delta > 0$ and any outgoing NS $\mathcal{H}$ (including v_c) in the future of $\mathcal{H}_3$, define the region Σ by

$$\Sigma: \qquad v_{\mathcal{H}} < v < 0,\ r < \delta,$$

where $v_{\mathcal{H}}$ is the advanced time of the end point of $\mathcal{H}$. Then, there is a metric $\tilde{g}_{\mu\nu}$ in Σ such that

1. at the boundary of Σ, $\tilde{g}_{\mu\nu}$ matches the Vaidya metric outside of Σ smoothly (C^1), and

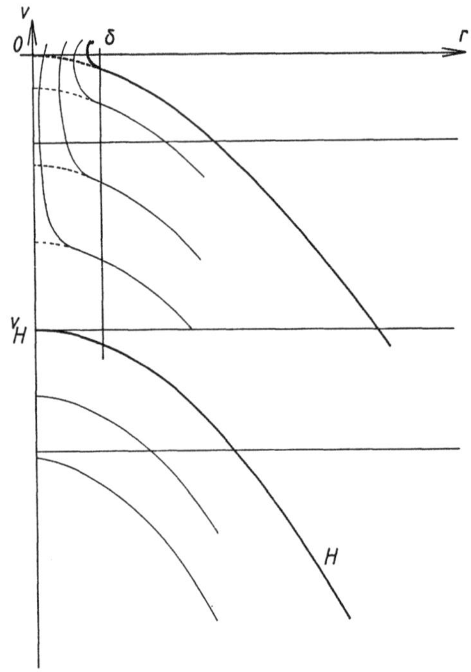

Fig. 6. Change of the geometry of null hypersurfaces, if the Vaidya model is modified inside of the region Σ given by $r < \delta$ and $v_H < v < 0$. The old event horizon is the boldface curve with the end point at O. In the modified spacetime, it is turned towards infinity inside Σ. Similarly, all outgoing null hypersurfaces which enter Σ are leaking out along the singularity, so that the new event horizon will be the null hypersurface H.

2. the event horizon corresponding to this new spacetime coincides with $\mathcal{H}$ outside of Σ.

The behaviour of the outgoing NS's for the new metric is illustrated by Fig. 6. We observe that the strongly contracting outgoing NS's begin to expand again after coming into the region Σ. Thus, one can ask the following question. As it is well-known (see, e.g. [8]), any null hypersurface which once contracts will keep contracting forever, if

1. the (classical) Einstein equations are satisfied, and
2. the WEC holds.

Are the assumptions 1 and 2 not natural to be imposed also in a neighbourhood of singularities? If yes, then the described change of the metric would be forbidden. However, these assumptions, together with the existence of trapped surfaces, will necessarily imply existence of singularities (singularity theorems); it is, therefore, something wrong with them. A more natural assumption is that some modification (like quantum gravity) of Einstein's theory near the singularities will be necessary so that the classical Einstein equations will be invalid there. Even if the equations happen to be valid, say, in some effective way, it is still not clear how the validity of WEC near the singularity is to be justified: it is violated already in a neighbourhood of such a low curvature structure like an apparent horizon (see the negative energy currents obtained in the previous section). Moreover, and this is an important point, the meaning of the theorem is not to propose $\tilde{g}_{\mu\nu}$ as a better and more natural metric than the original Vaidya one for a neighbourhood of the singularity. The *only* purpose of theorem 4 is to show that the position, and even the existence of the event horizon, in a spacetime with an evaporating black hole, does depend strongly on assumptions about the behaviour of the singularities and their neighbourhoods. This is a very surprising result: indeed, it could, in a rather pointed way, be described as "cosmic censorship upside down". Let us explain what the cosmic censorship is.

The so-called *cosmic censorship hypothesis* has been put forward by Penrose (see, e.g. [30]), and it is quite broadly accepted as a reasonable possibility. It states roughly the following. If the equations governing the behaviour of matter possess global solutions in the flat spacetime (coupling to gravity is switched off), then the system consisting of these equations with minimal coupling to gravity and of the Einstein ones will evolve sufficiently regular initial data in such a way that any possibly forming singularity will lie inside an event horizon and will, therefore, have no influence on the spacetime outside it. This means that a large part of the spacetime remains predictable even if we do not know what the singularity will do, because the event horizon remains in some bounded part of the space. The hypothesis is mathematically very interesting and its proof or disproof will without doubts be a great progress in our understanding of the mathematical structure of the classical Einstein equations. The cosmic censorship has, however, one aspect which seduces people sometimes to think that an effort to solve the mystery of collapse singularities is not justified, as the singularities will be hidden under event horizons: what happens inside black holes is not interesting, it is even not physical, because it is not observable. Our result contradicts just this, certainly false, aspect.

Another important point about our back reaction model and theorem 4 is the observation that the shift or removal of the event horizon does not lead to any change in the geometry or physics anywhere else than in Σ. Thus, for $v < 0$ and $r > \delta$, all observers will see always the same, independent of where the event horizon is. For $v > 0$, some differences between the Vaidya spacetime and the modified spacetime of theorem 4 could be due to possible radiation along the outgoing NS's coming out through the hole $v = 0$ and $r < \delta$. However, at the advanced time $v = 0$, practically all energy of the collapsing star is radiated away. If we assume energy conservation and non-existence of negative energy isolated objects, then only a negligible amount of energy remains to come out of this hole. It seems, therefore, that the presence and position of the event horizon is *irrelevant* to observational astrophysics. This point has its practical aspect: observable properties of black holes as derived from the Kerr-Newman model are independent of the fact that the Kerr-Newman spacetime contains an event horizon. The well-known results of black hole theory obtained as yet (as a representative literature, the book [1], the corresponding parts of the lectures by Proff. Börner and Straumann in this volume, etc., could be mentioned) are not affected by our results about event horizons.

The confusion about the role of the event horizon stems from the fact that the textbook definition of an event horizon is taken too literally. On one hand, this definition contains properties which have some physical importance, for example the strong time delay [28] (idealized to be infinite), on the other hand, there are some idealizations which make it easy to work with it mathematically, for example the asymptotical flatness or the stationarity leading to the existence of an event horizon. It seems that results like "loss of quantum coherence" and "non-unitarity of quantum gravity" [31] depend on the existence of event horizons; thus, they are based on assumptions which are not compelling.

Theorem 4 will be invalid in the case 1 ($m_0 > 0$). However, a modified theorem will be true in which the parameter δ satisfies:

$$\delta > 2m_0.$$

As far as $2m_0$ will be comparable to the Planck radius, the interpretation of the modified theorem 4 will remain the same as given above.

Finally, we come to the problem of positivity and conservation of energy in a gravitational collapse. Clearly, the positive energy theorems cannot be applied to the end stages of black hole evaporation: the DEC is violated by the strong negative energy currents and, moreover, there can be a singularity which is not surrounded by any outer trapped surface. Such singularities are present in the cases 2 and 3 ($m_0 \leq 0$). In the case 2, the final issue of the evaporation can even be an isolated object with an arbitrarily large negative energy, and so an arbitrary amount of energy could be radiated to infinity from the collapse region, similarly as in the Newton theory. Of course, the scenario of the case 2 must be completely false, if the theory is to remain self-consistent. Nevertheless, any attempt at the corresponding proof from first principles must be delayed until we understand more about Planck energy and Planck curvature physics.

Acknowledgement. The present author is thankful to B. Jensen for useful discussions, and to the organizers of the Schladming School for support and nice atmosphere. This work was partially supported by Schweizerischer Nationalfonds.

References

1. K.S. Thorne, R.H. Price, and D.A. Macdonald, ed.: "Black Holes: The Membrane Paradigm". Yale University Press, New Haven, 1986.
2. G.T. Horowitz, and M.J. Perry: Gen. Rel. Grav. 15, 1 (1983); W. Israel: Can. J. Phys. 64, 120 (1986).
3. E.H. Lieb, and W.E. Thirring: Ann. Phys. 155, 494 (1984).
4. D.N. Page: Phys. Rev. D13, 198 (1976).
5. J.R. Oppenheimer, and H. Snyder: Phys. Rev. 56, 455 (1939).
6. T. Pirani: in "Sources of Gravitational Radition". Ed. by L. L. Smarr. Cambridge University Press, Cambridge, 1979.

7. C. W. Misner, K. S. Thorne, and J. A. Wheeler: "Gravitation". Freeman, San Francisco, 1973.
8. S. W. Hawking, and J. F. R. Ellis: "The Large Scale Structure of Spacetime". Cambridge, Cambridge University Press, 1973.
9. R. Schoen, and S.-T. Yau: Commun. Math. Phys. 65, 45 (1979); Phys. Rev. Letters 43, 1457 (1979); Commun. Math. Phys. 79, 47 (1981); Commun. Math. Phys. 79, 231 (1981).
10. S. Deser, and C. Teitelboim: Phys. Rev. Letters 39, 249 (1977).
11. E. Witten: Commun. Math. Phys. 80, 381 (1981).
12. G. T. Horowitz: in "Asymptotic Behaviour of Mass and Spacetime Geometry". Ed. by F. J. Flaherty. Springer, Berlin, 1984.
13. R. Penrose: in "Battelle Rencontres". Ed. by C. DeWitt, and J. A. Wheeler. Benjamin, New York, 1968.
14. N. O. Murchadha, and J. W. York, Jr.: Phys. Rev. D10, 428 (1974).
15. A. Ashtekar: in "Asymptotic Behaviour of Mass and Spacetime Geometry". Ed. by F. J. Flaherty. Springer, Berlin, 1984.
16. O. Reula: J. Math. Phys. 23, 810 (1982).
17. T. Parker, and C. Taubes: Commun. Math. Phys. 84, 223 (1982).
18. O. Reula, and P. Tod: J. Math. Phys. 25, 1004 (1984).
19. S. W. Hawking: Commun. Math. Phys. 43, 199 (1975).
20. D. Boulware: Phys. Rev. D13, 2169 (1976).
21. N. D. Birrell, and P. C. W. Davies: "Quantum Fields in Curved Space". Cambridge University Press, Cambridge, 1982.
22. K. W. Howard: Phys. Rev. D30, 2532 (1984).
23. B. S. Kay, and R. M. Wald: in "An Italian Conference on General Relativity and Gravitational Physics". Ed. by U. Bruzzo, R. Cianci, and E. Massa. World Scientific, Singapore, 1987.
24. S. A. Fulling, M. Sweeny, and R. M. Wald: Commun. Math. Phys. 63, 257 (1978).
25. J. W. York, Jr.: Phys. Rev. D15, 775 (1985).
26. S. W. Hawking: Commun. Math. Phys. 18, 301 (1970).
27. P. Hajicek: Phys. Rev. D36, 1065 (1987).
28. J. W. York, Jr.: in "Quantum Theory of Gravity. Essays in Honor of the 60th Birthday of Bryce S. DeWitt". Ed. by S. M. Christensen. A. Hilger, Bristol, 1984.
29. P. Hajicek, and W. Israel: Phys. Letters 80A, 9 (1980).
30. F. J. Tipler, C. J. S. Clark, and G. F. R. Ellis: in "General Relativity and Gravitation", Vol. II. Ed. by A. Held. Plenum Press, New York, 1980.
31. S. W. Hawking: Phys. Rev. D14, 2460 (1976).

Quantum Gravity Motivated Computer Simulations[1]

Bernd A. Berg

Department of Physics and Supercomputer Computations Research Institute, The Florida State University, Tallahassee, FL 32306, USA

Abstract. Computer simulations are reviewed as a new tool to gain a better understanding of problems related to four-dimensional quantum gravity. There are two major directions: The more than twenty years old Regge calculus provides a discrete formulation of gravity that is well-suited for numerical investigations. Alternatively, conventional lattice gauge theory methods can be used to simulate certain field theoretical models of gravity.

1. Introduction

By comparison with most other scientific theories, quantum gravity is in a very special and somewhat unhappy situation. Due to lack of experimental guidance, the challenge is to develop quantum gravity purely by theoretical arguments. Although the task seems formidable, the situation is not so different from the historic one under which Einstein developed the general theory of relativity [1]. Disagreement of experimental data with Newton's theory of gravitation certainly played no role for the creation of general relativity. Instead, the major driving force was a theoretical one, namely the inconsistency between Newton's classical theory and special relativity.

Today's fundamental inconsistency is between quantum theory and general relativity. It is at the scale of the Planck length $l_{\rm P} \sim 10^{-33}$cm that the Heisenberg uncertainty principle combined with general relativity implies that quantum fluctuations are big enough to create microscopic black holes. Let us consider two points A and B, separated by a spatial distance Δx and a time difference Δt with ($\hbar = c = 1$)

$$\Delta t \; < \; \Delta x \; < \; l_{\rm P} \; \sim \; 10^{-33}\text{cm} \; .$$

The first inequality insures that A and B are outside each other's light cone. Local field theory states that experiments at A and B can be done independently of each other, no matter how close the points are. Yet, based on the

[1] Work in part supported by the Department of Energy under contract numbers DE-FG05-87ER40319 and DE-FC05-85ER250000.

uncertainty principle, we expect fluctuations of energy $\Delta E \sim (\Delta t)^{-1} \sim l_P^{-1}$. The Schwarzschild (i.e. black hole) radius associated with such a strong gravitational field is $R \sim G\Delta E > G/l_P$, where $G = l_P^2$ is Newton's constant. Thus we find the unreasonable result $R > \Delta x$ and the concept of local field theory is likely to break down at such distances. The correct physical theory may be non-local in character. Beyond this we would like the non-local theory to retain the good features of the usual continuum theory: Lorentz invariance, Poincaré invariance, gauge symmetries, unitarity and general coordinate invariance of general relativity. In addition the theory should contain a solution for divergence difficulties, such that physical quantities come out to be finite.

Unfortunately, presently none of the theoretical argument lines seems to lead to a really convincing conclusion, although there is a tendency towards string and superstring theories, preferred by a majority of workers in the field. Figure 1.1 gives a schematic indication of various theoretical offsprings. The diagram is mainly for illustrative purposes, completeness is not intended. The various options are distinct but not necessarily exclusive. Whereas the geometric formulation [1] and the field theory approach [2] are plainly equivalent on the classical level, it is less clear whether such an equivalence will continue to persist after quantization. Here and at each other branching point of the diagram it is a rather subjective decision how to continue. String theory or fundamental length, each in its own way, seem to overcome the problems with locality. For introductions see [3,4] and references as given there. However, these options do not necessarily rule out the more conventional quantum field theory approach a1–b1) and its various subcases.

It is not the intention of my lectures to present new arguments that may favour one of the approaches to quantum gravity over the others. Instead, I shall concentrate on the relatively new method of computer simulations as a tool to investigate models of quantum gravity. One may talk about a developing field of "experimental" mathematics and theoretical physics. Simulations are expected to become even more important, because the impressive progress in computer engineering still promises a doubling of available computer power every few years. My lectures emphasize the "engineering" point of view, this means they are intended to enable the interested student or researcher to write own simulation programs on gravity related problems of their interest. Presenting the existing, exploratory results, the lectures will carry us to the rather remote areas b1.1.2), b1.1.3) and a2.2.1.2) of the diagram. Of course none of the cases can make a serious claim to describe **the** theory of quantum gravity. My opinion is that such studies are useful nevertheless, because they investigate sophisticated mathematical models and any non-trivial information about such models is useful and may sooner or later find its application in science.

If quantum gravity is just another ("traditional") quantum field theory, it may be of crucial importance to understand the non-perturbative aspects of such a theory, similarly as it is a central question for the theory of strong interactions. To handle non-perturbative calculations one needs an appropriate regularization scheme. The regularization with a hypercubic lattice, pioneered by Wilson [5], has proven to be suitable for simulations of non-Abelian gauge

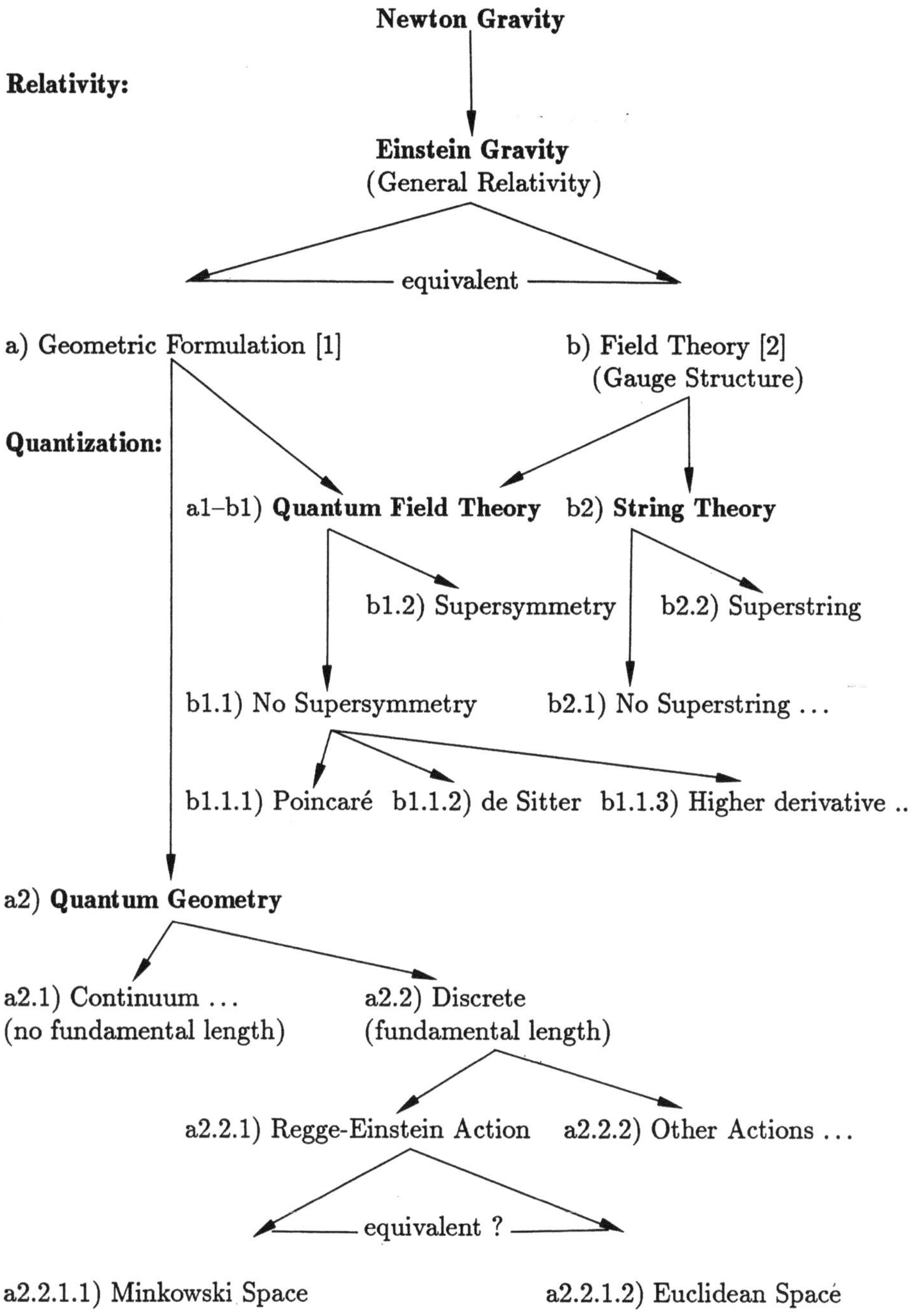
Newton Gravity
Relativity:
Einstein Gravity
(General Relativity)
equivalent
a) Geometric Formulation [1]
b) Field Theory [2]
(Gauge Structure)
Quantization:
a1–b1) Quantum Field Theory
b2) String Theory
b1.2) Supersymmetry
b2.2) Superstring
b1.1) No Supersymmetry
b2.1) No Superstring ...
b1.1.1) Poincaré
b1.1.2) de Sitter
b1.1.3) Higher derivative ...
a2) Quantum Geometry
a2.1) Continuum ...
(no fundamental length)
a2.2) Discrete
(fundamental length)
a2.2.1) Regge-Einstein Action
a2.2.2) Other Actions ...
equivalent ?
a2.2.1.1) Minkowski Space
a2.2.1.2) Euclidean Space

Fig. 1.1.

theories (see for instance [6]). It is in this sense that Smolin [7] has formulated de Sitter gravity on the lattice. In Section 6 of these lectures I will summarize simulations by Caracciolo and Pelissetto [8] that rely on this framework. This kind of simulation is most conventional in the sense that it closely resembles early work in numerical lattice gauge theory. Naturally the major problem is whether this lattice regularized theory admits, with proper choice of the integration measure, an ultraviolet fixed point which gives rise to a reparametrization invariant continuum limit. For asymptotically free theories such a problem can be addressed at the perturbative level; in other cases like standard de Sitter and Einstein gravity the problem is more difficult and indications can be obtained from numerical simulations. In the context of gravity an attractive alternative is provided by the Regge calculus [9] that yields directly a discrete, geometrical version of Einstein's action and is also suitable for more general actions. In the field theoretical approach the Regge calculus enters as an alternative regularization scheme and has been exploited by Hamber and Williams [10] to study higher derivative gravity numerically. Some of their results are reported in Section 5.

In the geometrical approach to quantum gravity the Regge calculus may well have a deeper meaning than just being another regularization scheme. In connection with the concept of a fundamental length we have a discrete space-time and the appealing aspects have, in particular, been emphasized by T.D. Lee and collaborators [4]. In that case the random lattice [11] emerges as a natural concept. The random lattice approach looks very appealing from the viewpoint of physical intuition. Unfortunately, it seems also to be the technically most involved approach as well. Structural results like existence of a mass gap (in the strong coupling limit), reflection positivity [12], connection between Euclidean and Minkowskian approach, etc. can hardly ever be handled. The hope of my simulation [13] is that qualitative features of a random lattice approach will survive the considerable technical simplifications gained by introducing an underlying regular, hypercubic structure.

Presently (supersymmetric) string theory is in the center of interest of numerous analytical investigations. However, although there are simulations of bosonic string theories [14], the connection of this work to gravity seems presently too remote to justify an inclusion in this review.

We conclude the introduction with a table of contents.

2. Regge-Einstein Action and Quantization

Consider the euclidean Einstein action

$$S_{\mathrm{E}} = \int d^4x \sqrt{g}\, R , \tag{2.1}$$

where R is the scalar curvature. Variation with respect to the metric leads to the classical equations of motion for the gravitational field in the vacuum

$$\delta S_{\mathrm{E}} = 0 \;\Rightarrow\; R_{\mu\nu} - \frac{1}{2} g_{\mu\nu} R = 0. \tag{2.2}$$

The euclidean formulation is conjectured [15] to be related to minkowskian gravity by analytic continuation. Such a relation is, however, by no means as well established as it is in case of relativistic field theory [12]. With the exception of some field theoretical models [16], the equivalence of euclidean gravity and physical gravity remains hypothetical. It is due to technical reasons (and rather unsatisfactory) that simulations were so far limited to euclidean gravity.

To see that the gravitational action (2.1) is unbounded (from above and below) consider the conformal transformation $\tilde{g}_{\mu\nu} = \Omega^2 g_{\mu\nu}$ where Ω is a positive function. The Einstein action transforms into

$$S_{\mathrm{E}} = \int d^4x \sqrt{g}\, (\Omega^2 R + 6 g^{\mu\nu} \partial_\mu \Omega \partial_\nu \Omega) , \tag{2.3}$$

which can be made arbitrarily large by choosing a rapidly varying conformal factor Ω [15].

It is well-known that one encounters two major difficulties when one attempts to quantize the Einstein theory of gravity. The field equations for the metric are derived from an unbounded action and the path integral seems therefore mathematically ill-defined. In addition the coupling constant in Einstein gravity (Newton's constant) has dimension of inverse mass squared (in units $\hbar = c = 1$), and this leads to a non-renormalizable quantum theory, as can be verified by doing explicit Feynman diagram perturbation theory [17].

A discrete version of Einstein's action has been given by Regge [9]. To describe Regge's action it is convenient to start with a simplicial lattice in flat four-dimensional euclidean space R_4. We then re-assign a new link length l_{ij} to each linked pair of sites in such a way that the pentahedral inequalities (which include the familiar triangular inequalities, the tetrahedral inequalities etc.) are fulfilled. The detailed geometry is outlined in Appendix B. As a result, in general the lattice does no longer fit in R_4 and the space defined by the lattice becomes curved. This space is called *Regge skeleton.*

The curvature of the Regge skeleton is concentrated on the $(d-2)$-simplices, that is the triangles t for $d = 4$. For each triangle t a deficit angle α_t is defined by

$$\alpha_t = 2\pi - \sum (\text{angles around } t \text{ in the plane perpendicular to } t) .$$

The Regge-Einstein action action is now given by

$$S = S_{\mathrm{RE}} = 2\sum_t \alpha_t A_t \ , \tag{2.4}$$

where α_t is the deficit angle and A_t the area of triangle t. The sum goes over all triangles of the skeleton. The most transparent derivation of (2.4) has been given by Friedberg and Lee [18]. For the four-dimensional Regge skeleton one has $R_{\mu\nu\lambda\sigma} = 0$ everywhere, except on the triangles. Let us consider the neighbourhood of triangle t. Using suitable local coordinates, the expression for $R_{\mu\nu\lambda\sigma}$ involves δ-functions in the two direction, say x_1 and x_2, perpendicular to t. By explicit calculation the equation

$$\sqrt{g}R \;=\; 2\delta(x_1)\delta(x_2)\,\alpha_t$$

can be derived. Inserting this equation into (2.1) yields (2.4). For a Regge skeleton the Einstein action and the the Regge-Einstein action are identical. Once the continuous space is replaced by the discrete skeleton, no further approximation is involved. In particular this implies that the Regge-Einstein action is as the Einstein action invariant under general coordinate transformations $x_\mu \to x'_\mu$.

In the path integral formulation a quantization of the Einstein action (2.1) would proceed by writing down the partition function

$$Z \;=\; \int D\mu(g)\; \mathrm{e}^{m_{\mathrm{P}}^2 S_{\mathrm{E}}} \ . \tag{2.5}$$

Immediately, this approach faces severe difficulties. On a perturbative level there is the well known fact [17] that the Einstein action is non-renormalizable. Even worse is that the unboundedness (2.3) of the action seems to render the partition function ill-defined. However, the latter argument is not entirely convincing as it neglects entropy effects that may come from the measure $D\mu(g)$. Equation (2.5) can be rewritten as

$$Z \;=\; \int_{-\infty}^{+\infty} dS\; n(S)\; \mathrm{e}^{m_{\mathrm{P}}^2 S} \tag{2.6}$$

and the possibility

$$n(S) \le \mathrm{e}^{-m_c^2 S} \quad \text{for} \quad S \to \infty \tag{2.7}$$

cannot be excluded *a priori*. In this case Z would exist at least for $m_{\mathrm{P}} < m_c$. The simulation [13] reported in Section 4 is motivated by investigating this question. Suitable for simulations is the discrete formulation of equation (2.5)

$$Z \;=\; \int D[\{l\}]\; F_\theta(\{l\})\; \mathrm{e}^{m_{\mathrm{P}}^2 S_{\mathrm{RE}}} \ , \tag{2.8}$$

where $F_\theta(\{l\}) \;=\; F_\theta(l_1, ..., l_N)$ takes care of the pentahedral constraints. It is a complicated function of the N links of the Regge skeleton. Even for a finite system the discrete Regge-Einstein action is unbounded (as well as the continuous Einstein action).

A conventional way to cope with the unboundedness of the Einstein action is to modify the gravitational action such that it becomes bounded, but the classical limit is not destroyed. Simulations of such actions are summarized in Sections 5 and 6. Although bounded, such actions tend to exhibit other problems (discussed later). The main difficulty in the present state is that no convincing arguments exist that lead to an unique choice of the action. In view of this situation simplicity and aesthetical beauty become the major guidelines. The first of those gives hope that at least some results will be obtained after all.

The choice of the measure $D\mu(g)$ is a delicate and essentially unresolved question [19]. For the exploratory numerical investigations, reported here, the choice will be essentially the simplest non-trivial measure. The hope is that qualitative features of the theory will be maintained under such an approximation that typically keeps only some of the invariances essential for general relativity.

3. The Monte Carlo Method

Originally the Monte Carlo (MC) method was introduced by Metropolis et. al. [20]. Here, I will give an introduction emphasizing practical aspects for actually doing calculations. For more detailed reviews the reader is referred to [21,22].

The typical task in statistical mechanics is to calculate expectation values

$$\overline{\mathcal{O}} = \langle \mathcal{O} \rangle = Z^{-1} \int \prod_{i=1}^{n} d\Phi_i \mathcal{O}(K) \exp\{-\beta S(K)\}. \tag{3.1}$$

The *partition function* Z is a normalization factor such that the expectation value $\langle \mathbf{1} \rangle = 1$. $S = S(K)$ is the action and a *configuration* K of the system is defined by

$$K = \{\Phi_1, ..., \Phi_n\}, \tag{3.2}$$

where n denotes the total number of variables. For simplicity I consider a finite discrete system. From the viewpoint of computer simulations this restriction is not very serious, because we are anyhow constraint to a finite value for n and a continuous variable can thought to be discretized in steps of the computer's digital accuracy. For a finite system we can count all configurations and label them by integers: K_k, $(k = 1, ..., N)$. For instance, $N = 2^n$ in case of Ising models (i.e. $\Phi_i = \pm 1$). With this notation equation (3.1) becomes

$$\overline{\mathcal{O}} = \langle \mathcal{O} \rangle = Z^{-1} \sum_k \mathcal{O}(K_k) \exp\{-\beta S(K_k)\}. \tag{3.3}$$

We like to attempt a numerical evaluation of (3.3). A naive sampling is certainly impractical. Already for the case of a two-dimensional Ising model on a 30×30 lattice 2^{900} configurations would contribute. The large number of configurations suggests a statistical treatment. Generating configurations randomly with a uniform distribution would, however, be completely inefficient. Most of the configurations have a very small Boltzmann factor $\exp\{-\beta S(K_k)\}$ and their contribution to the sum (3.3) is negligible, unless the number of such configurations (entropy) is very large.

Importance sampling generates configurations K_k with probability

$$P(K_k) \;=\; \text{const } e^{-\beta S_k} \quad \text{where} \quad S_k = S(K_k). \tag{3.4}$$

The constant is determined by the normalization condition $\sum_k P(K_k) = 1$. When configurations are generated with probability (3.4), expectation values become arithmetic averages:

$$\overline{\mathcal{O}} \;=\; \langle \mathcal{O} \rangle \;=\; \lim_{N\to\infty} \frac{1}{N} \sum_{k=1}^{N} \mathcal{O}(K_k). \tag{3.5}$$

Normally we cannot obtain directly configurations with probability (3.4) unless the problem is exactly solvable. But configurations with (3.4) as their equilibrium distribution can be generated by a ***Markov process***. The elements of the Markov chain are the configurations. They are sequentially generated such that each configuration only depends on the previous one. Assume that the configuration K_i is given. The transition probability to create the configuration K_j in one step from K_i is given by the matrix $W_{ji} = W(i \to j) \geq 0$, which is is required to satisfy the following properties:

i) *Ergodicity*:

$$e^{-\beta S_i} > 0 \;\text{ and }\; e^{-\beta S_j} > 0 \;\text{ imply :}$$

an integer number $k > 0$ exists such that $(W^k)_{ij} > 0$.

ii) *Normalization*:

$$\sum_j W_{ji} \;=\; 1.$$

iii) *Balance*:

$$\sum_i W_{ji} e^{-\beta S_i} \;=\; e^{-\beta S_j}.$$

Balance means: The Boltzmann weight factors form an eigenvector of the matrix $W = (W_{ji})$.

By definition, an ensemble is a collection of configurations such that to each configuration K_i a probability P^i is assigned and $\sum_i P^i = 1$. The ***Boltzmann ensemble*** E_{B} is defined to be the ensemble with probability distribution

$$P_{\mathrm{B}}^i \;=\; \text{const } e^{-\beta S_i}, \quad \sum_i P_{\mathrm{B}}^i = 1. \tag{3.6}$$

An *equilibrium ensemble* E_{eq} of the Markov process is defined by satisfying

$$WP_{\text{eq}} = P_{\text{eq}} \quad (\text{or in components}: P^j_{\text{eq}} = \sum_i W_{ji} P^i_{\text{eq}}).$$

The following result and the remark after equation (3.8) are central for the MC method.

Statement: Under the conditions i), ii) and iii) the Boltzmann ensemble is the **only** equilibrium ensemble of the Markov process.

My proof of this claim follows closely [22]. Let us first define a distance between ensembles. Suppose we have two ensembles E and E', each of which is a collection of many configurations. Denote the probability for configuration K_i in E by P^i and in E' by P'^i. We define the *distance* between E and E' as

$$||E - E'|| = \sum_i |P^i - P'^i|, \tag{3.7}$$

where the sum is over all possible configurations. Now suppose that E' resulted from the application of transition probability W to ensemble E. We can compare the distance of E' from the Boltzmann ensemble with the distance of E from the Boltzmann ensemble:

$$||E' - E_{\text{B}}|| = |\sum_j \sum_i W_{ji}(P^i - P^i_{\text{B}})| \qquad (\text{using balance})$$

$$\leq \sum_j \sum_i |W_{ji}(P^i - P^i_{\text{B}})| \qquad (\text{using the triangle inequality})$$

$$= \sum_i |P^i - P^i_{\text{B}}| = ||E - E_{\text{B}}|| \,. \tag{3.8}$$

The last line is obtained by making use of the normalization condition $\sum_j W_{ji} = 1$ and of $W_{ji} \geq 0$. This proves the statement.

Remark: We further conclude that the algorithm reduces the distance of an ensemble from the equilibrium Boltzmann ensemble, namely the inequality (3.8) is **strict** if W_{ji} never vanishes. This is easily seen: through the normalization condition $\sum P^i = \sum P^i_{\text{B}} = 1$ it follows that

$$P^k - P^k_{\text{B}} > 0 \implies P^l - P^l_{\text{B}} < 0 \quad \text{for some appropriate } l.$$

Consequently the contributions in $\sum_i W_{ji}(P^i - P^i_{\text{B}})$ flip sign and the equality in (3.8) can be excluded. Here we have used the (unrealistic) assumption that all matrix elements W_{ji} are greater than zero. However, this restriction is less severe than it looks. Ergodicity implies $(W^k)_{ji} > 0$ for k sufficiently large. As $(W^k)_{ji}$ can be used in equation (3.8) as well as W_{ji} it follows that the distance to the Boltzmann ensemble will be reduced by applying the Markov process sufficiently often. Obviously the approach to the Boltzmann distribution may become very slow, if a large value of k is needed until the transition amplitudes between all relevant configurations are non-zero.

There are many ways to construct a Markov process satisfying i), ii) and iii). In practice most MC algorithms are based on products of steps, each satisfying the condition of

iii') *detailed balance*:

$$W_{ji} e^{-\beta S_i} = W_{ij} e^{-\beta S^j} .$$

Using the normalization condition $\sum_i W_{ij} = 1$ detailed balance immediately implies balance iii).

Detailed balance still does not uniquely fix the transition probabilities W_{ji}. The *Metropolis algorithm* is a popular choice because of its computational simplicity in practical applications. Given an old configuration i, the Metropolis algorithm proposes a new configuration j with a probability density $f(j,i)$, $(\sum_j f(j,i) = 1)$. The new configuration is then accepted (or rejected) with probability

$$w_{ji} = \begin{cases} 1 & \text{for} \quad S_j < S_i \\ e^{-\beta(S_j - S_i)} & \text{for} \quad S_j > S_i \end{cases} \tag{3.9}$$

If the new configuration is rejected, the old configuration is kept. Consequently the Metropolis procedure gives rise to the transition probabilities

$$W_{ij} = f(j,i)\, w_{ji} \quad \text{for} \quad j \neq i$$

and

$$W_{jj} = f(j,j) + \sum_{j \neq i} f(j,i)\,(1 - w_{ji}) \quad \text{for} \quad j = i.$$

Obviously W_{ij}/W_{ji} satisfies balance, if

$$f(j,i) = f(i,j).$$

Otherwise the probability density $f(j,i)$ is unconstrained and can be chosen conveniently.

There is an amazing flexibility in the choice of the transition probabilities W_{ji}. In Section 3.1 I illustrate this for a simple example. Section 3.2 formulates the Metropolis algorithm for a Regge skeleton.

3.1 Example: Two-Dimensional $O(3)$ σ-Model

The two-dimensional $O(3)$ σ-model has been of recent interest because of its analogies with the four-dimensional Yang-Mills theory. The MC updating of this model is to some extent typical for conventional lattice gauge theory and of interest for Section 6 of these lectures. Vacuum expectation values are calculated with respect to the partition function

$$Z = \int \prod_j ds_j \; e^{-\beta S(\{s_j\})} . \tag{3.10}$$

The spins s_j are defined on the sites of a two-dimensional torus. They are normalized to $(s_j)^2 = 1$, and the measure ds_j is defined by

$$\int ds_j = \frac{1}{4\pi}\int_{-1}^{+1} d\cos(\delta_j)\int_0^{2\pi} d\phi_j. \tag{3.11}$$

The action is

$$S = \sum_l (1 - s_{i(l)} s_{j(l)}), \tag{3.12}$$

where the sum goes over all links l of the lattice and $i(l), j(l)$ are the endpoints of link l. Each spin interacts only with its nearest neighbors. We would like to change a single spin s, as depicted in Figure 3.1. The sum of its neighbors is

$$s_n = s_1 + s_2 + s_3 + s_4.$$

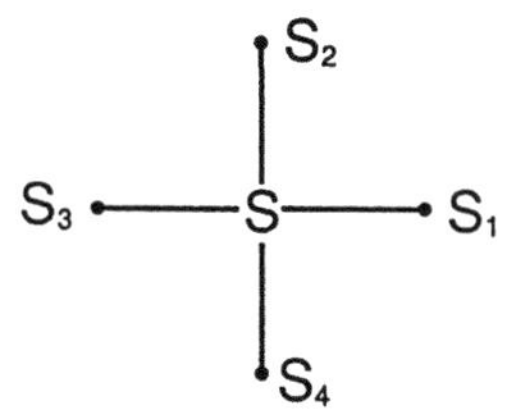

Fig. 3.1. Two-dimensional $O(3)$ σ-model, a single spin and its four neighbors.

Hence, the contribution of spin s to the action is $4 - ss_n$. We propose a new spin s' with the measure (3.11) (this defines the probability function f introduced above) and the Metropolis algorithm accepts or rejects the choice with probability

$$w(s \to s') = \begin{cases} 1 & \text{for} \quad s_n s' > s_n s \\ e^{-\beta(s_n s - s_n s')} & \text{for} \quad s_n s' < s_n s. \end{cases} \tag{3.13}$$

Repeating the Metropolis algorithm again and again for the same spin s leads (as has been proven in the previous Section) to the equilibrium distribution of this spin, which reads

$$P(s'; s_n) = \text{const } e^{\beta s_n s'} \quad \text{with} \quad \int P(s'; s_n) ds' = 1. \tag{3.14}$$

One would prefer to choose s' directly with the probability

$$W(s \to s') = P(s'; s_n) = \text{const } e^{\beta s' s_n}. \tag{3.15}$$

The algorithm, which creates this one variable distribution directly, is called the ***heat bath*** method. The heat bath algorithm is efficient, if the action is sufficiently simple to allow an explicit calculation of the probability $P(s'; s_n)$. Setting it up is an easy task in case of the $O(3)$ σ-model. Let

$$\alpha = \text{angle}(s', s_n), \quad x = \cos(\alpha) \quad \text{and} \quad \hat{s}_n = \beta |s_n|.$$

The Boltzmann weight becomes $\exp(x\hat{s}_n)$ and the normalization constant follows from

$$\int_{-1}^{+1} e^{x\hat{s}_n} dx = \frac{2}{\hat{s}_n} \sinh(\hat{s}_n) .$$

Therefore the probability is

$$P(s'; s_n) = P(x; \hat{s}_n) = \frac{\hat{s}_n}{2\sinh(\hat{s}_n)} e^{x\hat{s}_n} . \tag{3.16}$$

Computers are equipped with efficient pseudo random number generators, which generate uniformly distributed pseudo random numbers in the range $0 \le x_\omega < 1$. We have to convert a proposed pseudo random number x_ω into a choice of $x = \cos\alpha$. This is done by interpreting x_ω as the shaded area of Figure 3.2. It follows

$$x = \cos\alpha = \frac{1}{\hat{s}_n} \ln\left[\exp(+\hat{s}_n) - x_\omega \exp(+\hat{s}_n) + x_\omega \exp(-\hat{s}_n)\right] . \tag{3.17}$$

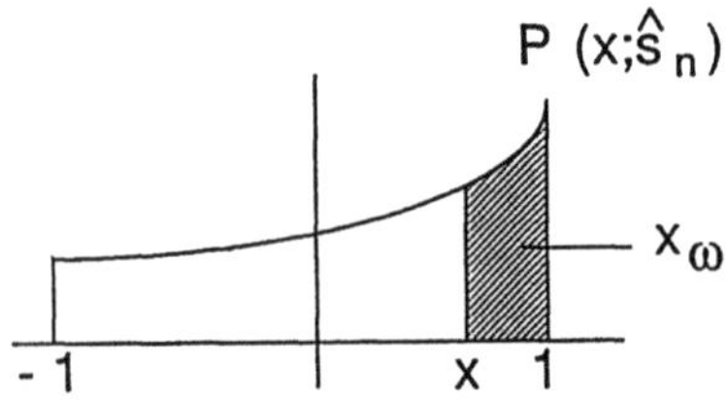

Fig. 3.2. Heat bath algorithm, interpretation of the random number $x_\omega \in [0, 1)$ as an area under the probability distribution.

Next we have to give s' a direction in the plane orthogonal to s_n . This is simply done by choosing a random angle β, uniformly distributed in the range $0 \le \beta < 2\pi$. $x = \cos\alpha$ and β completely determine s' with respect to s_n. Before finally storing s' in the computer memory, we have to calculate coordinates of s with respect to a cartesian coordinate system, which is globally defined for the whole lattice. The three components of s' with respect to this coordinate system are kept in the computer memory.

3.2 The Monte Carlo Algorithm for Regge Skeletons

In this Section the MC updating of a Regge skeletons is described in some details. The outlined procedure was used in [13].

Demands on computer memory are minimized by storing only the link lengths $\{l\}$. To update a link,

$$l \to l' , \tag{3.18}$$

we have:

1) To check whether the change $l \to l'$ is allowed geometrically.
2) To calculate the change for the Regge-Einstein action.

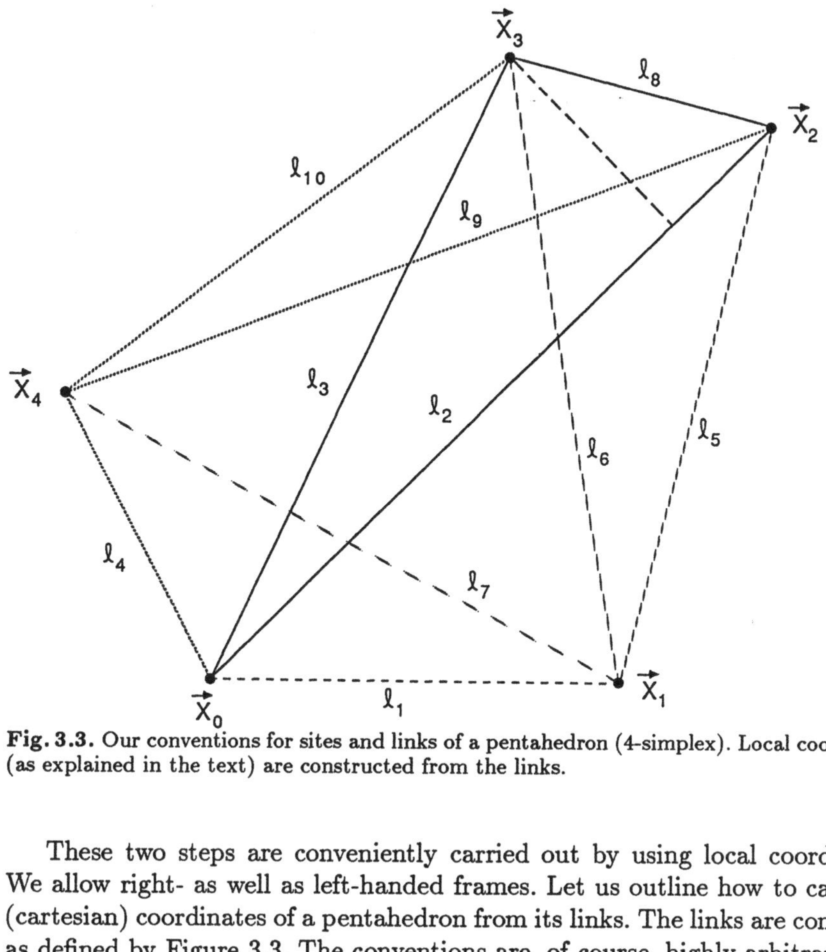

Fig. 3.3. Our conventions for sites and links of a pentahedron (4-simplex). Local coordinates (as explained in the text) are constructed from the links.

These two steps are conveniently carried out by using local coordinates. We allow right- as well as left-handed frames. Let us outline how to calculate (cartesian) coordinates of a pentahedron from its links. The links are connected as defined by Figure 3.3. The conventions are, of course, highly arbitrary.

First, consider the triangle with sites $\vec{x}_0, \vec{x}_2$ and $\vec{x}_3$. We are free to choose coordinates such that

$$\vec{x}_0 = (0,0,0,0),\ \vec{x}_2 = (x_2,0,0,0),\ \vec{x}_3 = (x_3,y_3,0,0); \qquad x_2, y_3 > 0\ . \tag{3.19}$$

This defines an orientation of the coordinate frame. Elementary geometry, see Figure 3.4, gives

$$x_2 = l_2, \quad x_3 = \frac{l_3^2 + l_2^2 - l_8^2}{2l_2}, \quad y_3 = \sqrt{l_3^2 - x_3^2}\ . \tag{3.20}$$

By adding another point, $\vec{x}_4$, we obtain a tetrahedron. We choose

$$\vec{x}_4 = (x_4, y_4, z_4, 0) \quad \text{with} \quad z_4 > 0 \tag{3.21}$$

and have to calculate x_4, y_4 and z_4. From the triangle $(\vec{x}_0, \vec{x}_2, \vec{x}_4)$ we get

$$x_4 = \frac{l_4^2 + l_2^2 - l_9^2}{2l_2}\ . \tag{3.22}$$

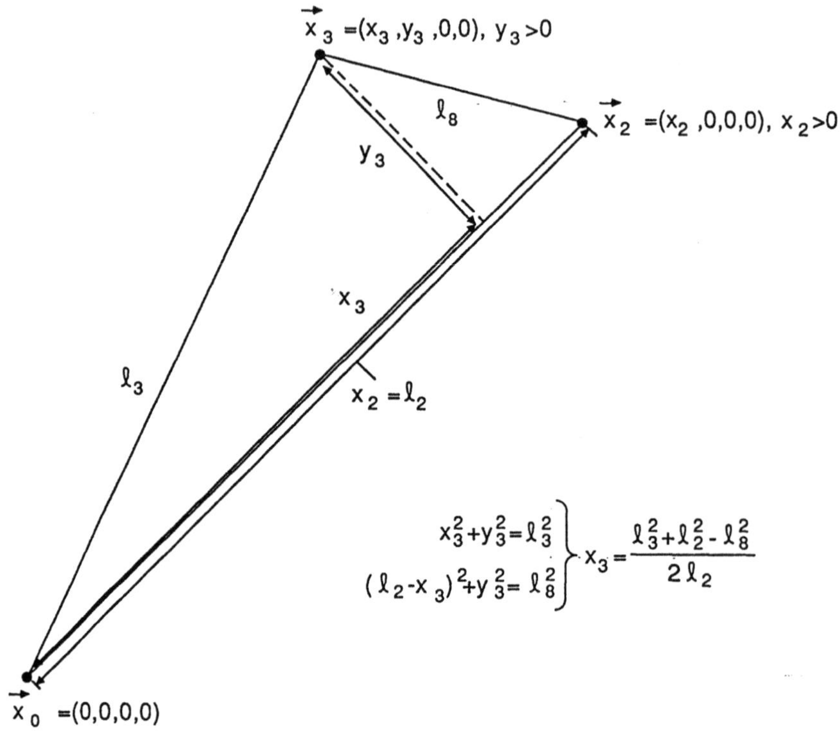

Fig. 3.4. Choice of coordinates and orientation for the first triangle.

To calculate y_4 and z_4 we regard the triangle $(\vec{x}_0, \vec{x}_3, \vec{x}_4)$ and define

$$a_4 = \frac{l_4^2 + l_3^2 - l_{10}^2}{2l_3} \quad \text{and} \quad \hat{a} = \frac{\vec{x}_3}{l_3} \ .$$

Using $a_4 = \vec{x}_4 \hat{a}$, $\hat{a} = (x_3/l_3)\hat{e}_x + (y_3/l_3)\hat{e}_y$ (with $\hat{e}_x$ and $\hat{e}_y$ unit vectors in x and y direction) and $y_4 = \hat{e}_y \vec{x}_4 = (l_3/y_3)(\hat{a}\vec{x}_4 - (x_3/l_3)x_4)$, we find

$$y_4 = \frac{l_3}{y_3}\left(a_4 - \frac{x_3}{l_3}x_4\right) \quad \text{and} \quad z_4 = \sqrt{l_4^2 - x_4^2 - y_4^2} \qquad (3.23)$$

Proceeding analogously we define the last coordinate point

$$\vec{x}_1 = (x_1, y_1, z_1, t_1) \quad \text{with} \quad t_1 > 0 \ , \qquad (3.24)$$

and consider the triangles $(\vec{x}_0, \vec{x}_2, \vec{x}_1)$ and $(\vec{x}_0, \vec{x}_3, \vec{x}_1)$. We obtain

$$x_1 = \frac{l_1^2 + l_2^2 - l_5^2}{2l_2} \ , \quad y_1 = \frac{l_3}{y_3}\left(a_1 - \frac{x_3}{l_3}x_1\right) \quad \text{with} \quad a_1 = \frac{l_1^2 + l_3^2 - l_6^2}{2l_3} \ . \qquad (3.25)$$

Finally we take the triangle $(\vec{x}_0, \vec{x}_4, \vec{x}_1)$ to define

$$b_1 = \frac{l_1^2 + l_4^2 - l_7^2}{2l_4} \quad \text{and} \quad \hat{b} = \frac{\vec{x}_4}{l_4} \ .$$

Using $b_1 = \vec{x}_1 \hat{b}$, $\hat{b} = (x_4/l_4)\hat{e}_x + (y_4/l_4)\hat{e}_y + (z_4/l_4)\hat{e}_z$, etc. gives

$$z_1 = \frac{l_4}{z_4}\left(b_1 - \frac{x_4}{l_4}x_1 - \frac{y_4}{l_4}y_1\right) \quad \text{and} \quad t_1 = \sqrt{l_1^2 - x_1^2 - y_1^2 - z_1^2}\,. \tag{3.26}$$

With help of these equation, step 1) of the MC procedure can easily be carried out. We choose the labelling such that the update corresponds to $l_1 \to l_1'$ in Figure 3.3. In our local coordinates the only coordinate point (vertex) that will be changed is is $\vec{x}_1 \to \vec{x}_1'$. Using equations (3.25) and (3.26) we calculate x_1', y_1' and z_1'. The proposed change is geometrically allowed, if

$$t_1'^2 = l_1'^2 - x_1'^2 - y_1'^2 - z_1'^2 > 0\,. \tag{3.27}$$

Otherwise the change is rejected. This procedure is sufficient to calculate the entropy of the Regge skeleton.

To simulate a gravitational action, the next step is to calculate the change of the action due to the link updating (3.18). With our local coordinate frames this is, in principle, straightforward. For each link we need a table that labels all triangles attached to it. Then, for each triangle we need a table that labels all pentahedra that contain it. By elementary geometry we calculate the angles cut out by each pentahedron in the plane perpendicular to the triangle. Typically this requires to calculate functions of the arccos-type. In this way we obtain the deficit angles as well as the areas. This procedure is mainly slow due to a considerable number of triangles attached to each single link and due to another large number of pentahedra that contain each triangle. Typically one ends up with the calculation of 240–480 arccos functions to update a single link.

To be concrete, let us discuss the Regge-Einstein action (2.4) with the scale-invariant measure as given by (4.3) and (4.4). We scan through the links of the lattice and propose new link lengths according to

$$l' = \exp(\varepsilon)\, l \quad \text{where} \quad \varepsilon \in [-a, a]. \tag{3.28}$$

In this way the scale invariant measure does not affect the detailed balance relation (3.9). For practical purposes the choice $a = 0.4$ has proven to be suitable. Of course, the new link is rejected if any of the pentahedral constraints are violated. Under the substitution (3.28) the action will change from S to S'. To obtain the correct value for S' the following procedure is advisable: a) Calculate the naive change of S, this means use the local coordinates and calculate the new areas and deficit angles for the triangles attached to the changed link; compare to the previous areas and deficit angles and denote the resulting new action by S_1'. b) Calculate the naive change in the volume, this means calculate the new volumes of the attached pentahedra and compare with the old volume. The result for the (total) new volume is denoted V_1'. Finally, the correct new action is obtained by rescaling:

$$S' = \lambda^2 S_1' \quad \text{where} \quad \lambda = \sqrt[4]{\frac{V}{V'}}\,. \tag{3.29}$$

S' is then compared to the previous action S and accepted or rejected according to the Metropolis algorithm (3.9). The unboundedness of the Regge-Einstein action is no obstacle to this procedure as all local changes are finite and well-defined. However, we would never reach equilibrium in case of an infinite partition function and expectation values would diverge as function of computer time.

4. A Simulation with the Regge-Einstein Action

This Section summarizes results of Ref. [13]. A four-dimensional model for discrete euclidean quantum gravity, based on the Regge-Einstein action (2.4) is investigated. The length scale is set by

$$V = \text{const}, \tag{4.1}$$

where V is the volume of the universe. More realistic models would introduce a second scale, for instance by coupling gravity to a Yang-Mills theory. The analog to the lattice spacing a in ordinary lattice gauge theory would be

$$l_0 = \sqrt[4]{v_0}, \tag{4.2}$$

where $v_0 = \langle v_\mathrm{p} \rangle$ is the expectation value for the volume of a single pentahedron. Due to $V = N_\mathrm{p} v_0$, where N_p is the total number of pentahedra, equation (4.1) implies that l_0 is constant configuration by configuration.

Two rather different scenarios are possible.

a) A fundamental length $l_0 = \text{finite} = O(l_\mathrm{P})$ exists.
b) A continuum limit $l_0 \to 0$ (in units l_P), similar as in ordinary quantum field theory, exists.

Attractive features of the fundamental length scenario have, in particular, been emphasized by T.D. Lee and collaborators [4,18]:

- Difficulties with locality at $l_\mathrm{P} \sim 10^{-33}$cm might be avoided.
- The theory might provide a natural ultraviolet cut-off. Renormalization of infinities may not be necessary.

Scenario a) is subject of this Section. In this scenario the Regge calculus, together with a random lattice [11] approach, may be of fundamental importance and it is notable that invariance under general coordinate transformations is a property of the discrete Regge-Einstein action. The more conventional scenario b) will be discussed in the next two Sections.

We are mainly interested in the question whether the discrete version of the partition function (2.8) exists for some non-zero Planck mass. To study this question numerically, we have to define a measure $D[\text{space}] = D[\{l\}]$. Let us consider a definite Regge skeleton. Scale transformations $l \to \lambda l$, $(\lambda > 0)$ leave the action invariant (in units l_0). This is an obvious consequence of the fact that the action scales with λ^2 due to the area of the triangles (the deficit angles are unchanged) and l_0 scales with λ and leads to the requirement that D[space] has to be invariant under the scale transformations $l \to \lambda l$. Otherwise the partition function Z is trivially ill-defined. For a sophisticated discussion of the measure see [24]. To find the (technically) simplest measure that is scale invariant we make the product ansatz

$$D[\text{space}] = \prod_l df(x_l) \prod_p F_\theta(x_p^1, \ldots, x_p^{10}). \tag{4.3}$$

In contrast to our previous notation we have denoted the link lengths by x_l (instead of l) to distinguish them clearly from the the link label l. The second product goes over all pentahedra, F_θ takes care of the pentahedral constraints and $x_p^1, \ldots, x_p^{10}$ are the ten links contained in the pentahedron p, see Figure 3.3. Our requirement of scale invariance of the measure $df(\lambda x_l) = df(x_l)$ immediately implies $f(\lambda x_l) = f(x_l) + f(\lambda)$ and therefore

$$f(x_l) = \text{const } \ln(x_l). \tag{4.4}$$

In principle a natural setting for the simulation would be the random lattice [11] and in a further step the the coordination number may become a random variable. Unfortunately such a skeleton is very difficult to deal with. In addition a single (rather small sized) random lattice would not exhibit the desired symmetry properties, but to sum over many such lattices would be required.

A more tractable possibility is to start with a hypercubical lattice, a lattice which is made topologically equivalent to the hypertorus by identifying opposite faces. Spurious finite size effects are avoided, since the boundary is formed by a replica of the same lattice. The number of edges can be easily increased, as the local incidence matrix is unchanged. Two such tesselations (model 1 and model 2) of the hypercubical lattice into pentahedra were investigated in Ref. [13] and the qualitative features turned out to be the same. They are defined in appendix B. Model 1 was first investigated by Roček and Williams [23] who showed that the lattice propagator, obtained from the second variation of the Regge action, agrees exactly with the continuum results in the weak field (small momentum) limit. Here, we consider the hypercubical model as a qualitative approximation to the random lattice. To get an impression about the accuracy that one might expect, we compare in Table 4.1 the average numbers of n-simplices per m-simplex for the random lattice with the corresponding numbers for a hypercubical lattice.

Numerical results are obtained by means of the MC procedure outlined in Section 3.3. It has to be remarked that the topology will not be changed by the allowed local fluctuations. However, it not even clear whether a summation over different topologies is desireable at all. In string theory there is a well-known example [25] where the partition function of such a summation does not exist, whereas it exists in a fixed topological sector (see for instance [26]). The present simulations [13] were done on rather small-sized 2^4 and 3^4 lattices. Vectorization on nowadays supercomputers would allow considerably larger lattices.

Table 4.1. Average number of n-simplices/m-simplex ($n > m$)

	0	1	2	3	4
0		30	150	240	120
1	37.8...		10	24	16
2	196.7...	10.4...		4.8	4.8
3	317.8...	25.2...	4.85...		2
4	158.9...	16.9...	4.85...	2	

Entries in the upper half plane are for the hypercubic model, whereas entries in the lower half plane are for the random lattice and are taken from Ref. [11]. (1,0)=(0,1), (2,0)=(0,2), (3,0)=(0,3) and (4,0)=(0,4) are the average numbers of links, triangles, tetrahedra and pentahedra per site, (2,1)=(1,2), (3,1)=(1,3) and (4,1)=(1,4) are the average numbers of triangles, tetrahedra and pentahedra per link etc.

Let us summarize the results now. First pure entropy

$$m_{\mathrm{P}}^2 = 0 \tag{4.5}$$

was investigated. Figure 4.1 exhibits how various quantities equilibrate after about 50 MC sweeps and Table 4.2 give the averages for various quantities of interest. Most notable is the negative expectation value for the scalar curvature (action density):

$$\langle S\rangle = \langle 2\alpha A_t\rangle = -0.560 \pm .006 \quad \text{(model 1 on a } 3^4 \text{ lattice).} \tag{4.6}$$

(Note that the present notation for the action S differs by a factor two from Ref. [13], see Appendix A.)

Table 4.2. Results [13] for entropy

	M1: 2^4	M1: 3^4	M2: 2^4	M2: 3^4
$\langle x_l\rangle$	2.8779 (17)	2.8651 (16)	2.8605 (16)	2.8557 (18)
$\langle A_t\rangle$	3.3814 (06)	3.3609 (23)	3.3571 (08)	3.3457 (11)
$\langle\alpha_t\rangle$	−0.0142 (6)	−0.0148 (4)	−0.0136 (6)	−0.0147 (6)
$\langle S\rangle = \langle 2\alpha_t A_t\rangle$	−0.566 (8)	−0.560 (6)	−0.676 (16)	−0.652 (20)

The hypercubical models 1 and 2 are marked by M1 and M2 respectively. Expectation values for various quantities and lattice sizes are given.

One question that has (as far as I know) no obvious solution is whether canonical or anomalous dimensions are realized, this means whether ratios of various dimensional quantities (average link length, average area of triangles,...) will stay constant (up to small finite size effects) under extension of the system size. For instance, it is well-known that the length of a random walk does not canonically scale proportional to the distance of its endpoints. In the present case the comparison of our 2^4 and 3^4 results clearly favours canonical dimensions. Finite size correction are even very small.

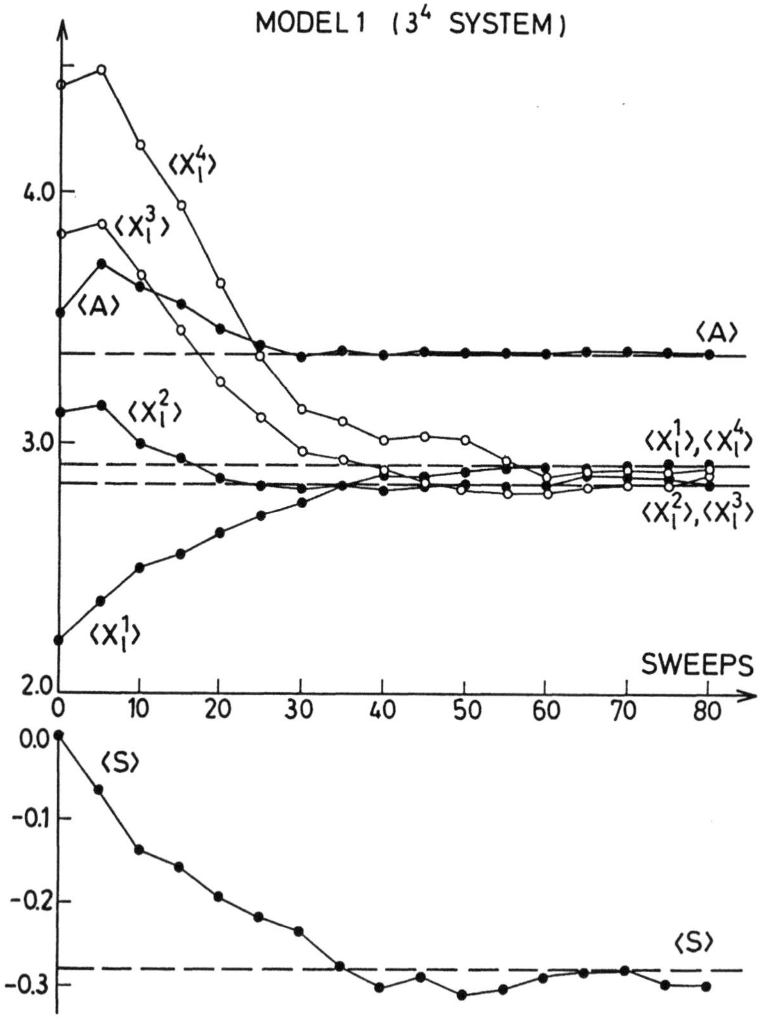

Fig. 4.1. First 80 sweeps, approach to equilibrium for various measured quantities (model 1, 3^4 system). Measurements are taken every sweep. Each dot represents an average over (the previous) five measurements. The solid lines are for guiding the eyes. Asymptotic averages are indicated by dashed, straight lines.

Next the question of a well-defined partition function for $m_{\mathrm{P}}^2 \neq 0$ was investigated on a 2^4 lattice. Figure 4.2 depicts the results of a long MC run ($\approx$ 60,000 sweeps) using various values for the Planck mass.
The data provide evidence for a first order phase transition that is tentatively estimated to happen in the range

$$0.02 < m_{\mathrm{P}}^2 < 0.04. \tag{4.7}$$

(Note that the present notation for the Planck mass squared differs by a factor two from Ref. [13], see Appendix A.) For a small enough Planck mass (in our

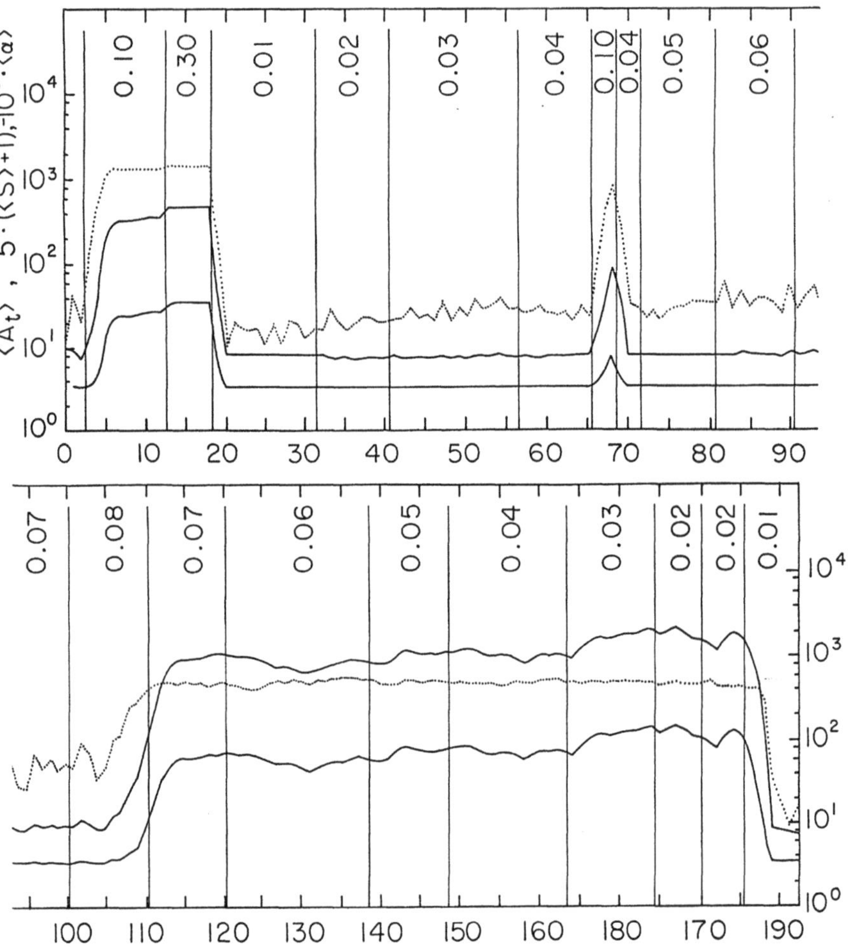

Fig. 4.2. m_P^2-dependence of various observables as defined in the text. The actual value of $2m_P^2$ is given in the upper part of the figure. For each observable 190 data points are connected by straight lines. Each data point is an average over 300 measurements. Data points are not drawn because of the limited resolution on the scale of this figure.

spurious units) we find an *entropy dominated phase* where the system fluctuates freely and all expectation values seem to be well-defined. Beyond a critical Planck mass energy (more precisely: euclidean action) dominates and the system ends up in various (irreproducible) metastable states. In this sense the energy dominated phase shows some similarities to spin glass models. Another remarkable property of the energy dominated phase is that, although the average action assumes a rather large positive value, the average deficit angle stays negative. In other words: the final action value is due to correlations between deficit angles and areas of triangles.

In conclusion, the partition function with the ordinary Regge-Einstein action seems to be well-defined for some range of the Planck mass m_P^2. This may be of importance for building a theory based on a fundamental length in connection with a random lattice. A crucial next question would be whether the entropy dominated phase may allow a Planck mass that is very large on the scale of usual hadronic masses. To address this question requires to introduce at least one further independent mass-scale. A simple way of doing this would be to investigate $SU(2)$ non-Abelian gauge theory on a Regge skeleton. One may then try to send the $SU(2)$ mass gap to zero while keeping the Planck mass fixed. In our approximation (without re-linking) the additional term to the action is known from the random lattice investigations [11] to be

$$S_{\mathrm{G}} = \frac{1}{g^2} \sum_{t_{ijk}} \lambda_{ijk} \mathrm{Tr}(1 - U_{ij} U_{jk} U_{ki}) \tag{4.8}$$

$$\text{with} \quad t_{ijk} \ \text{triangle} \quad \text{and} \quad \lambda_{ijk} = \frac{\text{area dual } t_{ijk}}{\text{area } t_{ijk}} \, .$$

A numerical investigation of such a theory is feasible as the additional updating would only approximately double the needed computer time. In such an approach D[space] and D[fields] would enter the functional integral in rather distinct, albeit aesthetically satisfactory, ways. Alternative approaches are outlined in the next two Sections.

5. Higher Derivative Gravity

Higher derivative gravity in terms of Regge's formulation has been discussed and simulated by Hamber and Williams [10]. This Section summarizes their results. The study of higher derivative gravity theories dates back more than thirty years, for a historical review see [27].

Within the framework of continuum local field theories one attempt to quantize gravity is by including in the action those terms that are generated by renormalization and to see whether the resulting action leads to a tractable and perhaps meaningful quantum theory. It turns out that two additional terms, involving fourth derivatives of the metric, need to be added to the Einstein action in order to obtain a bounded and perturbatively renormalizable theory [28]. Thus one is led to consider the extended gravitational action

$$S = m_{\mathrm{P}}^2 \int d^4x \, \sqrt{g} \left[R - l_{\mathrm{P}}^2 \left(\lambda + \frac{a}{4} R_{\mu\nu\rho\sigma} R^{\mu\nu\rho\sigma} + \frac{4b - a}{12} R^2 \right) \right] ,$$

with a cosmological constant term (proportional to λ) and two higher higher derivative terms (dimensionless coupling constants a^{-1} and b^{-1}) in addition to the usual Einstein term. Perturbation theory is usually performed around flat space, which requires $\lambda = 0$. One sets $g_{\mu\nu} = \eta_{\nu\mu} + h_{\mu\nu}$, where $\eta_{\nu\mu}$ is the usual Minkowski metric, and expands in powers of $h_{\mu\nu}$. If λ is nonzero, one has to expand around a solution of the classical equations of motion for higher derivative gravity with a λ term. Such a solution will be no longer constant over space time.

Similarly as for non-Abelian gauge theories the resulting theory can be made asymptotically free and is thus ultraviolet stable [29] (once the coupling constants are chosen with the right sign). The measure can play a delicate role in cancelling spurious divergences in loop diagrams and it has been argued that the quartic divergence for λ is absent to all orders of perturbation theory for an appropriate choice of the measure [29]. The unique local gauge invariant measure is then

$$du(g) = \prod_x g^{-(d+1)/2} \prod_{\mu \geq \nu} dg_{\mu\nu} \tag{5.1}$$

which is also scale invariant [30].

Unfortunately, it appears that the theory has potential problems with unitarity. The tree-level graviton propagator exhibits a massive ghost pole. Another severe problem is to achieve a natural explanation for the smallness of the cosmological constant,

$$\lambda < 10^{-122} m_{\mathrm{P}}^4, \quad \text{(i.e. in units of the Planck mass).}$$

It may be doubted that such questions can be conclusively answered within the context of weak coupling perturbation theory. Non-perturbative methods, similarly as for non-Abelian gauge theories may be needed. It is in this spirit that the authors of Ref. [10] investigate the discrete formulation of quantum gravity.

By the translation

$$\int d^4x \sqrt{g} \rightarrow \sum_{\text{pentahedra}} V_{\text{pentahedron}} = \sum_{\text{triangles } t} V_t$$

one obtains the lattice version of a cosmological constant term. Here $V_{\text{pentahedron}}$ is the volume of the corresponding simplex and and V_t is the volume associated with each triangle. V_t may be defined in a variety of ways, for instance using the dual lattice or using a baricentric subdivision. The latter is used for the numerical work reported in this Section. In the continuum limit, if it exists, a particular choice for V_t should become irrelevant. Similarly other terms relevant for higher derivative gravity have been obtained.

The full action of Ref. [10] is

$$\begin{aligned} S[l] &= m_{\mathrm{P}}^2 \sum_{\text{triangles } t} \left[2\alpha_t A_t - l_{\mathrm{P}}^2 \left(\lambda V_t + 4b \frac{A_t^2 \alpha_t^2}{V_t} \right) \right] \\ &- \frac{1}{3}(a-4b) \sum_{\text{sites } s} V_s \sum_{\text{triangles } t,t' \supset s} \epsilon_{t,t'} \left[\omega_{s,t} \frac{A_t \alpha_t}{V_t} - \omega_{s,t'} \frac{A_{t'} \alpha_{t'}}{V_{t'}} \right]^2 . \end{aligned} \tag{5.2}$$

Here $\epsilon_{t,t'}$ is equal to 1 if the two triangles t, t' have one edge in common and equal to -2 if they do not. The weight factors $\omega_{s,t}$ are defined by

$$\omega_{s,t} = \frac{V_t}{N_{s/t} V_s}$$

where $N_{s/t}$ is the number of sites per triangle. The action (5.2) is the analogue of the continuum action

$$S = m_{\mathrm{P}}^2 \int d^4x \sqrt{g} \left[R - l_{\mathrm{P}}^2 \left(\lambda + b R_{\mu\nu\rho\sigma} R^{\mu\nu\rho\sigma} + \frac{1}{2}(a - 4b) C_{\mu\nu\rho\sigma} C^{\mu\nu\rho\sigma} \right) \right],$$

with

$$C_{\mu\nu\rho\sigma} C^{\mu\nu\rho\sigma} = R_{\mu\nu\rho\sigma} R^{\mu\nu\rho\sigma} - 2 R_{\mu\nu} R^{\mu\nu} + \frac{1}{3} R^2 .$$

For their simulation Hamber and Williams use the measure

$$\int d\mu_\epsilon[l] = \prod_i \int_0^\infty \frac{dl_i^2}{l_i^2} F_\epsilon[l]. \tag{5.3}$$

Here $F_\epsilon[l]$ is the function of the edge lengths, that is equal to one when all simplicial inequalities are satisfied and, in addition, all links satisfy $|l| > \epsilon$. Otherwise $F_\epsilon[l]$ is zero. $F_{\epsilon=0}$ is identical with the previously (2.8,4.3) introduced F_θ. In the present context the measure (5.3) is attractive, since it is local and scale invariant as the continuum measure [30], and integrates directly over the elementary lattice degrees of freedom the conformal factor $\Phi_i = \ln(l_i/l_0)$. However, it is not identical to the continuum measure (5.1) and simplicity arguments remain a major reason for this choice.

As the aim is to obtain results for an asymptotically free field theory, a continuum limit has to be approached. It may be taken by requiring smoothness on the scale of the local lattice spacing

$$\left| \frac{A_t \alpha_t}{V_t} \right| \ll \frac{1}{\sqrt{V_t}} \quad \text{or equivalently} \quad \frac{A_t^2 \alpha_t^2}{V_t} \ll 1 .$$

All lengths can be expressed in terms of the unit length scale

$$l_0 = \sqrt{\frac{m_{\mathrm{P}}^2}{\lambda}} \sim 10^{61} l_{\mathrm{P}} , \tag{5.4}$$

then the partition function Z depends only on the dimensionless couplings m_{P}^4/λ, a and b. This rather unnatural equation illustrates, again, the problems with the smallness of the cosmological constant.

The numerical results of Ref. [10] were not obtained by means of the previously described MC method, but by using a time-discretized Langevin evolution equation

$$l_s^{(k+1)} = l_s^{(k)} \exp\left(-\varepsilon \frac{\delta S[l]}{\delta l_s^{(k)}} + \sqrt{2\varepsilon}\, \eta_s^{(k)} \right). \tag{5.5}$$

The stepsize ε was chosen between 10^{-3} and 10^{-4}, and the gaussian white noise satisfies

$$\left\langle \eta_p^{(k)} \right\rangle = 0, \quad \left\langle \eta_p^{(k)} \eta_q^{(l)} \right\rangle = \delta_{pq} \delta^{kl}.$$

Of course, $l_p^{(k+1)} = l_p^{(k)}$ whenever the pentahedral inequalities are violated.

For the simulation periodic N^4 lattices with $N = 2$ (240 links) and $N = 4$ (3840 links) were considered.

First, all coupling constants were set to zero and (as in part of Section 4) the entropy of the measure was investigated. To compute the averages, 1000 passes were performed on the 2^4 lattice, discarding the first 500. Expectation values for squared links $\sqrt{\langle l^2 \rangle}$ are (in naive units that assign link length 1 to a hypercube edge):

$$1.52,\ 1.50,\ 1.49 \text{ and } 1.50 \text{ where the initial link length are : } 1,\ \sqrt{2},\ \sqrt{3} \text{ and } 2. \tag{5.6}$$

This is consistent ($\sqrt{\langle l^2 \rangle} > \langle l \rangle$) with the results [13] reported in Section 4). Unfortunately, a direct comparison is presently not possible as $\langle l^2 \rangle$ was not measured in [13] and no results for $\langle l \rangle$ are reported in [10].

Other quantities measured in [10] are an average curvature $\mathcal{R}$ and an average curvature squared $\mathcal{R}^2$

$$\mathcal{R} = \langle l^2 \rangle \frac{\langle 2 \sum_t \alpha_t A_t \rangle}{\langle \sum_t V_t \rangle} \quad \text{and} \quad \mathcal{R}^2 = \langle l^2 \rangle^2 \frac{\langle 4 \sum_t \alpha_t^2 A_t^2 / V_t \rangle}{\langle \sum_t V_t \rangle}.$$

Also measured were the average deficit angle squared $\langle \alpha_t^2 \rangle$ and the average volume of a 4-simplex. For pure entropy the results are given in Table 5.1.

Table 5.1. Results [10] for entropy

$50\langle V_t \rangle / \langle l^2 \rangle^2$	$\mathcal{R}$	$\mathcal{R}^2$	$\langle \alpha_t^2 \rangle$
0.32 (2)	−10.36 (1.20)	58200 (600)	3.52 (9)

A measure of the "roughness" of space is given by $\mathcal{R}/\sqrt{\mathcal{R}^2} = 0.042 \pm 0.008$.

For various values of the coupling constants the numerical results are now summarized in Tables 5.2a and 5.2b. All these data are taken with $2m_{\mathrm{P}}^2 = 1$.

Table 5.2a. Results [10] with $m_{\mathrm{P}}^2 = 1,\ a = 4b = 0.005$

λ	$50\langle V_t \rangle / \langle l^2 \rangle^2$	$\mathcal{R}$	$\mathcal{R}^2$	$\langle \alpha_t^2 \rangle$
2.0	0.32 (2)	−0.612 (61)	8900 (220)	0.64 (12)
1.5	0.32 (2)	−0.420 (42)	8660 (260)	0.62 (11)
1.0	0.08 (2)	132 (22)	134000 (13600)	1.65 (30)
0.5	0.11 (2)	123 (14)	143000 (11600)	2.45 (33)

Table 5.2b. Results [10] with $m_P^2 = 1,\ a = 4b = m_P^2/2\lambda$

λ	$50\langle V_t\rangle/\langle l^2\rangle^2$	$\mathcal{R}$	$\mathcal{R}^2$	$\langle\alpha_t^2\rangle$
1.5	0.24 (4)	−0.044 (17)	496 (48)	0.021 (5)
1.0	0.18 (4)	−0.031 (44)	432 (45)	0.013 (3)
0.5	0.17 (3)	−0.030 (48)	208 (64)	0.006 (2)

Table 5.2a uses the rather small values $b = 0.00125$, $a = 4b$ and Table 5.2b relies on $b = m_P^2/8\lambda$, $a = 4b$. The values for the cosmological constant λ are given in the tables. The averages were obtained by doing 2000 passes each on the 2^4 lattice, discarding the first 1000 passes in the final averaging. Results on the 4^4 lattice turned out to be consistent, but rather inconclusive due to too low statistics. Table 5.2a indicates a first order phase transition between

$$\lambda = 1.0 \quad \text{and} \quad \lambda = 1.5\ . \tag{5.7}$$

Qualitatively this is consistent with the result (4.7).
The probability distribution of the curvature shows that the large $\mathcal{R}$ expectation value (at $\lambda = 0,5$) comes from long tails over a wide range in positive as well as negative directions. Altogether the speculative phase diagram in the $(m_P^2 - \lambda)$-plane is as indicated in Figure 5.1. This **assumes** that $b = 0.00125$ and $b = 0$ would give similar results.

6. De Sitter Lattice Gravity

The purpose of this Section is to review numerical results by Caracciolo and Pelissetto [8]. I closely follow their second paper and reproduce also parts of Ref. [7].

The de Sitter algebra generates the group of isomorphisms of spacetime with constant curvature (de Sitter space). The parameter l_U is the radius of curvature and will be identified with the radius of the universe. A formulation of gravity as a gauge theory of the de Sitter group, i.e. $SO(5)$ in the euclidean framework, has been proposed by MacDowell and Mansouri [31] (for a nice review see [7]). They introduce the connection

$$\rho_\mu(x) = \frac{1}{2} J_{ab}\omega_\mu^{ab} + P_a\tau_\mu^a$$

with $\mu = 1,\ldots,4$ and $a, b = 1,\ldots,4$, where ω_μ^{ab} is the spin connection and τ_μ^a the vierbeins. J_{ab} and P_a are generators of $SO(5)$. The action of the model is given by

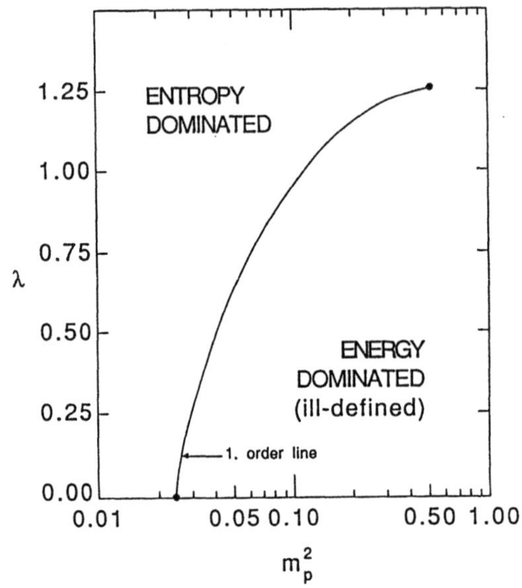

Fig. 5.1. Speculative phase diagram in the (m_{P}^2-λ)-plane. (For convenience a log-scale is used for m_{P}^2.)

$$S = \left(\frac{l_{\mathrm{U}}}{l_{\mathrm{P}}}\right)^2 \int d^4x \; \epsilon_{abcd} \, R^{ab}_{\mu\nu} R^{cd}_{\mu\nu} \tag{6.1}$$

where

$$R^{ab}_{\mu\nu} = \partial_\mu \omega^{ab}_\nu - \partial_\nu \omega^{ab}_\mu + \omega^{ac}_\mu \omega^{cb}_\nu - \omega^{ac}_\nu \omega^{cb}_\mu - \tau^a_\mu \tau^b_\nu + \tau^a_\nu \tau^b_\mu$$

is the curvature of the $SO(5)$ algebra, while

$$\hat{R}^{ab}_{\mu\nu} = R^{ab}_{\mu\nu} + \tau^a_\mu \tau^b_\nu - \tau^a_\nu \tau^b_\mu$$

is the usual curvature appearing in general relativity. It may be a little surprising that we have passed from a theory with two dimensional coupling constants, namely l_{P} and l_{U}, to a theory of one dimensionless coupling constant $g \sim 1/\beta \sim (l_{\mathrm{P}}/l_{\mathrm{U}})^2$. But actually this is the case. General relativity, with a cosmological constant, and not coupled to matter, is a function only of the dimensionless coupling g. In the MacDowell-Mansouri formalism the parameter l_{U} is introduced to make the correspondence between a dimensional gauge field and the metric, which is dimensionless and the cosmological constant can always be scaled into the metric such that the resulting theory is a function of g. This has been accomplished at the expense of attributing to the vierbein dimensions of a mass, and hence the inverse vierbein acquires the unconventional dimension of inverse mass. As a result of the presence of a field with a dimensionality inverse to the usual canonical the normal power counting argument breaks down and the fact that the coupling constant is dimensionless is of no help for the non-renormalizable gravity.

The translation of this theory on a hypercubic lattice has been given by Smolin [7]. This is very similar to Wilson's [5] regularization of non-Abelian gauge theories. With each link of the lattice on associates an element of the $SO(5)$ group

$$U_\mu = \exp\left(\frac{a}{2} J_{bc}\omega_\mu^{bc} + aP_c\tau_\mu^c\right) .$$

The ordered product of the link elements around an elementary plaquette is given by

$$U_{\mu\nu}^{AB}(n) = [U_\mu(n)U_\nu(n+\hat{\mu})^T U_\mu(n+\hat{\nu})^T U_\nu(n)]^{AB}$$

with $A, B = 1,\ldots,5$ $SO(5)$ indices and n denotes the vertex of an appropriate corner of the plaquette. The lattice action is

$$S = \frac{\beta}{4}\sum_{n\in Z^4}\sum_{\mu\nu\rho\sigma}{}' \epsilon^{\mu\nu\sigma\rho}\epsilon_{ABCD5}U_{\mu\nu}^{AB}(n)U_{\sigma\rho}^{CD}(n), \qquad \frac{\beta}{4} = \left(\frac{l_{\mathrm{U}}}{l_{\mathrm{P}}}\right)^2 . \tag{6.2}$$

Here $\sum'$ sums also planes in negative $\mu\nu$-directions in such a way that the action becomes compatible with the cubic invariance of the lattice. In the classical continuum limit, $a \to 0$, the action (6.1) is recovered. The continuum action is not invariant under $SO(5)$, but only under the $SO(4)$ subalgebra generated by the J_{ab}'s. The invariance under the P_a's is realized in a non-algebraic way through diffeomorphisms [31]. On the lattice the $SO(5)$ invariance is exactly maintained. But the reparametrization invariance is lost and one hopes to recover it in the (quantum) continuum limit. This problem is similar to the one with rotational invariance in lattice gauge theory.

In order to construct a quantum theory for the action (6.2) one has to specify the integration measure. Many proposals have been advanced for the continuum [19] and analogous measures can also be introduced on the lattice [32]. For the purposes of the numerical simulation the authors of Ref. [8] stay with the simplest choice, namely the $SO(5)$ Hurwitz [33] measure, although a more sophisticated measure may be relevant for the recovery of reparametrization invariance. Moreover, it has been shown [16] that a factor $sign|\det(\tau)|$ should be multiplied to the Lagrangian in order to achieve reflection positivity [12]. However, the calculation of the determinant would be rather computer time consuming and the hope of the reported simulation is that within the simplified model one may already acquire experience on the qualitative behaviour of the phases of the theory. The increased computer power of Class VII vector processors may allow to include the determinant as well as effects of a more realistic measure into the MC simulation.

As we have now defined the theory, we also have to introduce the lattice analogues of relevant observables. In [16] the quantity

$$T_\mu^b(n) = -\frac{1}{4a}\,\mathrm{Tr}\,[HP^b U_\mu(n)H^T U_\mu(n)]\; s_\mu \quad \text{where} \quad \begin{cases} +1 \text{ if } \mu \text{ is positive} \\ -1 \text{ otherwise} \end{cases}$$

was defined, with $H = \mathrm{diag}(1,1,1,1,-1)$. Its formal continuum limit is the vierbein and it has the advantage to be connected to a single site as far as the transformation properties are concerned. Under a gauge transformation one has

$$T_\mu^b(n) \rightarrow V_c^b(n) T_\mu^c(n).$$

From this quantity a gauge invariant matrix is built

$$G_{\mu\nu}(n) = \frac{1}{4}(T_\mu^a(n) + T_{-\mu}^a(n))(T_\nu^b(n) + T_{-\nu}^b(n))\delta_{ab}, \tag{6.3}$$

where we have symmetrized with respect to positive and negative directions in order to obtain a tensor with the correct transformations under the hypercubic group.

For $\beta \ll 1$ the theory allows a strong coupling expansion similar to that of ordinary lattice gauge theory. For $|\beta|$ sufficiently small the series is absolutely convergent and as consequence there is exponential clustering and a mass gap. For some Wilson loop expectation values the series has been worked out in the lowest few orders. This serves mainly for checking the correctness of the computer program.

Due to the compactness of the gauge group the lattice action is bounded. Therefore we can ask the question what are the configurations which realize the maxima of the action. In the weak coupling limit $\beta \rightarrow \infty$ they are the relevant configurations. The investigation of the weak coupling limit is somewhat involved as the model turns out to be frustrated, using the terminology of spin systems with antiferromagnetic interactions. The result is that for an Z_N^4 lattice with $N = even$ solutions with maximal action per site $S_{\max}/4! = 32$ can be found and the relevant link matrices are of form

$$\begin{pmatrix} z & 0 \\ 0 & n \end{pmatrix} \quad \text{with} \quad z \in O(4) \quad \text{and} \quad n = \pm 1.$$

For $n = +1$ we have $z \in SO(4)$, but maximum action solutions with $U_\mu^{55} = -1$ on closed hypersurfaces are also possible.

Now we discuss the numerical results of Ref. [8]. The Metropolis algorithm was used on lattices of size 4^4 and 8^4. Figure 6.1 reports the mean values of $S/4!$ per site at different β-values on the 4^4 lattice. The simulation starts with a random configuration at $\beta = 0.04$, after 1000 iterations β is increased by 0.01 and so on. Average values were taken after discarding 200 values to allow thermalization at the new value of β. When the system had settled in the weak coupling regime, β was decreased in steps of 0.01 until the system had come back to the strong coupling phase. Different experiments gave rise to different hysteresis behaviour. In all cases but one the system followed the behaviour corresponding to curve (a), while in one case the curve was (b), approaching the value predicted for the weak coupling regime. The system shows therefore at least three states and undergoes at least one phase transition. The large hysteresis is indicative for a first order transition. A measurement of 2×2 Wilson loops gave similar results. Figure 6.2 reports the average values on the same cycles for the metric $\frac{1}{4}\sum_{\mu=1}^{4} G_{\mu\mu}(n)$.

The different phases can be characterized by local order order parameters. The first order-parameter is $U_\mu^{55}(n)$ which should approach the values ± 1 and

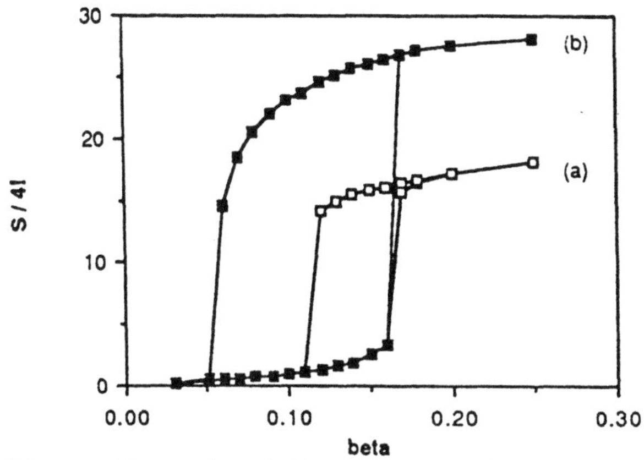

Fig. 6.1. Hysteresis cycle for the mean of $S/4!$ per site. Curve (a) refers to a cycle in which the system goes through a metastable state. Curve (b) refers, instead, to the cycle in which the cold state is close to the weak coupling result.

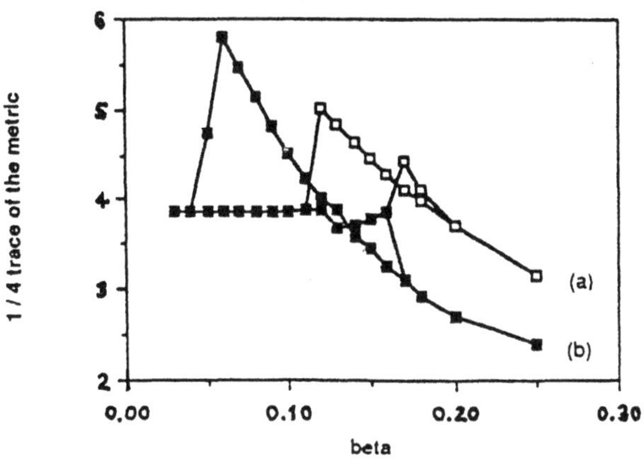

Fig. 6.2. Hysteresis cycle for the mean of 1/4 of the metric. The meaning of the labels (a) and (b) is as in Figure 6.1.

show freezing $SO(5) \rightarrow O(4)$ in the low-temperature regime. In the hot region U_μ^{55} is widely distributed in the interval $[-1, 1]$, symmetrically around its most probable value zero, see Figure 6.3a.
In the lower-action cold state the link variables try to belong to the $O(4)$ subgroup. The values $U_\mu^{55} = \pm 1$ are the most probable ones and occur with the same frequence, see Figure 6.3b.

In the observed higher-action cold state there is a complete ordering into $SO(4)$, see Figure 6.3.c.

A second order parameter is

$$P_{\mu\nu}(n) = \frac{\epsilon_{abcd} U_{\mu\nu}^{ab}(n) U_{\mu\nu}^{cd}(n)}{4 U_{\mu\nu}^{ab}(n) U_{\mu\nu}^{ba}(n)}$$

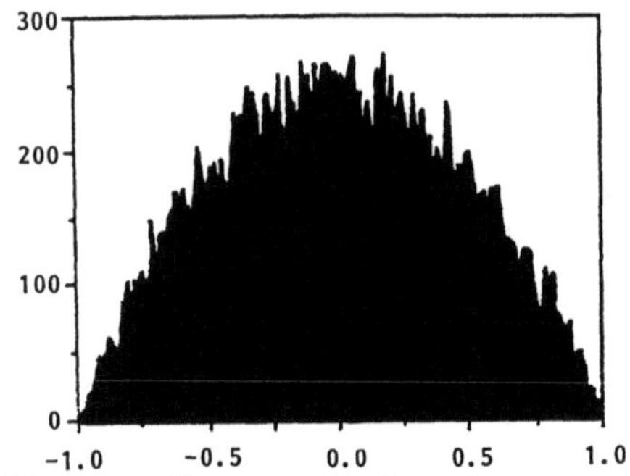

Fig. 6.3a. Histogram for the order parameter $U_\mu^{55}(x)$ mesured on a single configuration on a 8^4 lattice for $\beta = 0.06$.

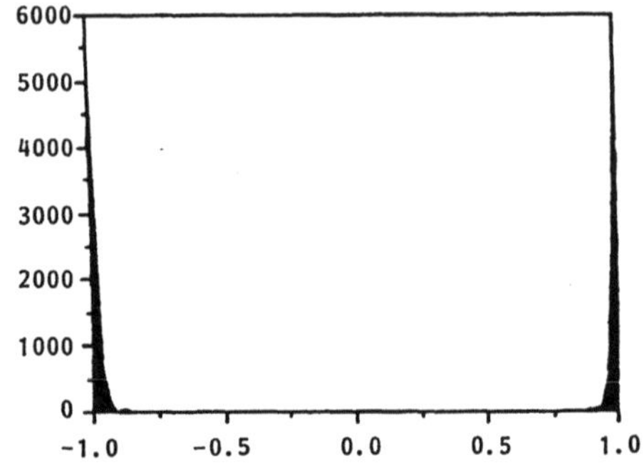

Fig. 6.3b. Histogram for the order parameter $U_\mu^{55}(x)$ mesured on a single configuration on a 8^4 lattice for the lower action cold state at $\beta = 2.0$.

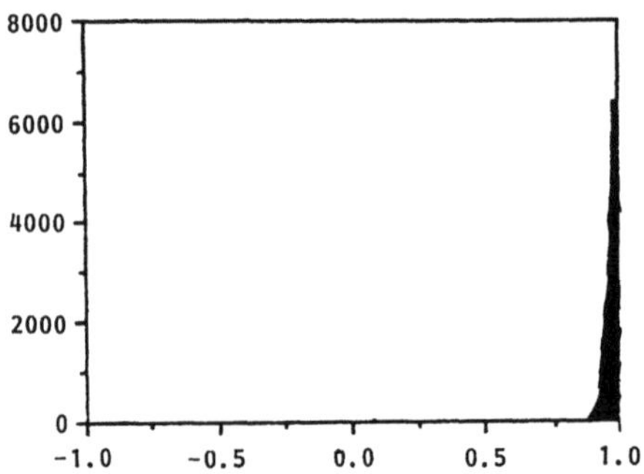

Fig. 6.3c. Histogram for the order parameter $U_\mu^{55}(x)$ mesured on a single configuration on a 8^4 lattice for the higher action cold state at $\beta = 2.0$.

and gives information about the topology of the space-time. The distribution of $P_{\mu\nu}$ is well peaked at zero in the hot phase, Figure 6.4. In both cold states it is peaked around ± 1. On a rather speculative level one may interpret the discrepancy as evidence for a highly degenerate vacuum with non-zero entropy and such a scenario would be reminiscent of the space-time foam picture by Wheeler and Hawking [34,15].

In order to discuss the relative stability of the different phases and the order of the transitions, the so called mixed phase method [35] was used. It puts into contact two half-systems in different phases. For several β-values these runs were performed on a 8^4 lattice. Figure 6.5 gives evidence for a first-order transition that occurs at

$$\beta_c = 0.08 \pm 0.01 \ . \tag{6.3}$$

At the most critical value the two subsystems do not mix at all because of a large latent heat. Possibly a second first order phase transition may happen around $\beta_c \approx 0.12$, but the evidence is not conclusive.

In conclusion, the main limitation of the model, as shown by the MC simulation, is that there is no way to define a continuum limit at β_c. The transition(s) which the system undergoes is (are) of the discontinuous type. This is amazingly similar to the result found by simulating a simplified random-lattice model, discussed in Section 4, where the original Regge-Einstein action is used. We should however keep in mind that the interpretational philosophy of these two approaches is quite different.

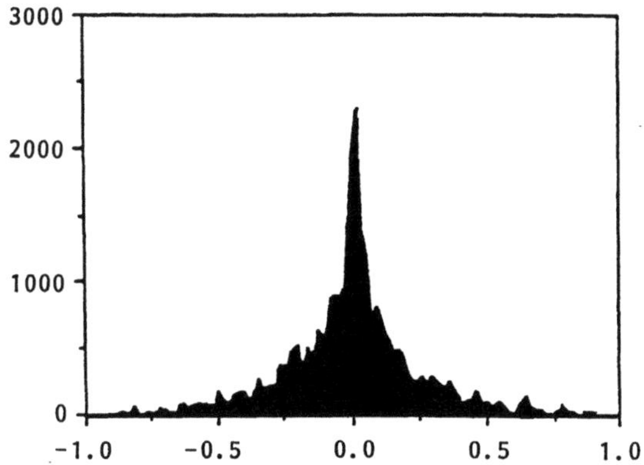

Fig. 6.4. Histogramm for the order parameter $P_{\mu\nu}(x)$ measured on a single configuration on a 8^4 lattice for $\beta = 0.06$.

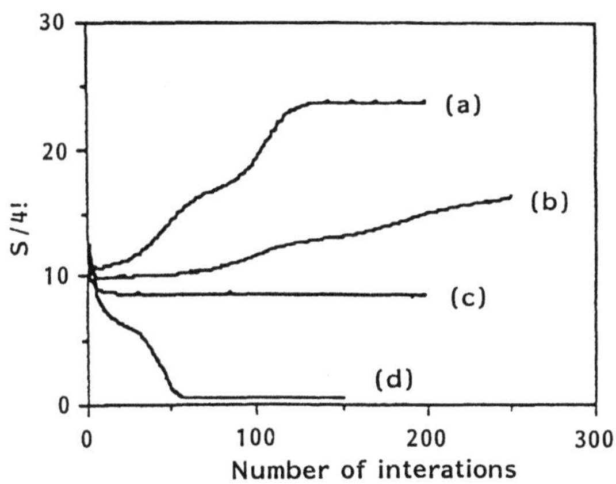

Fig. 6.5. MC time evolution for the mean value of $S/4!$ per site on a 8^4 lattice for a starting configuration which mixes the hot phases with a maximal action configuration for different values of β. (a) refers to $\beta = 0.10$, (b) refers to $\beta = 0.09$, (c) refers to $\beta = 0.08$ and (d) refers to $\beta = 0.07$.

7. Conclusions and Outlook

Simulations of mathematical models lead us to new physical as well as mathematical insight that may be relevant for the theory of quantum gravity. The major shortcoming at the present state of the art is, of course, that comparisons with experimentally measurable quantities seem to be out of reach. However, any new non-trivial information about complicated systems (like the investigated models) has normally turned out to be useful and sooner or later found its application in science. Therefore, the usefulness of such investigations might not just be limited to gravity. Spin glasses [36] exhibit amazing similarities to some features of the simulated models of quantum gravity. The phase structure of the various models is certainly interesting in its own and a new generation of simulations may clarify issues like for instance the phase diagram of Figure 5.1. Beyond, to study the coupled Yang-Mill-gravity model (4.8) seems to be worthwhile. Concerning the conventional approach [7,8]: such investigations may turn to models that are more realistic than the one presented in Section 5.

Acknowledgments. I would like to thank Prof. H. Mitter and the organizing committee for the kind hospitality extended to me at this pleasant winterschool at Schladming. Further I am indebted to Christian Lang for his encouragement to give these lectures.

Appendix A: Some Parameters and Constants

In this appendix we list some bare parameters.

a) Conventions in this review:

$l_{\mathrm{P}} = \sqrt{\frac{\hbar G}{c^3}} = 1.6 \times 10^{-33}\mathrm{cm}$: Planck length.

$m_{\mathrm{P}} = l_{\mathrm{P}}^{-1} = 1.2 \times 10^{19}\mathrm{GeV}$: Planck mass.

l_{U} : radius of the universe.

b) Conventions of the literature:

$\kappa = l_{\mathrm{P}}$: Planck length (Smolin [7]; Caracciolo and Pelissetto [8]).

$l = l_{\mathrm{U}}$: radius of the universe (Smolin [7]; Caracciolo and Pelisetto [8]).

$\kappa = \frac{1}{16\pi G}$, where G is the bare Newton constant (Hamber and Williams [10]).

$\mu = \sqrt{\kappa}$: Planck mass, $(\frac{1}{\sqrt{G}} = 1.2 \times 10^{19}\mathrm{GeV})$, (Hamber and Williams [10]).

m_{P}^2 (this review) $= m_{\mathrm{P}}/2$ (Berg [13]): Planck mass squared.

S (this review) $= 2S$ (Berg [13]): Regge-Einstein action.

Appendix B: Basic Geometry

Standard notation is established by Coxeter's [37] book *Regular Polytopes*.

a) Simplices

Examples of simplices are sites (0-simplices), links (1-simplices), triangles (2-simplices), tetrahedra (3-simplices) and pentahedra (4-simplices).

(n–1)-simplex: Dimension: $d = n - 1$; n vertices (sites).

Let $k \in \{1, \ldots, n\}$. The $(n-1)$–simplex contains $\binom{n}{k}$ $(k-1)$–simplices.

This means:

1) # Sites : $\binom{n}{1} = n$ (obvious). Example, pentahedron : 5.

2) # Links : $\binom{n}{2} = \frac{n(n-1)}{2}$. Example, pentahedron : 10.

The first site is connected to the $(n-1)$ others sites, the second site is connected to the $(n-2)$ remaining sites etc. . This gives $(n-1)+(n-2)+\ldots+1 = \frac{n(n-1)}{2}$ links. Of course this is the number of ways to choose two different objects out of n.

3) # Triangles : $\binom{n}{3}$. Example, pentahedron : 10.

4) # Tetrahedra : $\binom{n}{4}$, #pentahedra : $\binom{n}{5}$, etc. .

b) The d-Dimensional Hypercube

By Z_n^d we denote the hypercubic N^d lattice with periodic boundary conditions. The relevant properties of the d-dimensional hypercube are summarized in the following. By # simplices we denote how often a certain kind of simplex is contained in the d-dimensional hypercube.

Sites: The sites of the hypercube may be labelled by the coordinates

$$(k_1, k_2, \ldots, k_d) \text{ with } k_i \in Z_2 = \{0, 1\}. \tag{B.1}$$

Therefore, # sites: 2^d.

Links: At each site (0,0,...,0) there are d positive directions, hence d links in these directions. By symmetry, we have d links attached at each site of the hypercube. Each link connects two sites.

Therefore, # links: $d\ 2^{d-1}$.

The general case, k-dimensional hypercubes:. ($k \leq d$)

$$\# \; k - \text{dimensional hypercubes} : \binom{d}{k} \; 2^{d-k}.$$

Proof: At each site there are $\binom{d}{k}$ ways to pick out k directions out of d. Each k dimensional cube connects 2^k sites. We allow $k \in \{0, 1, \ldots, d\}$. For instance,

$$k = 0 : \; \# \text{ Sites} : \binom{d}{0} 2^d \text{ (see above).}$$

$$k = 1 : \; \# \text{ Links} : \binom{d}{1} 2^{d-1} \text{ (see above).}$$

$$k = 2 : \; \# \text{ Plaquettes} : \binom{d}{2} 2^{d-2}.$$

$$k = 3 : \; \# \text{ Cubes } (= 3 - \text{dimensional hypercube}) : \binom{d}{3} 2^{d-3}$$

Examples :

Cube (= 3-dim. hypercube)	4-dimensional hypercube
$2^3 = 8$ sites	$2^4 = 16$ sites
$\binom{3}{1} 2^2 = 12$ links	$\binom{4}{1} 2^3 = 32$ links
$\binom{3}{2} 2^1 = 6$ plaquettes	$\binom{4}{2} 2^2 = 24$ plaquettes
	$\binom{4}{3} 2^1 = 8$ cubes

c) Tesselations of Hypercubes

We outline two simple methods to decompose the d-dimensional hypercube into a number of d-simplices. These decompositions define model 1 and model 2 in Section 4.

Model 1: This construction relies on drawing appropriate diagonals into the 2, 3,..., d-dimensional hypercubes. The coordinates of the 2^d sites of the d-dimensional hypercube are given by equation (B.1). We will proceed by induction. In two dimensions the square plaquette is cut into two triangles (2-simplices) by adding the link $l = \{(0,0),(1,1)\}$. Let us consider the d-dimensional hypercubical lattice. The boundary of the d-dimensional hypercube consists of $2d$ $(d-1)$-dimensional hypercubes and d of the $(d-1)$-dimensional hypercubes contain the vertex (0,...,0). (The other d of the $(d-1)$-dimensional hypercubes contain the vertex (1,...,1).) We assume that each of the $(d-1)$-dimensional hypercubes is partitioned into $(d-1)!$ $(d-1)$-simplices. The assumption is obviously true for the 3–dimensional case (where the $(d-1)$-dimensional hypercubes are plaquettes). Let us regard the d $(d-1)$-dimensional hypercubes containing the vertex (0,...,0). Out of each $(d-1)$-simplex contained in one of these hypercubes we construct a d-simplex by by adding the site (1,...,1) to the $(d-1)$-simplex. In that way we obtain a partition of the d–dimensional cube into $d(d-1)! = d!$ d-simplices. Obviously, these simplices fill up the whole d–dimensional hypercube. Result: We have obtained a tesselation of the d–dimensional cube into $d!$ d-simplices.

Remark: It is easy to see that all these d-simplices are isomorphic. Therefore, each has the volume $v = 1/d!$.

Example: 4-dimensional hypercubical lattice.

Each hypercube is partitioned into 24 pentahedra. Further we find: 84 tetrahedra, 110 triangles and 65 links contained in each hypercube. In our MC procedure we have to update links. We have to avoid double counting as links are typically contained in several hypercubes. As there is a one-to-one correspondence between hypercubes and sites, all different cases are covered by only considering links connected to the site (0,0,0,0). There are 15 such links:

- 1-4: Four links along an edge of the hypercube (each contained in $2^3 = 8$ hypercubes). Link length 1.
- 5-10: Six links along a diagonal of a plaquette of the hypercube (each contained in $2^2 = 4$ hypercubes); link length $\sqrt{2}$.
- 11-14: Four links along a 3–dimensional cube which is part of the boundary of the hypercube (each of these links is contained in two hypercubes); link length $\sqrt{3}$.
- 15: The link $l = \{(0,0,0,0),(1,1,1,1)\}$ along the hypercube diagonal (contained in precisely one hypercube); link length $2 = \sqrt{4}$.

All links contained in a hypercube sum up to $4 \cdot 8 + 6 \cdot 4 + 4 \cdot 2 + 1 = 65$.

Model 2:. This model is obtained by omitting the hypercube diagonal of the previous construction and adding a *new vertex* (site) in the center of the hypercube. 2^d new links are added by connecting the new site with all the corners of the hypercube. In four dimensions we end up with 30 (instead of 15) links to update per hypercube. The first 14 links are identical with the previous construction and 16 new links connect to the additional vertex. The new links are of length 1.

For both models it is now rather easy to generate a table that specifies the triangles that contain a certain link. Similarly we need another table specifying the pentahedra that contain a certain triangle.

References

1. A. Einstein, Grundzüge der Relativitätstheorie, Vieweg & Sohn, Braunschweig 1965
2. S. Weinberg, Gravitation and Cosmology: Principles and Applications of the General Theory of Relativity, John Wiley & Sons, New York 1972
3. M. Green, Europhysics News, 18 (1987) 14
4. T.D. Lee, Difference equations as the Basis of Fundamental Physical Theories, symposium in honor of G.C. Wick, Pisa, Italy, October, 1984
5. K. Wilson, Phys. Rev. D10 (1974) 2445
6. C. Rebbi (editor), Lattice Gauge Theories and Monte Carlo Simulations, World Scientific, Singapore 1983
7. L. Smolin, Nucl. Phys. B148 (1979) 333
8. S. Caracciolo and A. Pelisetto, Phys. Lett. B193 (1987) 237 and Preprint, Pisa, SNS 6/87
9. T. Regge, Nuovo Cimento 19 (1961) 558
10. H. Hamber and R. Williams, Phys. Lett. 157B (1985) 368 and Nucl. Phys. B269 (1986) 712
11. N.H. Christ, R. Friedberg and T.D. Lee, Nucl. Phys. B202 (1982) 89; B210 [FS6] (1982) 337
12. K. Osterwalder and R. Schrader, Commun. Math. Phys. 31 (1973) 83 and 42 (1975) 281; K. Osterwalder and E. Seiler, Ann. Phys. (NY) 110 (1978) 440
13. B. Berg, Phys. Rev. Lett. 55 (1985) 904 and Phys. Lett. 176B (1986) 39
14. Proceedings of the 1987 Seillac conference on lattice gauge theory: contributions by J. Ambjørn, A. Krzywicki and A. Migdal.
15. S. Hawking, in General Relativity - an Einstein Centenary Survey, S. Hawking and W. Israel (editors), Cambridge University Press 1979
16. P. Menotti and A. Pelissetto, Phys. Rev. D35 (1987) 1194
17. G. 't Hooft, in Recent Developments in Gravitation, Cargèse Lecture notes (1978) and references given therein
18. R. Friedberg and T.D. Lee, Nucl. Phys. B242 (1984) 145; G. Feinberg, R. Friedberg, T.D. Lee and H.C. Ren, Nucl. Phys. B245 (1984) 343
19. H. Leutwyler, Phys. Rev. 134 (1964) 1155B; E. Fradkin and G. Vilkovisky, Phys. Rev. D8 (1973) 424; L. Faddeev and V. Popov, Sov. Phys. Usp. 16 (1974) 777; V. De Alfaro, S. Fubini and G. Furlan, Nuovo Cim. 57B (1980) 227 and 76A (1983) 365; K. Fujikawa, Nucl. Phys. B226 (1983) 437
20. S.N. Metropolis, A.W. Rosenbluth, A.H. Teller and E. Teller, J. Chem. Phys. 21 (1953) 1087
21. K. Binder (editor), Applications of the Monte Carlo Method in Statistical Physics, Springer-Verlag 1984
22. M. Creutz, Quarks and Gluons on a Lattice, Cambridge University Press, 1983
23. M. Roček and R. Williams, Z. Phys. C21 (1984) 171
24. H. Cheeger, W. Müller and R. Schrader, Commun. Math. Phys. 92 (1984) 405
25. D. Weingarten, Phys. Lett. 90B (1980) 280; T. Eguchi and H. Kawai, Phys. Lett. B110 (1982) 143
26. B. Baumann and B. Berg, Phys. Lett. 164B (1985) 131
27. N.H. Barth and S.M. Christensen, Phys. Rev. D28 (1983) 1876
28. K.S. Stelle, Phys. Rev. D16 (1977) 953
29. E. Fradkin and A.A. Tseytlin, Nucl. Phys. B201 (1982) 469
30. N.P. Konopleva and V.N. Popov, Gauge Fields, Harwood Academic, New York 1979

31. S.W. MacDowell and F. Mansouri, Phys. Rev. Lett. 38 (1977) 739 and 38 (1977) 1375 (E)
32. P. Menotti and A. Pelisetto, Nucl. Phys. B288 (1987) 813
33. E. Wigner, Group Theory and its Applications to the Quantum Mechanics of Atomic Spectra, Academic Press, New York 1959;
A. Hurwitz, Nachr. Gött. Ges. Wissensch. (1897) p.71
34. J.A. Wheeler, in Relativity, Groups and Topology, B.S. and C.M. DeWitt (editors), Gordon and Breach, New York 1963
35. M. Creutz, L. Jacobs and C. Rebbi, Phys. Rev. D20 (1979) 1915
36. G. Parisi, in Unification of the Fundamental Particle Interactions, S. Ferrara, J. Ellis and P. van Nieuwenhizen (editors), Plenum Press, New York 1980
37. H. Coxeter, Regular Polytopes, Dover Publications, New York 1973

This article was processed by the author using the TEX Macropackage from Springer-Verlag.

Neutrino Oscillations and Their Damping

Leo Stodolsky

Max-Planck-Institut für Physik und Astrophysik
Werner-Heisenberg-Institut für Physik
P.O.Box 40 12 12, D-8000 München, Fed. Rep. of Germany

The neutrino oscillations of which you have been hearing a lot about may seem a bit mysterious, especially to those not accustomed to the problem of particle oscillations. Therefore, I would like to present an easy way of seeing what they're about.

We have to do with a phenomenon of linear combinations of two states, quite analogous to the two states of a photon, left and right circularly polarized, or the two states of a spin, up or down. Quantum mechanically all the various kinds of two state systems are equivalent, so all we really have to understand is one and we have understood them all.

1. Particle Oscillations

We will take, then, a spin (1/2) as our model, since its behavior is easy to visualize under various influences. One begins describing a spin by naming the two basis states. For an ordinary spin they are up and down, usually refered to the z axis. For neutrinos the two basis states are ν_e, the particle emitted in neutron decay, and ν_μ the particle emitted in $\pi \rightarrow \mu\nu_\mu$. We use the word "particle " loosely, not necessarily implying thereby that the object in question is an eigenstate of propagation in vacuum, a state of definite mass.

The fact that the neutrino is emitted either as ν_e or ν_μ means that in this picture they are emitted either with "spin" up or spin down. In matrix notation we have

$$|\nu_e\rangle = \begin{pmatrix} 1 \\ 0 \end{pmatrix}$$

$$|\nu_\mu\rangle = \begin{pmatrix} 0 \\ 1 \end{pmatrix}$$

This "spin" is of course not the usual spin. It is in a theoretical two-dimensional space called "flavor space".

Next, an ordinary spin (taking it to have a magnetic moment) can be influenced by an external magnetic field. In general, the spin will tend to precess around the magnetic field direction. If we describe the spin by a polarization vector $\boldsymbol{P}$ the orientation of $\boldsymbol{P}$ sweeps out a cone in space (Fig. 1). There are,

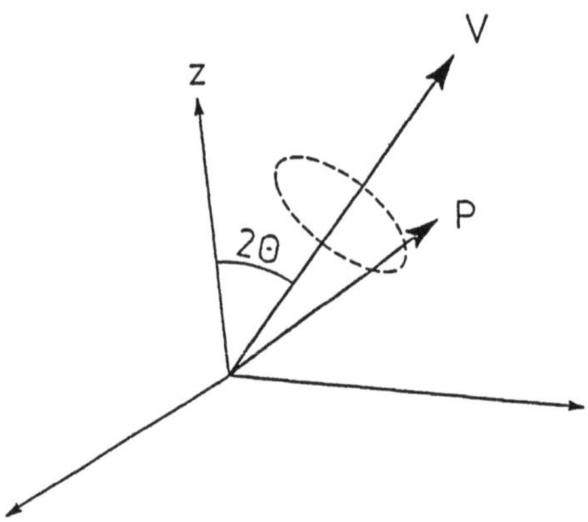

Fig. 1. Relationship of $\boldsymbol{P}$ and $\boldsymbol{V}$ in the absence of damping. The z or flavor axis is vertical so that $\boldsymbol{P}$ "up" represents a ν_e and $\boldsymbol{P}$ "down" a ν_μ. $\boldsymbol{P}$, representing the density matrix, precesses around $\boldsymbol{V}$ which represents the influence of the mass mixing matrix and the index of refraction of the medium.

however, two stationary states or eigenstates for the spin; the states where the spin points up or down with respect to the magnetic field $\boldsymbol{B}$. The eigenstates have different energy , which leads to the precession when we start the system in a state which is not either parallel or anti-parallel to the field.

This situation is described by the Schroedinger equation, $i\dot{\psi} = H\psi$ where the ψ is a two component wavefunction, and

$$H = \boldsymbol{B} \cdot \boldsymbol{\sigma},$$

with $\boldsymbol{\sigma}$ the Pauli matrices. We explicitly reserve the possibility that the field $\boldsymbol{B}$ is time dependent in direction and magnitude, in which case the spin in general will undergo various complicated precessions and wobblings.

Now, just as the spin may not be in a stationary state, even with a constant field, so in general the neutrino state at emission will not be in a stationary state. That is, there may be something playing the role of the magnetic field, which will cause a precession in flavor space. This something in particle physics is a possible "mass matrix" for the neutrino, tending to "mix" the two flavor states. Since any 2×2 matrix can be expanded in terms of the Pauli matrices, the Schroedinger equation for the flavor space is governed by a Hamiltonian

$$H = \boldsymbol{\sigma} \cdot \boldsymbol{V}$$

with $\boldsymbol{V}$ the analogy to the magnetic field. A possible constant unit matrix has been absorbed into the definition of H. The vector $\boldsymbol{V}$ is given as follows: If the two neutrino states of definite mass have masses m_1, m_2, the length given by

$$V = (m_1^2 - m_2^2)/2E$$

where E is the neutrino energy. If these two states are given by

$$cos\theta \; |\nu_e\rangle + sin\theta \; |\nu_\mu\rangle$$

and

$$cos\theta\ |\nu_\mu\rangle - sin\theta|\ \nu_e\rangle$$

the angle that $\boldsymbol{V}$ makes with the z axis is 2θ:

$$\text{angle}(V - z) = 2\theta$$

We now study the polarization vector $\boldsymbol{P}$ describing the flavor spin state of the neutrinos. If the neutrino is in a pure state, that is, described by a single wavefunction quantum mechanically, the three components of $\boldsymbol{P}$ are the expectation values of the corresponding $\boldsymbol{\sigma}$. If the system is not in a pure state it is an average over such expectation values. By definition, when the neutrino is emitted its polarization vector $\boldsymbol{P}$ in flavor space points up or down along the z axis. As time goes on $\boldsymbol{P}$ will precess around $\boldsymbol{V}$, leading to a projection along the opposite direction of z, in other words, a component of the other flavor. The precession rate is given by the length of $\boldsymbol{V}$, the mass difference squared. The 1/E factor in the precession rate is a time dilatation factor since the neutrino is moving with nearly the speed of light. If $\boldsymbol{V}$ is big, the precession is rapid, and if θ is big, a large component of the opposite flavor develops.

The equation governing the time development of $\boldsymbol{P}$ is

$$\frac{d\boldsymbol{P}}{dt} = \boldsymbol{V} \times \boldsymbol{P}$$

2. Effects of a Medium

So far our discussion has been for propagation in vacuum. If the propagation is in a medium, the neutrino will have an index of refraction, implying a further contribution to $\boldsymbol{V}$.

If the interaction with the medium is flavor conserving, which to good accuracy seems to be the case, the new contribution to $\boldsymbol{V}$ must be along the z axis, and its magnitude can be calculated in the standard electroweak theory in terms of the densities of the left-handed particles with electron number and muon number [1]:

$$V^{\text{med}} = V_z^{\text{med}} = -G\sqrt{2}(N_e^L - 1/2N_{\nu_e}^L) - G\sqrt{2}(N_\mu^L - 1/2N_{\nu_\mu}^L)$$

An important point to notice here is that the medium induced $\boldsymbol{V}$, which in the spin analogy plays the role of an additional applied magnetic field, varies with density of the medium.

We are now in the position to understand subtle effects, like that of Michaeev and Smirnov [2] for the solar neutrino problem. We need only to visualize the behavior of a spin precessing around a magnetic field. If the density of the medium seen by the neutrino varies as it passes through some medium, say the sun, we have a time varying "magnetic field".

In general this is a complicated problem without any general explicit solution [3]. But various special cases are easy to understand.

First, obviously, there is the case of a uniform medium. Here we just add the constant $\boldsymbol{V}^{\text{med}}$ to the vacuum $\boldsymbol{V}$ to get a new constant vector. If $\boldsymbol{V}^{\text{med}}$ is much bigger than the vacum $\boldsymbol{V}$, we get a big resultant vector essentially along the z axis, hardly any precession, and the result that the medium tends to suppress mixing [4].

3. Adiabatic Case

An important situation with a varying medium is the "adiabatic" case where the external field varies slowly, that is, where the density of the material through which the neutrino is passing varies slowly. In this case we have the phenomenon of gyroscopic stability: if the applied field is changing slowly compared to the precession rate of the spin, then the spin will tend to follow the applied field, and stay lined up with it. In quantum mechanical language, if the Fourier components associated with the varying of the applied field are small compared to the energy splitting of the two instantaneous spin eigenstates, then the spin will stay in the eigenstates.

Say that we start with spin up and the external field also pointing roughly up. If the field begins to vary slowly and gradually swings around so that it ends up pointing down, without having passed through zero along the way (i.e. it always has a non-zero horizontal component), then the spin will follow the field and also end up pointing down. In flavor space such a procedure turns a ν_e into a ν_μ. This is the Michaeev-Smirnov trick. The slowly varying field is given by the slowly varying density of the sun as the neutrino comes out from the center, and the relevant parameters need only be in the right broad range of values so that the "spin" goes from up to down.

Note, however, that it is important that the field varies from up to down and not vice versa. This requirement turns out to mean that the ν_μ -like neutrino should be heavier than the ν_e -like neutrino, which is fine since it accords with our prejudices.

By using the above formulas, the known weak interaction constants and the density of the sun, it turns out that typical values of the neutrino mixing parameters for which the trick works are mixing angles like 10^{-3} and mass squared differences of 10^{-4}eV^2, values which seem quite possible.

It is also clear from this picture that we can get Michaeev-Smirnov effects for other problems than just simple $\nu_e - \nu_\mu$ mixing. All that is necessary is a vacuum V and a medium V which point in opposite directions and that one goes smoothly over into the other. One, may, for example, have the mixing of the ordinary neutrino with a "sterile" neutrino .

We can also guess what happens under other conditions. If we depart from the adiabatic case, say if the density varies too fast, we can see that the spin will not be able to follow and will not be inverted, instead undergoing some complicated gyrations.

Of course, for detailed answers one can solve the problem numerically by computer, and for a V, which goes smoothly from one value to another, so that it can be approximated by the tanh-function, Nötzold [3] has shown how to find analytic answers.

4. Damping of Particle Oscillations

Thus far our polarization vector $\boldsymbol{P}$, as it moves, maintains a constant length. If we generalize our problem beyond that of the treatment of pure states, to those including statistical mixtures, there is the possiblility that the vector $\boldsymbol{P}$ shrinks. In the problem of the propagation of a beam, as for neutrinos leaving the sun or a reactor, it is usually sufficient just to think of the neutrino wavefunction, i.e pure states. But in problems where we are concerned with the development of a whole statistical ensemble, as say in the early universe or in a supernova, we must deal with the density matrix of the neutrinos, and if the state is not describable by a single wavefunction, P will be less than one.

In that case, we find [1] that the equation describing the precession of the spin is generalized to

$$\frac{d\boldsymbol{P}}{dt} = \boldsymbol{V} \times \boldsymbol{P} - D\boldsymbol{P}_T$$

where the new term on the right leads to a shrinking of $\boldsymbol{P}$. The index on $\boldsymbol{P}$ means transverse to the flavor (z) axis and the "damping parameter" D can be calculated from the basic scattering amplitudes. It represents, roughly speaking, the effect of collisions with the medium in breaking the coherence of the wavefunction which is necessary to oscillations, and is on the order of the scattering rate of the particle on the medium. The solutions to this equation can lead to various interesting effects, like inhibition or shifting of the oscillations. A full analysis of all these interesting possibilities still remains to be done.

Acknowledgements. I would like to thank R. Golub for a discussion of neutron physics suggesting the "spin" analogy.

Further Reading

For some of the other topics covered in the lectures see my reviews in:

- Neutrinos, Dark Matter and Cryogenic Detection in MPI-PAE/PTh 2/88.
- A Unified View of the Macro- and the Micro Cosmos, A. De Rújula, D.V. Nanopoulos and P.A. Shaver; eds. World Scientific, Singapore (1988).
- Comments on Nuclear and Particle Physics 18 (1988) 157–173.
- Low Temperature Detectors for Neutrinos and Dark Matter, K. Pretzl, N. Schmitz and L. Stodolsky; eds. Springer-Verlag (1988).

References

1. L. Stodolsky, Phys. Rev. D 36 (1987) 2273.
2. S.P. Michaeev and A.Yu. Smirnoff, Nuovo Cim. C9 (1986) 17;
   H.A. Bethe, Phys. Rev. Lett. 56 (1986) 1305, for detailed studies see
   W. Haxton, Phys. Rev. D 35 (1984) 2352.
3. D. Nötzold, Phys. Rev. D 36 (1987) 1625.
4. L. Wolfenstein, Phys. Rev. D 20 (1979) 2634.

This article was processed by the author using the TEX Macropackage from Springer-Verlag.

Phase Transitions as the Origin of Large Scale Structure in the Universe

Neil Turok [1]

NASA/Fermilab Astrophysics Center
Fermi National Accelerator Laboratory, Batavia, Illinois 60510, USA

Abstract

A review of the formation of large scale structure through gravitational growth of primordial perturbations is given. This is followed by a discussion of how symmetry breaking phase transitions in the early universe might have produced the required perturbations, in particular through the formation and evolution of a network of cosmic strings.

1. Very Large Scale Structure and Gravity

The standard big bang model of the early universe is both simple and successful. It assumes that at early times the universe was a homogeneous and isotropic thermal bath of all the particle species seen today. From this and Einstein's equations it correctly predicts the Hubble expansion, microwave background radiation and the abundances of the light elements (for a review, see [1]).

Nevertheless it is obviously only a first approximation – the universe is very clearly not homogeneous and isotropic today and so there must have been some inhomogeneity at early times. In fact, as we shall see in this Section, a number of recent observations point to the existence of structure in the universe on scales much larger than the distances matter has moved (under gravity) since the big bang. This is exciting – the new very large scale observations are giving us a more or less direct 'window' on the primordial fluctuations.

Where could the large scale structure in the universe have come from?

A beautiful idea has emerged from fundamental particle physics in the last few years. It is based on the notion of spontaneous symmetry breaking, which has emerged as fundamental in unified theories of particle interactions. What could be simpler than that the mechanism which broke the symmetry of particle interactions also broke the spatial symmetry of the universe?

---

1 Address after September 1st 1988: Joseph Henry Laboratory, Princeton University, Princeton NJ 08544, USA

This happens through the dynamics of the symmetry breaking process, as I shall discuss in Section 2. At very high temperatures the symmetry is unbroken but as the universe cools and expands a phase transition occurs, forming defects as the universe goes out of thermal equlibrium which survive to much later epochs and seed the formation of structure.

One particular class of defects shows particularly interesting behavior in an expanding universe. Cosmic strings [2] are linelike defects formed in symmetry breaking phase transitions, and have the property that they evolve into a 'scaling solution', where the strings remain a fixed fraction of the total density throughout the radiation dominated era. This results in a highly predictive theory of galaxy formation, an alternative to the more popular theory based on quantum fluctuations produced during a period of inflation. I will concentrate on cosmic strings here but will also compare and contrast the two scenarios in Section 2.

Of course we still do not know what the right fundamental theory is. Some grand unified theories predict strings and others do not. Over the last few years there has been increased hope that superstring theories may offer an 'ultimate' unified theory, including gravity. Some 'superstring-inspired' models also predict cosmic strings, but so far these are so contrived that it is hard to take them seriously.

1.1 Observations and Motivations

Cosmology is one of the most rapidly growing areas in physics for two very good reasons.

First the observational situation is qualitatively better than just a few years ago and improving all the time. Advances in telescope technology have made large scale complete surveys of the three dimensional galaxy distribution feasible. There are also now very stringent and improving limits on the microwave background anisotropy and gravity wave background, which put strong constraints on the magnitude of the primordial inhomogeneities – typically the magnitude of the density fluctuation must have been less than a part in ten thousand.

What makes the large scale observations interesting is that the distribution of galaxies is very clearly not random on very large scales. Early three dimensional surveys revealed giant 'filaments' (roughly linear overdense regions in the galaxy distribution) about $200h_{50}^{-1}$ Mpc long and $10h_{50}^{-1}$ Mpc across [3], large 'voids' $120h_{50}^{-1}$ Mpc in diameter [4]. A complete 'slice of the universe' out to $300h_{50}^{-1}$ Mpc shows the galaxy distribution to be 'frothy', i.e. many galaxies lying on the surfaces of almost empty 'bubbles' $40-60h_{50}^{-1}$ Mpc across [5], while deep 'needle' surveys out to further than $2000h_{50}^{-1}$ Mpc have shown a shell-like distribution of galaxies on a scale of as much as $200h_{50}^{-1}$ Mpc [11]. The rich clusters of galaxies, called Abell clusters [12], are (roughly) defined observationally to be regions $3h_{50}^{-1}$ Mpc in radius containing more than 50 bright galaxies. These seem to be significantly clustered on scales of at least $200h_{50}^{-1}$ Mpc [13]. For comparison the present Hubble radius H_0^{-1} is $6000h_{50}^{-1}$ Mpc, the mean galaxy separation (i.e. the inverse cube root of the number density) is $10h_{50}^{-1}$ Mpc [14] and the mean Abell cluster separation is $110h_{50}^{-1}$ Mpc [13]. h_{50} is of course Hubble's constant H_0 in units of $50\,\mathrm{km\,s^{-1}}$, observationally $1 < h_{50} < 2$.

Statistics to describe these structures are still being developed – the simplest is the 2-point correlation function $\xi(r)$, defined to be the excess probability over random of finding a galaxy at a distance r from another galaxy [7]. There are different ideas relating the primordial fluctuations to the final $\xi(r)$ we see today.

We now believe that most of the mass in the universe is dark. There is no strong reason why the shining matter distribution should exactly trace the dark matter distribution. What processes could end up making the matter in one part of the universe shine and leave another part dark? One popular idea is that of *threshold biasing* – that shining matter might only be formed at the *peaks* of the underlying matter distribution [21, 22]. With Gaussian noise, the peaks are more correlated than the underlying distribution, and this is used crucially to get the standard inflation plus CDM model (which I shall call the ICDM model) to agree with the galaxy-galaxy correlation function and pair velocity distribution. However this idea, at least as applied in the ICDM model, does not appear to work particularly well as far as fitting large scale structure [79].

Another idea, and one that is at least partly realised in the cosmic string scenario, is that the distribution of galaxies and clusters may reflect a perturbation distribution with a *scale invariant* correlation function, i.e. objects of mean separation d have a correlation function $\xi(r/d)$ with ξ the same function for all objects. This is consistent with the observational fact that clusters poorer than Abell clusters [16], the richer Abell clusters [13] and even superclusters [15] have clustering apparently consistent with a single scale invariant correlation function. In fact ξ is approximately equal to $0.2(r/d)^{-2}$. Written the same way galaxies appear slightly *more* correlated: $\xi_{gg} \approx (r/d)^{-2}$ [18], consistent with the view that gravity has enhanced the correlations on small scales.

Statistics on these observations are still not good and the ideas for explaining them clearly preliminary, but the observations are improving all the time and becoming increasingly constraining. Clearly data like this are going to play a crucial role in deciding which theories of the origin of large scale structure are viable.

The fact that the distribution of galaxies does not necessarily reflect the distribution of the mass (and in particular the dark matter) in a simple way represents a profound problem for matching cosmological theories to observation. Which is why recent determinations of the actual peculiar velocity field (the velocities of galaxies relative to the overall Hubble flow) are of such great importance. These seem to indicate a very large scale flow field centred on a region ('the great attractor') $80h_{50}^{-1}$ Mpc from us with magnitude about $500\,\mathrm{km\,s^{-1}}$ at our galaxy [17]. I will discuss below how this velocity field is related, in the linear regime, to the mass perturbation. If these observations are correct, they practically rule out the ICDM model.

So what if there is large scale structure in the universe? The most striking thing about it is that matter could not have moved these distances under gravity since the big bang. In the linear regime (and on these scales the structure is not yet nonlinear) there is a precise relation between the distance a particle has moved relative to the Hubble flow since the big bang δr and its present peculiar velocity δv: $\delta r = H_0^{-1}\delta v$ (see below). Now galaxies are very rarely observed to have velocities of greater than $600\,\mathrm{km\,s^{-1}} = 2 \cdot 10^{-3}c$ relative to the observed structures. This means they can have moved no more than $12h_{50}^{-1}$Mpc since the big bang. So when we see much larger structures we are seeing the primordial density perturbation field itself! This is an encouraging thought.

The second good reason for a growing interest in cosmology is that only in the last few years have any well-defined ideas as to the *origin* of large scale structure come forward. In particular cosmic strings and quantum fluctuations in inflationary models are well defined theories with clear predictions. For example, the quantum fluctuation scenario with hot dark matter is close to being ruled out by its excessively large predicted anisotropy in the microwave background [19].

1.2 Dark Matter

The greatest single obstacle to the interpretation of the observations of large scale structure is the by now firmly established fact that most of the matter in the universe is dark (for a nice review, see [9]). The shining matter – gas and stars – is only the 'tip of the iceberg'. The best evidence for dark matter comes from observations of the rotation curves of spiral galaxies [20]. To keep a particle in orbit around a galaxy requires an acceleration of v^2/r where v and r are the orbital velocity and radius. Assuming that this is provided by the gravitational pull of the mass $M_{<r}$ inside the orbit one finds $M_{<r} = (v^2 r)/G$. Observations indicate that v^2 remains constant with increasing r out to distances several times the optical radius of a galaxy. Thus most of the mass is dark. Estimates of the mass of apparently virialised clusters of galaxies confirm this – and extrapolating the observed mass to light ratios in clusters to the whole universe leads to an estimate of Ω typically between 0.1 and 0.3. This may well be an underestimate, particularly if galaxies form preferentially in denser regions like clusters, making the mass to light ratio larger outside clusters.

Nucleosynthesis calculations [8] in fact indicate that even the *baryons* are mostly dark! They yield the bounds $.04 < \Omega_{baryons} h_{50}^2 < 0.14$ whereas the observed $\Omega_{luminous}$, i.e. the density in stars and gas is less than 0.02. So we can't even see the baryons!

However things are worse than this, according to most theorists. There is a strong prejudice that $\Omega = 1$. This is because in a matter or radiation dominated universe $\Omega = 1$ is an unstable fixed point of Einstein's equation. If Ω deviates from 1 then the deviation grows rapidly in time. Given that Ω is today fairly close to unity it would seem crazy that we should apparently be here precisely around the time when it started to deviate from unity. Inflation 'solves' this problem (called the flatness problem) – during a period of inflation $\Omega = 1$ becomes an *attractor* in Einstein's equation so that in a typical region of the universe which underwent inflation Ω would still be very close to 1.

1.3 Newtonian Approximation

For many purposes the Newtonian approximation to Einstein's equation is entirely adequate. In particular it is a good approximation when dealing with scales (a) far inside the horizon so velocities are small and (b) where gravity is fairly weak so the dimensionless gravitational potential is much less than 1. It is also a lot simpler than full relativistic perturbation theory and brings out the main points about the growth of density perturbations under gravity.

So let us begin with Newton's equations. A particle at 'physical' coordinate $\boldsymbol{r}(t)$ obeys

$$\ddot{\boldsymbol{r}} = -\boldsymbol{\nabla}_{\boldsymbol{r}}\Phi \tag{1}$$

where the Newtonian potential Φ is given by

$$\boldsymbol{\nabla}_{\boldsymbol{r}}^2\Phi = 4\pi G\rho \tag{2}$$

with ρ the density. Before discussing perturbations, let us describe the homogeneous matter dominated background solution to (1) and (2). We set $\rho = \rho_b$, independent of $\boldsymbol{r}$, and find in spherical coordinates about some point chosen as the origin (where $r = \Phi = 0$),

$$\begin{aligned}\Phi &= \frac{2\pi G\rho_b r^2}{3} \\ \ddot{\boldsymbol{r}} &= -\frac{4\pi G\rho_b \boldsymbol{r}}{3} \\ \rho_b r^3 &= const\end{aligned} \tag{3}$$

the last equation being matter conservation. Integrating we find

$$\frac{\dot{r}^2}{2} - \frac{4\pi G\rho_b r^2}{3} = const \tag{4}$$

and choosing the constant to be zero (i.e. Ω to be 1) we find

$$\begin{aligned} r &\propto t^{\frac{2}{3}} \\ \rho_b &= \frac{1}{6\pi G t^2}\end{aligned} \tag{5}$$

the standard Friedmann matter dominated universe. This solution is true for every particle, and we write it as $\boldsymbol{r} = a(t)\boldsymbol{x}$ where $a(t)$ is the scale factor. We can choose $a(t)$ equal to 1 initially in which case $\boldsymbol{x}$ is the initial coordinate of the particle (the 'comoving coordinate'). The Newtonian approximation gives the correct result as a consequence of homogeneity and Birkhoff's theorem [6] – it is only the contribution of the matter *inside* a spherical shell which is important, and this shell may be chosen small enough so that fields are weak and velocities small so Newton's equations apply. By homogeneity the same rate of expansion must apply on all scales.

1.4 Linear Perturbations in Cold Dark Matter

For cold dark matter we also have mass conservation in a comoving volume

$$\int \rho dV_{comoving} = const \; . \tag{6}$$

Now imagine we give each particle a small perturbation at some early time and follow it thereafter

$$\boldsymbol{r} = a(t)(\boldsymbol{x} + \boldsymbol{\psi}(\boldsymbol{x}, t)) \tag{7}$$

$\boldsymbol{\psi}$ is called the comoving displacement. Mass conservation tells us the resulting density perturbation: if the initial unperturbed density is $\rho_i = \rho_b a^3$ then

$$\begin{aligned} \rho_i d^3\boldsymbol{x} &= \rho_b a^3 d^3\boldsymbol{x} \\ &= \rho(\boldsymbol{r}) d^3\boldsymbol{r} \end{aligned} \tag{8}$$

so the total density

$$\begin{aligned} \rho(\boldsymbol{r}) &= \frac{a^3 \rho_b}{Det(\frac{\partial \boldsymbol{r}}{\partial \boldsymbol{x}})} \\ &\approx \rho_b(1 - \boldsymbol{\nabla}_x \cdot \boldsymbol{\psi}) \end{aligned} \tag{9}$$

where we expand the determinant to first order in the perturbation using $Det(1 + A) = 1 + Tr(A) + o(A^2)$. Thus the fractional density perturbation is given by

$$\begin{aligned} \frac{\delta\rho}{\rho}(\boldsymbol{r}) &\equiv \frac{\rho(\boldsymbol{r}) - \rho_b}{\rho_b} \\ &= -\boldsymbol{\nabla}_x \cdot \boldsymbol{\psi} \; . \end{aligned} \tag{10}$$

At this point it is convenient to Fourier transform $\boldsymbol{\psi}(\boldsymbol{x})$ and decompose it into two pieces;

$$\begin{aligned} \boldsymbol{\psi}(\boldsymbol{k}) &\equiv \int d^3\boldsymbol{x} e^{i\boldsymbol{k}\cdot\boldsymbol{x}} \boldsymbol{\psi}(\boldsymbol{x}) \\ &\equiv \boldsymbol{\psi}_{\parallel}(\boldsymbol{k}) + \boldsymbol{\psi}_{\perp}(\boldsymbol{k}) \; . \end{aligned} \tag{11}$$

$\boldsymbol{\psi}_{\parallel}$ (parallel to $\boldsymbol{k}$) is the *compressional* piece and $\boldsymbol{\psi}_{\perp}$ (perpendicular to $\boldsymbol{k}$) is the *rotational* piece. Now from (2) and (9) only the *compressional* piece causes a perturbation in the potential $\delta\Phi$. Fourier transforming and solving (2) we find

$$\boldsymbol{\nabla}_r \delta\Phi = -4\pi G \rho_b a \boldsymbol{\psi}_{\parallel} \tag{12}$$

where to lowest order we use $\boldsymbol{\nabla}_r = a^{-1}\boldsymbol{\nabla}_x$. Now (1) decomposes into

$$\begin{aligned} \ddot{\boldsymbol{\psi}}_{\parallel} + 2\frac{\dot{a}}{a}\dot{\boldsymbol{\psi}}_{\parallel} &= 4\pi G \rho_b \boldsymbol{\psi}_{\parallel} \\ \ddot{\boldsymbol{\psi}}_{\perp} + 2\frac{\dot{a}}{a}\dot{\boldsymbol{\psi}}_{\perp} &= 0 \; . \end{aligned} \tag{13}$$

These equations have a simple interpretation. The left hand side describes a test particle moving in comoving coordinates – setting it to zero results in

$\dot{\boldsymbol{\psi}} \propto a^{-2}$, so the peculiar velocity (the physical velocity $\dot{\boldsymbol{r}}$ minus the Hubble flow $\frac{\dot{a}}{a}\boldsymbol{r}$) $\boldsymbol{v}_p = a\dot{\boldsymbol{\psi}} \propto a^{-1}$. In comoving coordinates a test particle slows down. This is all that happens to the rotational piece. The right hand side of (13) describes the feedback of the compressional perturbation on itself – the gravitational instability first discovered by Jeans. If the background were not expanding, $\boldsymbol{\psi}$ would grow exponentially but in a matter dominated universe with $a \propto t^{2/3}$ we find

$$\begin{aligned} \ddot{\boldsymbol{\psi}}_{\|} + \frac{4}{3t}\dot{\boldsymbol{\psi}}_{\|} &= 4\pi G\rho_b\boldsymbol{\psi}_{\|} = \frac{2}{3t^2}\boldsymbol{\psi}_{\|} \\ \ddot{\boldsymbol{\psi}}_{\perp} + \frac{4}{3t}\dot{\boldsymbol{\psi}}_{\perp} &= 0 \\ \rightarrow \boldsymbol{\psi}_{\|} &= \boldsymbol{A}(\boldsymbol{x})t^{2/3} + \boldsymbol{B}(\boldsymbol{x})t^{-1} \\ \boldsymbol{\psi}_{\perp} &= \boldsymbol{C}(\boldsymbol{x})t^{-1/3} + \boldsymbol{D}(\boldsymbol{x}) \end{aligned} \tag{14}$$

with $\boldsymbol{A}, \boldsymbol{B}, \boldsymbol{C}$, and $\boldsymbol{D}$ arbitrary functions of x. From (10) $\delta\rho/\rho$ behaves just as $\psi_{\|}$, so in the growing mode it grows like $a(t)$.

More generally the right hand side of (13) has an extra term $-a^{-1}\nabla_r\Phi_s(\boldsymbol{r})$ if there is a source driving the perturbation producing a potential Φ_s. For example for a point mass it reads $-Gm\boldsymbol{x}/(a^3x^3)$. Likewise in the equation for $\delta\rho/\rho$ the source term is $4\pi G\rho_s(\boldsymbol{x})$, where $\rho_s(\boldsymbol{x})$ is the density of the source. Notice that with cold dark matter the evolution is *purely local* in comoving coordinates.

In general if one has a theory for the origin of the perturbations one wants to evolve the perturbations through the radiation era into the matter era. Now the density ρ_b on the right hand side of (13) is only the *matter* density, at early times much less than the total density. Setting it to zero and using $a \propto t^{\frac{1}{2}}$, the radiation dominated result, one finds the solution $\psi = E + F\ln t$ with E and F constants, i.e. slow logarithmic growth. In fact solving (13) through the transition one finds at late times ($a \gg a_{eq}$, the scale factor at equal matter and radiation density)

$$\boldsymbol{\psi} = \boldsymbol{A}(\boldsymbol{x})\left(1 + \frac{3}{2}\frac{a}{a_{eq}}\right) \tag{15}$$

where $\boldsymbol{A}(\boldsymbol{x})$ is the perturbation at early times [7]. Thus there is very little growth in the radiation era and one can think of the perturbation growth as beginning at a_{eq}.

1.5 Local Nonlinearity "doesn't matter"

Whilst these equations are correct only in the linear regime they can be used perfectly well on a surface surrounding a nonlinear mass concentration as long as the density perturbation is always small *on the surface.* If one has a comoving surface S surrounding a volume V then the change in the volume enclosed by S is

$$\delta V = a^3 \int_S \boldsymbol{\psi} \cdot \boldsymbol{dS} \tag{16}$$

and the change in the mean density ρ inside S defined by $\rho_{av}(V+\delta V) \equiv \rho_i V_i = \rho_b V$ where $V = V_i a^3$. Now we find

$$\begin{aligned}\frac{\delta\rho}{\rho} &\equiv \frac{(\rho_{av} - \rho_b)}{\rho_b} \\ &= -\frac{\delta V}{V} \\ &= -\frac{\int_S \boldsymbol{\psi} \cdot \boldsymbol{dS}}{V_i}\end{aligned} \tag{17}$$

which of course grows just as (14). Thus linear theory has a wide range of applicability.

1.6 Peculiar Velocities

Linear theory also gives information about peculiar velocities in a very direct way – we see from $\boldsymbol{v}_p = a\dot{\boldsymbol{\psi}}$ and (14) that peculiar velocities grow as $t^{\frac{1}{3}}$ in the compressional mode. We also see using $\boldsymbol{\psi} \propto a$ in the growing mode that $\boldsymbol{v}_p = \dot{a}\boldsymbol{\psi} = H\delta\boldsymbol{r}$, the relation used in the introduction.

In particular geometries v_p is related to $\delta\rho/\rho$ – using (17) we find that defining the Hubble flow $v_H = Hr$ we get

$$\begin{aligned}\frac{v_p}{v_H} &= \frac{\delta\rho}{\rho} && \text{for planar collapse} \\ &= \frac{1}{2}\frac{\delta\rho}{\rho} && \text{for cylindrical collapse} \\ &= \frac{1}{3}\frac{\delta\rho}{\rho} && \text{for spherical collapse}\end{aligned} \tag{18}$$

Observations of peculiar velocities such as those recently reported [17] thus give a unique window on $\delta\rho/\rho$. If they are correct, we are moving in part of a very roughly spherical collapse pattern centred on the 'great attractor', 80 h_{50}^{-1} Mpc away from us. Our velocity towards it is approximately $500\,\mathrm{km\,s^{-1}}$. This corresponds to a value of $\delta\rho/\rho$ of approximately 0.4 on this scale.

1.7 Sources

As discussed above gravitational sources (like loops of cosmic string) add an additional driving term to the right hand side of (13). It may then be solved fairly simply with a Green's function [65], [70]. Since the equation is completely local $\delta\rho/\rho\,(x)$ simply evolves in proportion to $\delta\rho_S\,(x)$, the source density. In linear theory therefore a point mass always remains a point mass – its mass simply grows as $a(t)$. In the linear regime (13) with a source which 'turns on' in the matter era at t_i is easily solved with a Green's function

$$\begin{aligned} \delta\rho/\rho(x,t) &= \int_{t_i}^{t} dt' G(t,t') 4\pi G\rho_S(\boldsymbol{x},t') \\ G(t,t') &= \frac{3}{5}(t^{2/3}t'^{1/3} - t^{-1}t'^2) \end{aligned} \tag{19}$$

Extended sources like large cosmic string loops trace out an extended 'trumpet-shaped' structure in comoving coordinates leading to a density contrast of the same pattern [70], quite similar to the giant 'filament' mentioned above [3]!

1.8 Hot Dark Matter

If the dark matter particles have appreciable thermal velocities at t_{eq}, when perturbations start to grow, then the above approach must be modified – it is hard to follow the particles! Instead one must follow the particle trajectories in phase space and I will discuss this in the next Section.

However the main effect is simple to understand. Roughly speaking, particles move a distance $d \approx vt$ in an expansion time, where v is their velocity. Structure on smaller scales than this is washed out by particle diffusion ('free streaming') while structure on larger scales evolves unaffected, just as with cold dark matter. This length scale corresponds to a mass scale, often called the 'neutrino Jeans mass'. Since peculiar velocities are damped as we discussed above, the thermal velocity v of the particles decays as a^{-1}. The comoving scale $d_c = d/a \propto t^{-\frac{1}{3}} \propto a^{-\frac{1}{2}}$ corresponds to a mass scale $M_J \propto a^{-\frac{3}{2}}$. With neutrinos for example, of mass 25 h_{50}^2 eV such as would be required to make $\Omega = 1$ while respecting the nucleosynthesis bound, M_J is of order $10^{15} M_\odot$ at matter-radiation equality and about $10^9 M_\odot$ today.

1.9 Linearising Liouville

Clustering of hot dark matter can be treated quite precisely by linearising the equation for the phase space density, the Liouville equation. As before I will remain strictly Newtonian.

The phase space density f is defined so that the number of particles in the phase space volume $d^3\boldsymbol{r}d^3\boldsymbol{p}$ at time t is $f(\boldsymbol{r},\boldsymbol{p},t)d^3\boldsymbol{r}d^3\boldsymbol{p}$. Thus the ordinary density is given by $\rho = m\int d^3\boldsymbol{p}\ f$, where m is the mass of the particles. Conservation of particle number yields the continuity equation

$$\frac{\partial f}{\partial t} + \boldsymbol{\nabla}\cdot(f\boldsymbol{v}) = 0 \tag{20}$$

where the divergence and velocity are in the full phase space: $\boldsymbol{\nabla} = (\boldsymbol{\nabla}_r, \boldsymbol{\nabla}_p)$, $\boldsymbol{v} = (\dot{\boldsymbol{r}}, \dot{\boldsymbol{p}})$. Recalling Liouville's theorem that $\boldsymbol{\nabla}\cdot\boldsymbol{v} = \boldsymbol{\nabla}_r\cdot\dot{\boldsymbol{r}} + \boldsymbol{\nabla}_p\cdot\dot{\boldsymbol{p}} = \boldsymbol{\nabla}_r\cdot\boldsymbol{\nabla}_p H - \boldsymbol{\nabla}_p\cdot\boldsymbol{\nabla}_r H = 0$ where the total Hamiltonian is H, we get

$$\frac{\partial f}{\partial t} + \boldsymbol{v}\cdot\boldsymbol{\nabla} f = 0 \tag{21}$$

which is the statement that the function f is constant along particle trajectories. We will return to this in the next Section. Next we use the fact that $(\dot{\boldsymbol{r}}, \dot{\boldsymbol{p}}) = (\boldsymbol{p}/m, -m\nabla_r \Phi)$ with Φ the Newtonian potential. It is convenient to change coordinates to $\boldsymbol{x} = \boldsymbol{r}/a, \boldsymbol{q} = a\boldsymbol{p} - m a \dot{\boldsymbol{r}}$, comoving position and peculiar momentum, and $d\tau = dt/a$, conformal time. This yields

$$\frac{1}{a}\frac{\partial f}{\partial \tau} + \frac{\boldsymbol{q}}{ma^2} \cdot \nabla_x f - m\ddot{a}\boldsymbol{r} \cdot \nabla_q f - m\nabla_x \Phi \cdot \nabla_q f = 0 \tag{22}$$

(Φ is independent of $\boldsymbol{q}$). Here $\ddot{a}\boldsymbol{r} = \frac{d^2 a}{dt^2}\boldsymbol{r} = -\nabla_x \Phi_0$ from Newton's equations for the background as discussed above, where Φ_0 is the zeroth order part of the potential. The homogeneous, zeroth order solution to this is

$$\begin{aligned} f_0 &= \frac{2}{e^{q/(T_\nu a)} + 1} \\ \rightarrow \rho_0 &= \frac{m}{a^3} \int d^3\boldsymbol{q} \frac{2}{e^{q/(T_\nu a)} + 1} = \frac{2mT_\nu^3}{\pi^2}\eta_3 \end{aligned} \tag{23}$$

where $\eta_3 \approx 1.803$ and $T_\nu \propto a^{-1}$ is the neutrino temperature, equal to $(4/11)^{\frac{1}{3}}$ the temperature of the microwave background radiation. We use the relativistic phase space density of one species of neutrino (and antineutrino) when they decoupled as initial conditions, and this is preserved as the universe expands even as the neutrinos go out of thermal equilibrium and go nonrelativistic, as we learn from (22). As an exercise the reader can calculate the required mass of the neutrino needed to make $\Omega = 1$, stated earlier.

It is fairly straightforward to linearise and solve (23) through the matter-radiation transition including sources just as for cold dark matter [24, 25]. The interested reader may refer to [26] for a pedagogical account.

1.10 Phase Space Constraints for Hot Dark Matter

The fact that for neutrino (and more generally fermionic) dark matter the initial phase space density (23) has a maximum, namely $\frac{1}{2}$ per neutrino or antineutrino, has a nice consequence first pointed out by Tremaine and Gunn [27]. Because it is constant along particle trajectories, it cannot increase beyond this maximum. This puts a strong constraint, valid even in the nonlinear regime, on the clustering of neutrinos, relevant to the issue of whether neutrinos can form the dark matter in galactic halos.

Crudely, the phase space density corresponding to a virialised lump of N neutrinos with volume r^3 and momentum p_ν is just $Nr^{-3}p_\nu^{-3}$ and by the constraint, this cannot be larger than one. However for a virialised lump we have $v^2 \approx GM/r \approx GNm/r$. Solving for N and substituting back into the previous formula we obtain $m_\nu^4 > 1/(vr^2 G)$ (I use units where $\hbar$ is 1). A more detailed treatment [27] yields the bound

$$m_\nu > 30\text{eV} \left(\frac{100 \text{ km s}^{-1}}{\sigma}\right)^{\frac{1}{4}} \left(\frac{15 \text{ kpc}}{r_c}\right)^{\frac{1}{2}} \tag{24}$$

assuming the collapsed object is an isothermal sphere with core radius r_c and one dimensional velocity dispersion σ. Observations of dwarf galaxies with dark matter halos can therefore put strong constraints on the neutrino scenario.

1.11 Summary

In this Section I have given a flavour of the new observations and what is encouraging about them. I have also given a brief introduction to the kind of calculation necessary to follow the gravitational clustering of matter around primordial perturbations. In the next Section I will give an account of one theory as to where these perturbations came from, the theory of cosmic strings.

2. Phase Transitions, Defects and Fluctuations

In this Section I will discuss the physics of symmetry breaking phase transitions predicted to occur in grand unified theories in the very early universe [28, 29, 2]. The discussion will necessarily be qualitative and crude – it is difficult to do precise calculations in this area! Nevertheless according to our present view the precise details are not crucial in most cases. This is because the phase transitions concerned happened very early in the life of the universe, typically at times of 10^{-34} seconds. Transitions producing domain walls or monopoles would have been disastrous fairly quickly – the defects quickly coming to dominate the universe. Those producing string, however, which I shall concentrate on as the most interesting case, lead to a 'scaling solution' for the string network, in which the strings remain a fixed fraction of the total energy density in the universe. As I shall discuss, we have good reason to believe that the late time evolution of the scaling solution is rather insensitive to the precise details of the initial conditions.

For the moment I will ignore the possibility that the phase transitions were inflationary – indeed it is very difficult to arrange for inflation in a gauge symmetry-breaking transition – but will return to the question of how cosmic strings relate to inflation later on.

2.1 A Physical Picture of Phase Transitions

According to our current and successful understanding of high energy physics, the underlying field theory describing nature has a high degree of symmetry but most of this symmetry is broken at low energy.

As the simplest example of how this works, consider the theory of a scalar field ϕ invariant under $\phi \to -\phi$. If we give the field a potential $V(\phi)$ as in Figure 2.1 then the potential shares this symmetry, so the classical ground

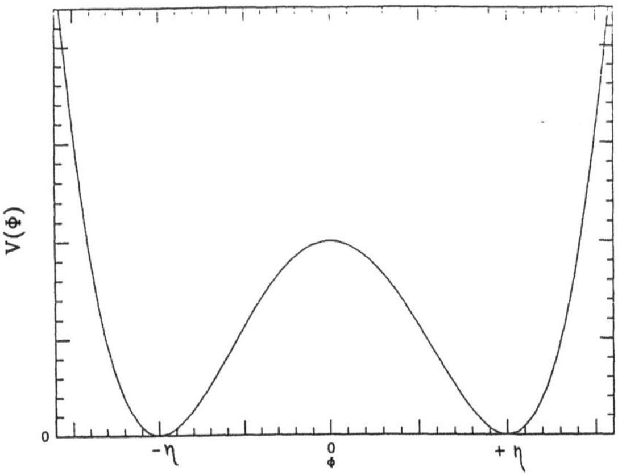

Fig. 2.1. The potential $V(\Phi) = \frac{\lambda}{8}(\Phi^2 - \eta^2)^2$.

state of the theory is degenerate – in vacuo ϕ can sit at + or at −. The chosen configuration breaks the symmetry[2].

At high temperature, however, the symmetry is 'restored' simply because thermal excitations are energetic enough to 'kick' the field ϕ over the potential barrier. Any masses become irrelevant at high temperatures and so by dimensions one must have $<\phi>^2 \approx T^2$. Thus ϕ explores a wide range of the potential. As the system cools the field explores a smaller and smaller range until at low temperatures it flops down into one of the two minima. In some parts of space it chooses the + minimum and in other parts the − minimum. Because of this defects are inevitably formed – in going from a + region to a − region ϕ must cross over the potential barrier in $V(\phi)$. Where it does so there is a localised lump of potential and gradient energy – a *domain wall* separating the + and − regions. It is important to realise that these defects are, at low temperatures, a *nonequilibrium* phenomenon – they are Boltzmann suppressed. To understand them one needs to understand how the fields go out of equilibrium as the universe expands.

According to the most naive picture, at high T the system is characterised by thermal oscillations with $k \approx T$. Shorter wavelength modes are suppressed. Thus the crudest picture of the ϕ field is that it is smooth on a scale of the thermal wavelength. On might therefore expect that the size of the + and − domains would, as long as one stayed in equilibrium, be just this. However what is really relevant is not the shortest wavelength modes – these oscillate vigorously and simply average out to zero as the system cools. Rather it is the *long* wavelength modes, with their more sluggish oscillations, which determine how the system goes out of equilibrium.

---

[2] This idea is *classical* – quantum mechanically of course tunnelling can occur between the + and − states. The true ground state is a symmetric combination of + and −. In field theory, however, in macroscopic volumes the tunnelling rate is very tiny so the classical picture is reasonable – there is simply a *classical* probability of being in + or −.

2.2 Gaussian Approximation

Let us try to understand this in a little more detail. To begin with assume that Φ has a long wavelength component $\bar{\Phi}$ which we shall initially treat as constant. We write $\Phi = \bar{\Phi} + \delta\Phi$ where $\delta\Phi$ is the short wavelength component. I will be more precise about the division into short and long wavelengths later. We then calculate the partition function for Φ by first summing over the fluctuations $\delta\Phi$ about $\bar{\Phi}$ in the Gaussian approximation (i.e. keeping only quadratic terms in $\delta\Phi$ in the energy).

The small fluctuations are just massive fields whose mass is determined by the value of $\bar{\Phi}$. For $V(\Phi) = \frac{\lambda}{8}(\Phi^2 - \eta^2)^2$ (shown in Figure 2.1) the mass squared of $\delta\Phi$ is $m^2 = V''(\bar{\Phi}) = \frac{\lambda}{2}(3\bar{\Phi}^2 - \eta^2)$. In general if Φ is a Higgs field it gives a mass to other fields as well – in particular fermion and gauge fields. Their contribution may be included in exactly the same way.

Now recall that a massive boson field has a partition function

$$
\begin{aligned}
Z_b = Tr(\mathrm{e}^{-\beta H}) &= \sum_{\{n_k\}} \mathrm{e}^{-\beta \sum_k E_k} \\
&= \prod_{\boldsymbol{k}} \sum_{n_k=0}^{\infty} \mathrm{e}^{-\beta n_k \omega_k} = \prod_{\boldsymbol{k}} \frac{1}{(1-\mathrm{e}^{-\beta\omega_k})}
\end{aligned}
\tag{25}
$$

where $\boldsymbol{k}$ labels each momentum mode and $n_{\boldsymbol{k}}$ are the occupation numbers for each mode. ω_k is just $\sqrt{\boldsymbol{k}^2 + m^2}$, and $\beta = 1/T$. For fermions one has instead

$$
Z_f = \prod_{\boldsymbol{k}} (1 + \mathrm{e}^{-\beta\omega_k}) \tag{26}
$$

The free energy contributed by a massive field is (recall $Z \equiv \mathrm{e}^{-\beta F}$)

$$
\begin{aligned}
F_b &= \frac{V}{\beta} \int d^3k \ln(1 - \mathrm{e}^{-\beta\omega_k}) \\
&\approx V\left(-\frac{\pi^2 T^4}{90} + \frac{m^2T^2}{24} + \ldots\right) \qquad \frac{m}{T} \ll 1
\end{aligned}
\tag{27}
$$

at high temperature. A Dirac fermion contributes

$$
\begin{aligned}
F_f &= 4\frac{V}{\beta} \int d^3k \ln(1 + \mathrm{e}^{-\beta\omega_k}) \\
&\approx V\left(-\frac{7\pi^2 T^4}{180} + \frac{m^2T^2}{12} + \ldots\right) \qquad \frac{m}{T} \ll 1
\end{aligned}
\tag{28}
$$

similarly. Here V is the volume: we replace $\sum_{\boldsymbol{k}}$ by $V\int d^3k \equiv V\int d^3k/(2\pi)^3$. The high T expansion is obtained by scaling out the T dependence and Taylor expanding the integrand in m/T about $m/T = 0$.

It is easily seen that (27) and (28) are dominated by modes with $k \approx T$, just as we expected above. Thus for $m/T \ll 1$ we can regard (27) and (28) as the contribution to the full partition function of only summing over the high k

modes, leaving the low k modes (i.e. $k \ll T$) not summed over. We then have an effective theory for the low k modes with a potential energy given by adding the contributions of (27) and (28) to the zero temperature potential for $\bar{\Phi}$

$$\begin{aligned} V_{eff}(\bar{\Phi}) &\approx V(\bar{\Phi}) + m^2(\bar{\Phi})\frac{T^2}{24} \\ &\approx \frac{\lambda}{8}\bar{\Phi}^4 - \frac{\lambda}{4}\eta^2\bar{\Phi}^2 + \frac{\lambda}{16}\bar{\Phi}^2 T^2 \\ &\approx \frac{\lambda}{8}(\bar{\Phi}^2 - \eta_{eff}^2)^2 \end{aligned} \tag{29}$$

ignoring $\bar{\Phi}$ independent terms. Here

$$\begin{aligned} \eta_{eff}^2 &= \eta^2\left(1 - \frac{T^2}{T_c^2}\right) \\ T_c &= 2\eta\ . \end{aligned} \tag{30}$$

To repeat, the extra temperature dependent terms come from summing over the short wavelength fluctuations about the long wavelength 'background'. This has a very important effect – at high temperature it changes the sign of the $\bar{\Phi}^2$ term and pushes the minimum to $\bar{\Phi} = 0$, the 'unbroken' phase. The transition occurs when the $\bar{\Phi}$ field becomes effectively massless, at $T = T_c = 2\eta$, called the critical temperature. This is shown in Figure 2.2. If the Φ field couples to other fields then it also gives them a mass and their contribution may be included in V_{eff} similarly. For example Φ generally gives gauge fields a mass squared $\approx g^2\Phi^2$ giving a correction term in (29) of $\approx g^2\Phi^2T^2$ and we find $T_c \approx \sqrt{\lambda/g^2}\eta$.

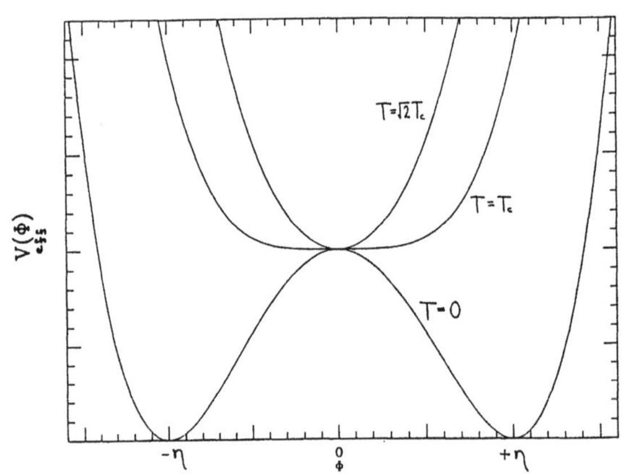

Fig. 2.2. The effective potential $V_{eff}(\Phi)$ for the long wavelength modes, at $T = 0$, $T = T_c$, and $T = \sqrt{2}T_c$.

What is the cause of this 'restoration' of symmetry? In the partition function we are really finding the most likely configuration for the system plus heat bath. In this context this amounts to minimising the free energy with respect to $\bar{\Phi}$.

The most important effect of $\bar{\Phi}$ being nonzero and giving the other modes a mass is to cut down the number of states available (at given T) to that mode. This is an *entropy* effect, and at high T as we have seen it tends to favour $\bar{\Phi} = 0$.

One other point must be made. In the Gaussian approximation the exact expressions in (27) and (28) do not make sense for all $\bar{\Phi}$: for $\bar{\Phi}^2 < \eta^2/3$ we have $m^2 = \frac{\lambda}{2}(3\bar{\Phi}^2 - \eta^2) < 0$. This means that ω_k and hence F are not even real! This is clearly a problem with the approximation. Nevertheless the qualitative behaviour exhibited by the small m/T expansion is quite believable.

The function $V_{eff}(\bar{\Phi})$ is often called the finite temperature effective potential. In quantum mechanics, by considerations similar to those I explained in the context of symmetry breaking, this should never have two minima – $\bar{\Phi}$ can tunnel through from one minimum to another. However we are interested in an evolving macroscopic system, and we are approximating it as a *classical* $\bar{\Phi}$ plus a quantum fluctuation. Ignoring quantum tunnelling should be reasonable and so $V_{eff}(\bar{\Phi})$ is the relevant quantity. However as far as I am aware none of this has been made rigourous and I would certainly encourage those of you feeling queasy with it to try and do better yourselves [30]!

2.3 Spatial Distribution of the Fields near T_c

The interpretation of $V_{eff}(\bar{\Phi})$ as an effective potential for the long wavelength modes means that we can use it to discuss their distribution at high T. Clearly $\bar{\Phi}$ will not be spatially constant at finite T. To understand its spatial distribution consider a massive field in equilibrium. As we approach T_c from above $\bar{\Phi}$ looks like a massive field with m decreasing to zero. For T below T_c, in the neighbourhood of the minima $\bar{\Phi} = \pm\eta_{eff}$ of $V_{eff}(\bar{\Phi})$, $\bar{\Phi}$ looks like a massive field with mass squared $m^2_{eff} = V''_{eff}(\bar{\Phi}) = \lambda\eta^2_{eff} = \lambda\eta^2(1 - (T/T_c)^2)$. Now in thermal equilibrium a massive field has spatial correlation function

$$< \Phi(r)\Phi(0) >= \frac{Tr(\Phi(r)\Phi(0)e^{-\beta H})}{Tr(e^{-\beta H})} . \tag{31}$$

The trace over all states is conveniently calculated in an occupation number basis just as the partition function in (25) was. With the mode expansion for Φ one finds

$$< n_k \mid \Phi(r)\Phi(0) \mid n_k >= \int \frac{d^3k}{2\omega_k} e^{i\mathbf{k}\cdot\mathbf{r}}(2n_k + 1) \tag{32}$$

and so summing over all states

$$< \Phi(r)\Phi(0) >= \int \frac{d^3k}{2\omega_k} e^{i\mathbf{k}\cdot\mathbf{r}} \left(\frac{2}{e^{\beta\omega_k} - 1} + 1\right) \tag{33}$$

the second piece being the usual vacuum piece. For $r \gg m^{-1} \gg T^{-1}$ the integral in (33) is dominated by long wavelength modes and gives

$$< \Phi(r)\Phi(0) >\approx \frac{T}{4\pi r} e^{-mr} . \tag{34}$$

This tells us that $\bar{\Phi}$ is uncorrelated on scales larger than $\xi \equiv m_{eff}^{-1}$, which is called the correlation length. If we defined $\bar{\Phi}_R$ as the average of $\bar{\Phi}$ over a region of size $R \gg \xi$ then it is straightforward to check that (34) implies the same result for $< \bar{\Phi}_R^2 >$ as that obtained if $\bar{\Phi}_R$ were the result of averaging over $(R/\xi)^3$ volumes within which $\bar{\Phi}$ took on randomly chosen values $\pm\sqrt{m_{eff}T}$. This serves as a rough picture of the distribution of $\bar{\Phi}$. Near the transition point the magnitude of the fluctuation, $\sqrt{m_{eff}T}$, is much less than T but the range, m_{eff}^{-1}, is much greater than T^{-1}. The energy in a typical fluctuation $\approx$ *potential* $\times$ *volume* $\approx m_{eff}^2 \Phi^2 \xi^3 \approx T$, as expected.

In fact if the system remains in equilibrium the correlation length goes to infinity at T_c, when $m_{eff} = 0$. In any real system with a finite rate of cooling, this never occurs. Instead the maximum correlation length is determined by the rate of cooling. I leave it as an exercise to find the maximum value of ξ by using the equilibrium value, with $T \propto a^{-1}$ and demanding that its rate of increase $\dot{\xi}/\xi$ not exceed the Hubble expansion rate $\dot{a}/a$.

2.4 Out of Equilibrium: Defect Formation

As the universe cools below a symmetry breaking phase transition, even though V_{eff} has the symmetry breaking form thermal fluctuations are initially still large enough to erase whole domains – in the example above a whole + domain may be 'flipped' to a – domain. Domains are really 'frozen in' very soon after the temperature falls below the energy required to kick them over the potential barrier. This is because below this temperature 'flipping' becomes exponentially (Boltzmann) suppressed, and the finite expansion rate of the universe means that it rapidly becomes negligible. The temperature at which domains are still frequently 'flipped' is given by

$$T \approx \xi^3 \Delta V_{eff} \approx \xi^3 \frac{\lambda}{8} \eta_{eff}^4 \tag{35}$$

being the energy required to take a domain ξ^3 over the potential barrier separating the two minima of V_{eff}. Using the equilibrium expressions right down to this temperature we find

$$\begin{aligned} T &= \frac{\eta}{8\sqrt{\lambda}} \left(1 - (T/T_c)^2\right)^{\frac{1}{2}} \\ \rightarrow T^{-1} &= \sqrt{\frac{64\lambda}{\eta^2} + \frac{1}{T_c^2}} \approx T_c \end{aligned} \tag{36}$$

for small λ. This temperature, the temperature of domains freezing in, is called the *Ginzburg* temperature T_G. Below this temperature the domains and defects rapidly go out of thermal equilibrium [2].

At T_G we have $\xi = m_{eff}^{-1} = \eta_{eff}^{-1}/\sqrt{\lambda} \approx 1/(8T_c\lambda)$ which for λ not too small and $T_c \approx m_{Gut} << m_{Planck}$ is smaller than the horizon scale at this time. For very small λ the assumption of thermal equilibrium would be dubious. However λ must be quite small for the small m/T expansion to be reasonable.

I have given a very simplistic account of what is a very complex subject. For one recent attempt to do better see [31].

2.5 Defects and Topology

In the previous Section we discussed the simplest example of a symmetry, the discrete Z_2 symmetry $\Phi \rightarrow -\Phi$. In unified theories we are interested in more complicated symmetries including continuous symmetries. Nevertheless defect formation can be understood in a simple and general way.

Say we have a theory invariant under a symmetry group G which is spontaneously broken by a Higgs field Φ to a subgroup H, defined as $H = \{g \in G : g\Phi = \Phi\}$. Now the whole theory and in particular the Higgs potential $V(\Phi)$ is invariant under $\Phi \rightarrow g\Phi$. Thus if we have one point η on the minimum V_{min} of V we can obtain a whole set of values of Φ degenerate in energy by applying elements of G to η. In fact if there is no 'accidental' degeneracy in the theory the resulting set will be the entire space V_{min}. V_{min} is not, however, equal to G because if we apply two different group elements g and g' we may obtain the same point on V_{min}. This happens if $g\Phi = g'\Phi$ which is true iff $g^{-1}g' \in H$ so that g and g' are related by right multiplication in H . V_{min} is therefore obtained by *identifying* elements in G related by right multiplication in H – it is a *coset space*

$$V_{min} = \frac{G}{H} \; . \tag{37}$$

Thus the topology of V_{min} is simply a group theoretic entity, not depending on details of the Higgs potential or its coupling constants. In our simple example, $G = Z_2$ was broken to $H = 1$ by Φ and so from (37) $V_{min} = Z_2$, the two points $+$ and $-$.

As another simple example, consider a theory with a continuous $G = U(1)$ broken by a complex scalar field Φ transforming under $g = e^{i\theta}$ as $\Phi \rightarrow e^{iQ\theta}\Phi$. Here $H = 1$ so from (37) $V_{min} = U(1)$ which is just a circle. The corresponding quartic potential, $V = \lambda(|\, \Phi \,|^2 - \eta^2)^2$ is the 'Mexican hat' potential. As a more complicated example if $G = SO(3)$ is broken by a 3-component vector $\hat{\Phi}$ to a subgroup $H = SO(2)$ (rotations about the axis defined by $\hat{\Phi}$) then we find $V_{min} = SO(3)/SO(2) = S_2$, a two-sphere.

The topology of V_{min} has very important physical consequences. In the 'domain' picture explained above, $\bar{\Phi}$ falls onto different points of V_{min} in different domains. Moving through space, Φ will wind around V_{min} and this will in general result in defects just as in Section 2.1. In the simplest case, V_{min} has two points $+$ and $-$. In going along any path in space from a domain where $\Phi = +$ to one where $\Phi = -$, Φ has to climb over the potential barrier in between. This means that somewhere there must be a localised wall of energy density separating the two domains, a 'domain wall'. In the second example above, as we traverse a loop in space Φ will in general wind around the circle V_{min}. This means that somewhere inside the loop Φ must vanish. In fact Φ must vanish somewhere on every surface bounded by the loop. Where it vanishes, there is a nonzero energy density. The lowest energy configuration compatible with Φ winding around V_{min} on the loop is a *line* of energy density – a string – passing through the loop. Neither walls nor strings can have edges or ends – they must either be closed or infinite. In the third example above, traversing a two-sphere in space one would find in general that Φ would wind around the S^2 of V_{min}. This results in a localised *point* defect – a magnetic monopole (see Figure 2.3).

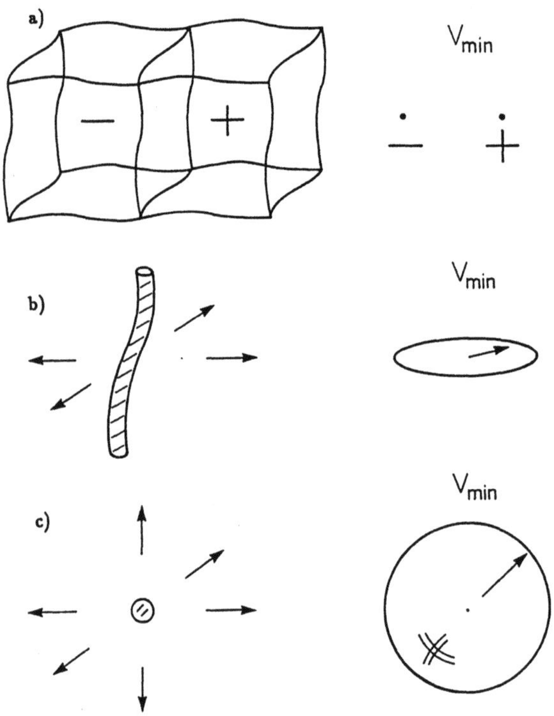

Fig. 2.3. The three simplest types of defects. On the left is the spatial distribution of the fields. Arrows show the location of the fields on the minimum of the potential V_{min}, which is shown on the right.
a) domain wall
b) string
c) magnetic monopole

2.6 A Useful Theorem

A useful theorem in the case of strings is that the group of loops in $V_{min} = G/H$ not continuously deformable into one another (the homotopy group $\pi_1(G/H)$) is given by

$$\pi_1\left(\frac{G}{H}\right) = \pi_0(H) \tag{38}$$

in the case where G is simply connected. $\pi_0(H)$ is just the number of disconnected components of H. A pictorial 'proof' of this statement is given in Figure 2.4. This formula is still useful when G is simple but not simply connected because in this case one has $G/H = \bar{G}/\bar{H}$ where $\bar{G}$ is the simply connected covering group of G and $\bar{H}$ is the subgroup of $\bar{G}$ leaving Φ invariant. For example $SO(3)$ can be broken by *two* 3-vectors to 1. This produces strings because the simply connected covering group of $SO(3)$ is $\bar{G} = SU(2)$ and it would be broken to a subgroup $\bar{H} = Z_2$, because the 3-vectors are in the adjoint representation of $SU(2)$, and are left invariant by the two elements of its centre. Thus we would have $\pi_1(V_{min}) = Z_2$. These strings are more complicated than the $U(1)$ strings discussed above – in this case a string running in one direction is topologically equivalent to one running in the opposite direction. Strings of this type form in many simple grand unified theories, for example in the symmetry breaking transitions

$$\begin{aligned} SO(10) &\rightarrow \text{`}SU(5) \times Z_2\text{'} \\ E(6) &\rightarrow \text{`}SO(10) \times Z_2\text{'} \end{aligned} \tag{39}$$

the first being mediated by a Higgs **126** and the second by a **351**. The quotation marks indicate that the full global structure of H is more complicated, as explained in [32].

In low energy models 'inspired' by superstrings cosmic strings of the simplest $U(1)$ type as well as those of the more complicated Z_n type occur [33].

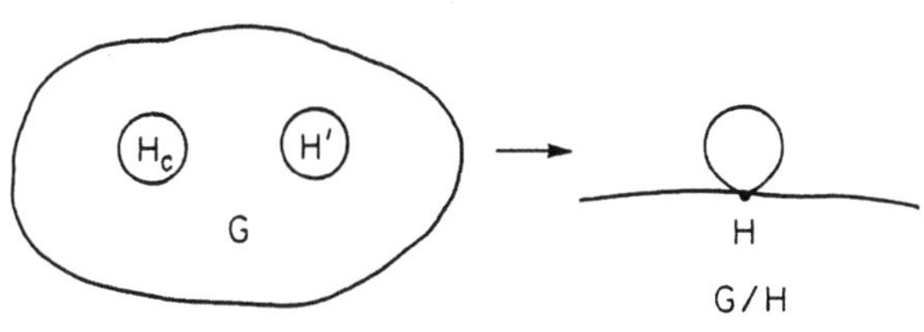

Fig. 2.4. A pictorial demonstration that $\pi_1(G/H) = \pi_0(H)$ when G is simply connected. On the left, H has two components, the component connected to the identity, H_c, and the disconnected component H'. When G/H is constructed from G one identifies all the elements of H with the identity. This can be pictured as squeezing all the elements of H together, forming a 'tunnel' as shown on the right. This clearly has a noncontractible loop.

As an excercise I leave the reader to construct the 'proof' analogous to that in Figure 2.4 that $\pi_2(\frac{G}{H}) = \pi_1(H)$ assuming $\pi_2(G)$ and $\pi_1(G)$ are trivial. The latter are always true for simple Lie groups. Since H must ultimately contain the $U(1)$ of electromagnetism, if the grand unified theory is based on a simple group then magnetic monopoles are inevitable. This is the basis of the famous 'magnetic monopole problem' [34].

2.7 Cosmic Strings

Strings are the most interesting topological defects because their existence is compatible with the observed universe and may even explain its large scale structure. Their possible existence in relativistic field theories was first pointed out by Nielsen and Olesen [45]. They are simple generalisations of the flux lines found in superconductors. For a review, see [35].

The microscopic structure of a string is illustrated in Figure 2.5 for the simplest case, a $U(1)$ string. Outside the string, Φ winds around V_{min} as we go around the string. At the centre of the string Φ goes to zero. The 'magnetic' gauge field of the $U(1)$ group that is broken in forming the string is nonzero and points along the string in one direction in the minimum energy configuration. The energy per unit length μ of a static straight string is obtained from the energy functional

$$\begin{aligned} \mu \equiv E/L &= \int d^2x[(\boldsymbol{D}\Phi)^2 + \frac{1}{2}\boldsymbol{B}^2 + V(\Phi)] \\ &= 2\pi\eta^2 f(\frac{\lambda}{g^2}) \;\approx\; m_{Gut}^2 \\ V(\Phi) &= \frac{\lambda}{2}(\mid \Phi \mid^2 - \eta^2)^2 \\ \boldsymbol{D}\Phi &= (\boldsymbol{\nabla} - ig\boldsymbol{A})\Phi \end{aligned} \tag{40}$$

where f is a slowly varying function equal to unity when $\lambda = g^2$ [38, 39]. These strings generally carry no electric charges and their strongest coupling to other matter is through their gravitational field. The gravitational coupling is thus the only parameter relevant to the (ordinary) cosmic string theory of galaxy formation. It is given by the dimensionless number $G\mu \approx (m_{Gut}/m_{Planck})^2 \approx 10^{-6}$ typically in those grand unified theories predicting strings.

In some theories the string can couple to long range fields other than gravity. *Global* strings occur where the $U(1)$ symmetry broken by Φ is not gauged. Here there is a massless Goldstone boson which couples to the string and is radiated from it. This kind of string occurs in axion models [40]. *Superconducting* strings form in theories where Φ couples to another electrically charged field χ in such a way as to force χ to be nonzero on the string. These strings may carry a net electric charge or current [41]. With the assumption of a primordial magnetic field (which is hard to justify) these strings are the basis for the OTW scenario of explosive galaxy formation [42]. I will not discuss this scenario in these lectures.

2.8 Simulating String Formation

Cosmic string formation was originally understood numerically by simply throwing down values for the Higgs field on V_{min} at random on a lattice of domains of size ξ, with a prescription for smoothly varying the Higgs field phase in going from one domain to the next. This is of course a crude model and

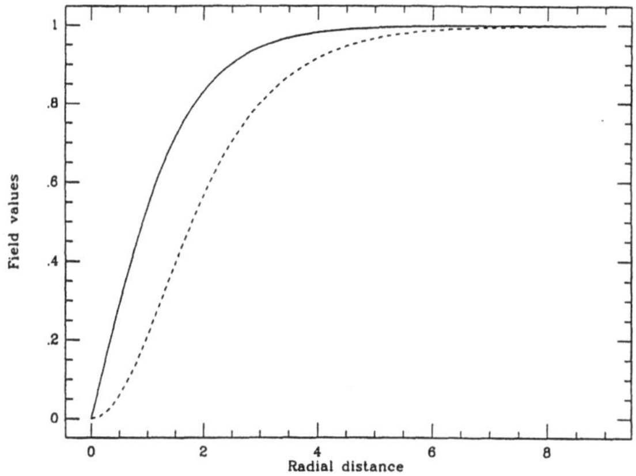

Fig. 2.5. The profile of a cosmic string. The solid line is $|\Phi|/\eta$ where the angular dependence (in cylindrical coordinates) is simply $\Phi = |\Phi| e^{i\theta}$. The dashed line is $A_\theta gr$, with g the gauge coupling. The gauge potential A_θ asymptotically tends to $1/(gr)$ so the total 'magnetic' flux down the string is $\int \boldsymbol{B} \cdot d\boldsymbol{S} = \int \boldsymbol{A} \cdot d\boldsymbol{l} = 2\pi/g$. The radius is in units of $1/(g\eta)$, and the solution is shown for $\lambda = g^2$, the case where the Higgs particle mass equals the gauge particle mass in the broken phase.

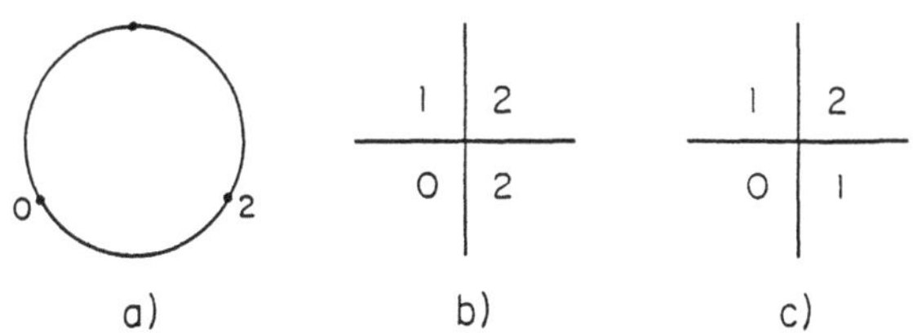

Fig. 2.6. The approximation used in simulations of the formation of the simplest 'U(1)' cosmic strings.
a) The space V_{min} (the circle) is approximated by 3 points. Each domain is assigned a value 0, 1 or 2, corresponding to one of these points. The rule for the phase of the Higgs field is to move along the shortest path on V_{min} when going from one domain to the next.
b) An edge with four surrounding domains – this edge is a string.
c) This edge is not a string.

ignores any interactions. The scale ξ is the correlation length at the Ginzburg temperature as discussed in Section 2.4.

In more detail, for $U(1)$ strings V_{min} is approximated by three points assigned the values 0, 1 or 2. For a cubic lattice of domains, any edge is surrounded by four domains. The rule for going from one domain to the next is that Φ takes the shortest path along V_{min}. This is illustrated in Figure 2.6. The reader can easily check that at any vertex on the lattice the same number of strings leave as enter. With periodic or reflecting boundary conditions this means that the strings have no ends – they are all in the form of closed loops.

The numerical simulations first done by Vachaspati and Vilenkin [56] have been checked by many people on a variety of lattices and are in good agreement [49, 46, 52]. Most (about 80%) of the string is in strings as large as the box in which the simulation is performed ('infinite' strings). The remainder is in the form of closed loops. These have a *scale invariant* distribution – the number of loops of size (a *course* measure like r.m.s. radius) greater than R per unit volume, $n_>(R) \propto R^{-3}$. The term *scale invariant* simply means that the result depends only on R and not on the domain size ξ. Dimensional analysis then determines the power law. Both the 'infinite' strings and the loops are in the form of Brownian random walks. Recently David Mitchell and myself have shown how these results may be obtained by counting states in the quantised closed bosonic string, an intriguing connection [99].

A picture of the string network at formation is shown in Figure 2.7.

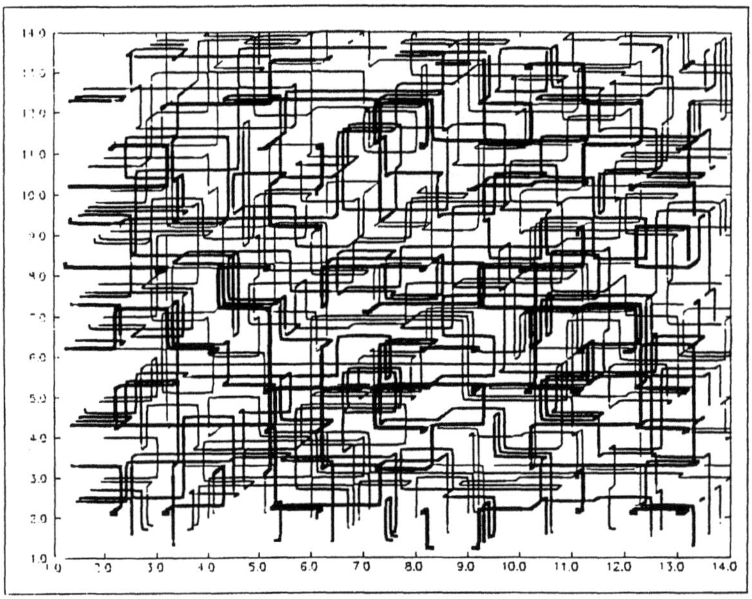

Fig. 2.7. A box of strings at formation. The strings are formed on a cubic lattice with the algorithm described in Figure 2.6. The box is tilted and the string width decreases away from the nearest side of the box to give an impression of perspective. Periodic boundary conditions are used, so all of the string is in the form of closed loops – most of it is actually in a single very long loop.

2.9 Evolution of Strings

After the strings form we have to evolve them. At first sight this appears horribly complicated – they are inherently nonlinear entities. However, two effects straighten out the strings very rapidly. In the early stages (until the temperature falls to $\approx (G\mu)^{\frac{1}{2}} m_{Gut}$) the strings are heavily damped by collisions with particles [35]. Secondly the expansion of the universe stretches the string out. The typical radius of curvature of the string increases rapidly while the width of the string remains fixed.

Quite quickly it becomes a very good approximation to treat the strings as infinitely thin relativistic lines or 'Nambu' strings, the action for which is simply the area of the two dimensional worldsheet they sweep out in spacetime [43]

$$S = -\mu \int dA = -\mu \int d^2\sigma \sqrt{-\gamma}$$
$$\gamma = det(\gamma_{\mu\nu}) \qquad \gamma_{\mu\nu} = \partial_\mu x^a(\sigma)\partial_\nu x^b(\sigma) g_{ab}(x(\sigma)) \tag{41}$$

where $g_{ab}(x)$ is the metric of the background spacetime. $x^a(\sigma)$ are the four spacetime coordinates of the string worldsheet which is parametrised by the two worldsheet coordinates $\sigma^\mu = (\tau, \sigma)$. The string worldsheet has one spacelike coordinate σ and one timelike coordinate τ.

Let me explain how (41) arises. For a $U(1)$ theory producing strings the action is

$$S = \int d^4y \sqrt{-g} \left((D\Phi)^2 - V(\Phi) - \frac{1}{4}F^2 \right)$$
$$V(\Phi) = \frac{\lambda}{2}(|\Phi|^2 - \eta^2)^2 \tag{42}$$
$$D_\mu \Phi = (\partial_\mu + igA_\mu)\Phi$$
$$F_{\mu\nu} = \partial_\mu A_\nu - \partial_\nu A_\mu$$

We seek an approximate classical solution to the full field theory equations in the form of a curved and moving string. Locally it should look like a Lorentz boosted version of the static cylindrically symmetric solutions obtained by minimising the energy functional (40). It should also be built around a curved worldsheet $x^a(\sigma)$.

First we imagine such a worldsheet is given and construct two normal vectors to it, $n^a_A(\sigma)$, $A = 1, 2$. These obey $n^a_A x^b_{,\mu} = 0$ and are orthonormal $n^a_A n^b_B g_{ab} = -\delta_{AB}$. Now for any point in space y^a closer to the string than its radius of curvature R we can associate two worldsheet coordinates σ^μ and two radial coordinates ρ^A uniquely

$$y^a = x^a(\sigma) + \rho^A n^a_A(\sigma) \; . \tag{43}$$

This is illustrated in Figure 2.8. We will use this relation to change variables from the four y^a coordinates to the four $\xi^\alpha = (\sigma^\mu, \rho^A)$ coordinates. Our ansatz for the field configuration corresponding to this worldsheet is then $\Phi(y) = \Phi_s(\rho)$ and $A^a(y) = n^a_A(\sigma)A^A_s(\rho)$ where the subscript s refers to the static cylindrically symmetric solution. The reader can check that this ansatz gives the exact solution for a straight moving string (i.e. just the Lorentz boosted version of the static straight solution). We expect that it will provide us with an approximate solution if the radius of curvature of the string R is much greater than its width W. Now we simply change coordinates in the action (42) from y^a to ξ^α. The Jacobian is most easily calculated by noting that

$$\sqrt{-g} det\left(\frac{\partial y}{\partial \xi}\right) = \sqrt{-det\left(g_{ab}\frac{\partial y^a}{\partial \xi^\alpha}\frac{\partial y^b}{\partial \xi^\beta}\right)} \tag{44}$$

and

$$M_{\alpha\beta} = g_{ab}\frac{\partial y^a}{\partial \xi^\alpha}\frac{\partial y^b}{\partial \xi^\beta} = diag(\gamma_{\mu\nu}, -\delta_{AB}) + o(\rho/R) \tag{45}$$

where worldsheet derivatives of x^a are down by order ρ/R. Thus $detM = det\gamma$ to order ρ/R. We also have to the same order $(D\Phi)^2 = M^{\alpha\beta}D_\alpha\Phi D_\beta\Phi = -(D_A\Phi_s)^2$ and $F^2 = F_s^2$, so when the $d^2\rho$ integral is performed the Lagrangian reduces to minus the energy per unit length, μ. The terms of order ρ are down by W/R after integrating ρ across the string profile, where W is the string width. Thus to lowest order we find the action (41). For loops seeding galaxies R is of the order of 10 pc and W is of order $m_{gut}^{-1} \approx 10^{-33}$ cm so W/R is indeed pretty small! However, in the very small regions of the string worldsheet like 'cusps' (see below), where the worldsheet becomes highly curved, we expect the corrections to be important.

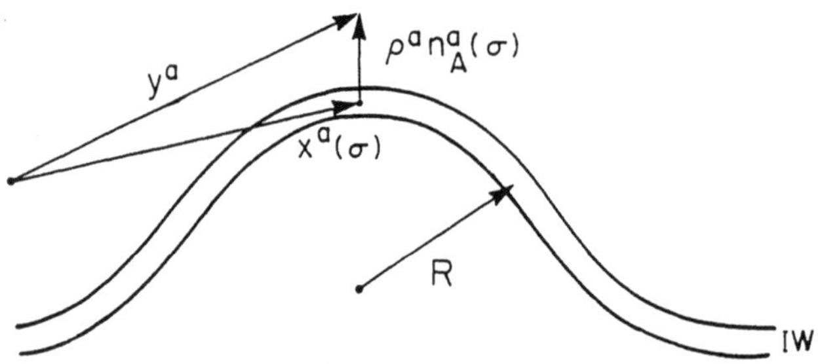

Fig. 2.8. Coordinates used in the derivation of the Nambu action. It is assumed that the radius of curvature of the string, R, is much greater than its width W.

In fact the $o(W/R)$ corrections to (41) vanish by symmetry – the first non-trivial corrections arise at order $(W/R)^2$. They are

$$\Delta S = \mu \int d^2\sigma(\alpha K^2 - \beta\omega^2) \tag{46}$$

where K is the extrinsic curvature of the worldsheet and ω the 'twist' along the string [44]. α and β are positive constants of order W^2. The sign of the extrinsic curvature term corresponds to a *negative* rigidity – cosmic string likes to be crinkled rather than straight. This might lead to interesting effects in the late time evolution of loops, where presumably it will be counteracted by the gravitational radiation acting to reduce high frequency modes on the string.

The nicest feature of the Nambu action is that it is completely geometrical. The parameter μ does not enter into the equations of motion at all. The motion of the string just depends on the background spacetime metric. Waves propagate along the string with the speed of light.

Of course our discussion breaks down where two strings collide. In this case one has to solve the full nonlinear field equations. However it really is a very local process (the string fields approach the vacuum exponentially rapidly outside the string) and can be done perfectly well in flat spacetime. Numerical calculations by Shellard [36] (for global strings) and Matzner [37] (for local

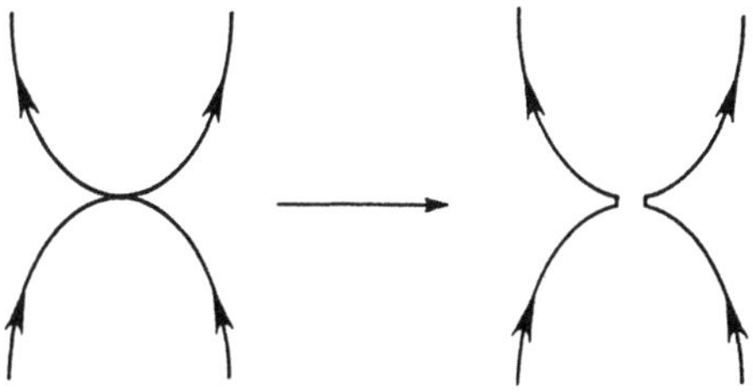

Fig. 2.9. When two strings collide they reconnect the other way.

strings) show that when two strings collide they always reconnect the other way (see Figure 2.9) for velocities less than .95 in the centre of mass frame, i.e. essentially always. This is a very nice result because the string interactions are therefore also fixed and cannot be adjusted. For strings with more complicated internal structure like superconducting strings this may not be the case.

2.10 Free Strings in an Expanding Universe

We know from present observation that at early times Ω was very close to unity and that the universe was radiation dominated and quite homogeneous. Thus the background metric in which we have to evolve the strings is completely specified. The metric is

$$ds^2 = dt^2 - a^2(t)d\boldsymbol{x}^2 = a^2(\eta)(d\eta^2 - d\boldsymbol{x}^2) \tag{47}$$

with $a(t) \propto t^{\frac{1}{2}} \propto \eta$ and η is the *conformal* time, in which the metric is conformally related to the Minkowski metric. The action (41) is invariant under *reparametrisations* of the string worldsheet. This means that to obtain a well defined initial value problem one has to choose a gauge. A convenient gauge is

$$\begin{aligned} x^0(\tau,\sigma) &= \tau \\ \partial_\tau \boldsymbol{x} \cdot \partial_\sigma \boldsymbol{x} &= 0 \end{aligned} \tag{48}$$

(i.e. choosing τ to be the time and choosing σ at each instant so that the motion of the point labelled by σ is perpendicular to the string). Upon varying (41) one finds in this gauge the equations [47]

$$\begin{aligned} \ddot{\boldsymbol{x}} + 2\frac{\dot{a}}{a}\dot{\boldsymbol{x}}(1-\dot{\boldsymbol{x}}^2) &= \frac{1}{\epsilon}\partial_\sigma\left(\frac{\boldsymbol{x}'}{\epsilon}\right) \\ \dot{\epsilon} &= -2\frac{\dot{a}}{a}\epsilon\dot{\boldsymbol{x}}^2 \\ \epsilon &= \sqrt{\frac{\boldsymbol{x}'^2}{1-\dot{\boldsymbol{x}}^2}} \end{aligned} \tag{49}$$

where dots refer to τ derivatives and primes to σ derivatives. The quantity ϵ has a simple meaning – the string energy per parameter σ length (see Section 2.14 below) is given by $\mu a(\eta)\epsilon$. The numerator in ϵ is just the comoving length of the string per unit σ, corresponding to a potential energy μ times (physical length). The denominator contributes the kinetic energy of the string.

In (49) there are two competing forces. On the left hand side is a damping term due to the expansion of the universe. On the right hand side is the curvature term which accelerates the string. The result of the competition between these terms is that for curvature scales R on the string much *smaller* than the Hubble radius $R_H = H^{-1} = a/\frac{da}{dt}$ the curvature wins and the string evolves as in flat spacetime. One finds in the case [47]

$$E \approx const \rightarrow \rho \propto a^{-3} \qquad R \ll R_H . \tag{50}$$

Highly curved strings behave like *matter*. For curvature scales much *greater* than the Hubble radius the damping term wins so $\dot{\boldsymbol{x}} \approx 0$ and the string is *conformally* stretched, its shape remaining fixed but its length growing like a. Thus

$$E \propto a \rightarrow \rho \propto a^{-2} \qquad R \gg R_H . \tag{51}$$

Either of these two behaviours would be cosmologically disastrous – the string would quickly come to dominate the radiation background (for which $\rho \propto a^{-4}$). However this is not the end of the story. Loops smaller than the Hubble radius are not a problem because they slowly radiate themselves away into gravity waves for which $\rho \propto a^{-4}$. As we shall see the long strings which would cause problems if there were no string-string interactions gradually chop themselves up into loops which then radiate themselves away.

2.11 Inside the Horizon: Free Strings in Flat Spacetime

In flat spacetime the string equations are particularly simple and can be solved analytically as follows [100, 48]. From (49) we have $\dot{\epsilon} = 0$. If we initially parametrise the string so that $\epsilon(\sigma)$ is equal to 1 along the string then it will remain so for all time. We then have

$$\begin{gathered} \ddot{\boldsymbol{x}} = \boldsymbol{x}'' \\ \dot{\boldsymbol{x}} \cdot \boldsymbol{x}' = 0 \qquad \dot{\boldsymbol{x}}^2 + \boldsymbol{x}'^2 = 1 . \end{gathered} \tag{52}$$

The general solution is given by

$$\begin{gathered} \boldsymbol{x} = \frac{1}{2}(\boldsymbol{a}(\sigma + \tau) + \boldsymbol{b}(\sigma - \tau)) \\ \boldsymbol{a}'^2 = \boldsymbol{b}'^2 = 1 . \end{gathered} \tag{53}$$

Here $\boldsymbol{a}'$ and $\boldsymbol{b}'$ are arbitrary curves on the suface of a unit sphere. For a closed loop in its centre of mass frame $\boldsymbol{a}'$ and $\boldsymbol{b}'$ are in addition constrained to have $\int d\sigma \boldsymbol{a}' = \int d\sigma \boldsymbol{b}' = 0$. Some simple nonintersecting Fourier mode solutions [55] are shown on Figure 2.10 (courtesy A. Stebbins).

It is easy to check that a closed loop has period $L/2$ where L is the parameter length in this gauge. Some of the trajectories in Figure 2.10 have 'cusps' – points where $\mid \boldsymbol{x}' \mid = 0$ and the string reverses back on itself. From (52) these points

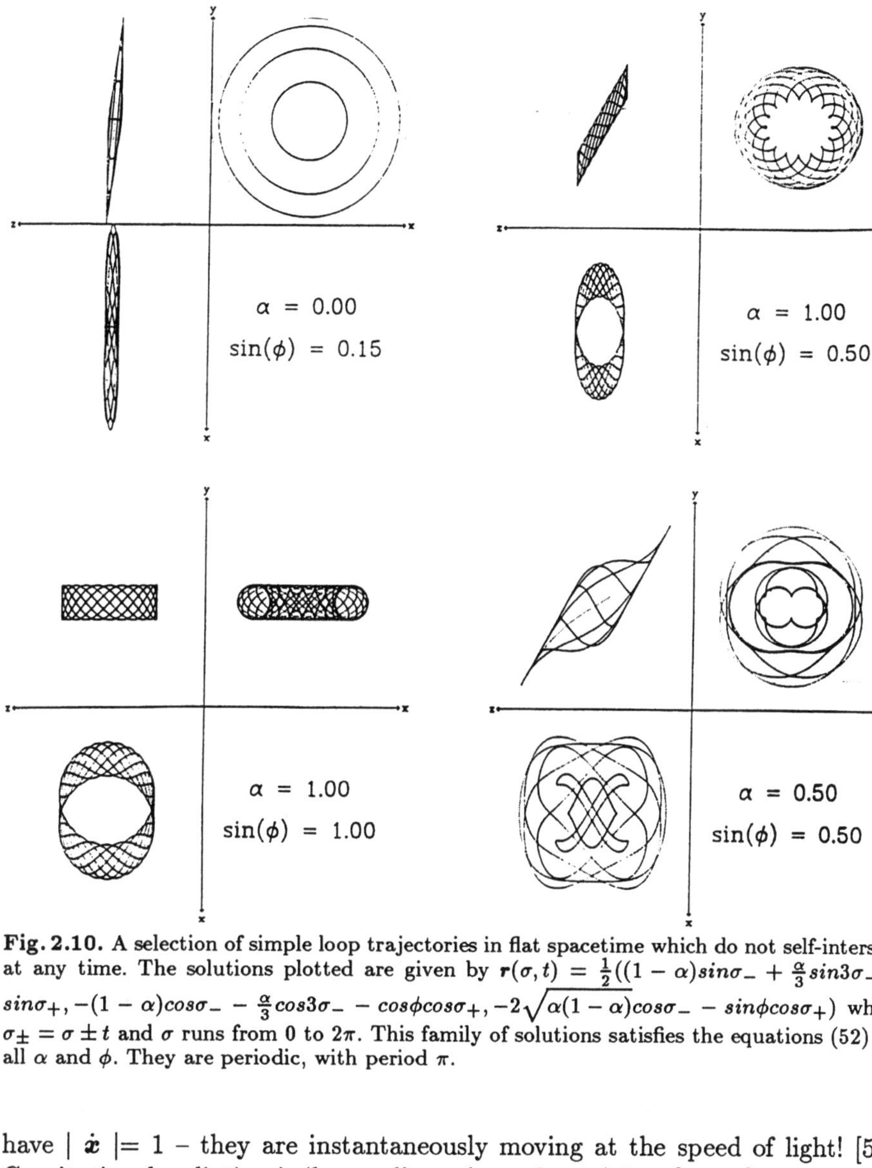

Fig. 2.10. A selection of simple loop trajectories in flat spacetime which do not self-intersect at any time. The solutions plotted are given by $\boldsymbol{r}(\sigma, t) = \frac{1}{2}((1-\alpha)sin\sigma_- + \frac{\alpha}{3}sin3\sigma_- + sin\sigma_+, -(1-\alpha)cos\sigma_- - \frac{\alpha}{3}cos3\sigma_- - cos\phi cos\sigma_+, -2\sqrt{\alpha(1-\alpha)}cos\sigma_- - sin\phi cos\sigma_+)$ where $\sigma_\pm = \sigma \pm t$ and σ runs from 0 to 2π. This family of solutions satisfies the equations (52) for all α and ϕ. They are periodic, with period π.

have $|\ \dot{\boldsymbol{x}}\ | = 1$ – they are instantaneously moving at the speed of light! [55]. Gravitational radiation is 'beamed' out from these points [55, 58].

2.12 Interacting String Networks in an Expanding Universe: the Scaling Solution

Now we proceed to the full problem of the evolution of an interacting string network in an expanding background. We simply take the initial configuration as shown in Figure 2.7 and evolve it with the equations (49) and the rule for interactions that whenever two strings cross we reconnect them the other way as in Figure 2.9.

A. Albrecht and myself have been developing a numerical code to do this for several years [49]. We have recently made qualitative improvements over our early code and brought the results to what (for us!) is a rather convincing state. Bennett and Bouchet [52] have also developed a code and our latest code agrees at least in part with theirs.

A picture of part of one of our simulations is shown in Figure 2.11. We begin with initial conditions as in Figure 2.7 in a 24^3 or 30^3 box, with 10 points on the string for every initial 'correlation length' ξ. We then evolve (49) numerically. The only parameter to vary in the simulation is the ratio of ξ to the Hubble radius $R_H \equiv a/\frac{da}{dt} = 2t$. The initial string density is approximately $\rho_s = \mu/\xi^2$, so starting at large or small R_H/ξ corresponds to starting at high or low string density relative to the background radiation density ($\rho_b = \rho_{radiation} = 3/(32\pi G t^2) = 3/(8\pi G R_H^2)$).

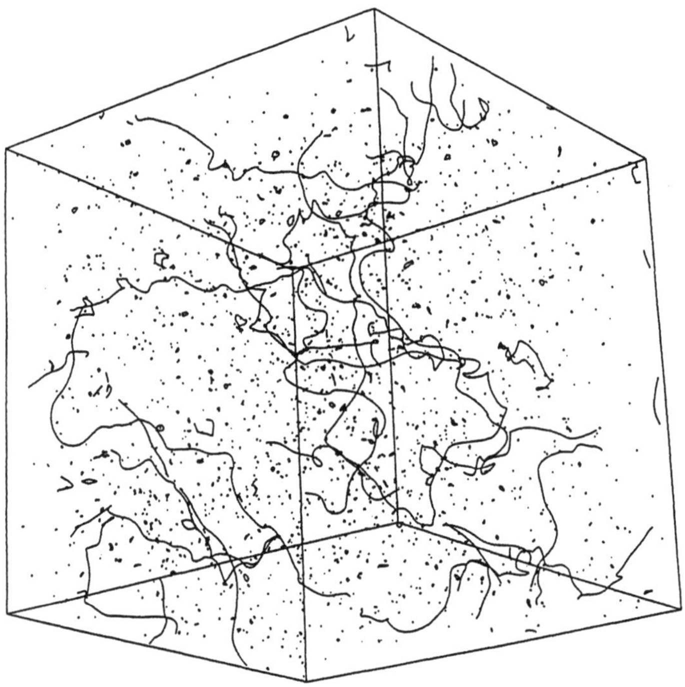

Fig. 2.11. A string network in a radiation dominated universe. The cube shown is half a Hubble radius on each side and the scale factor of the universe has doubled since the beginning of the simulation. The picture is from a forthcoming paper by Albrecht and myself.

It is useful to separate the strings into 'long string', strings longer than the Hubble radius, and 'loops', strings shorter than the Hubble radius. The precise dividing line is not very important.

Figure 2.12 shows how the density in long string (defined as loops for whom E/μ is greater than $2R_H$) behaves for differing initial string densities. A good way to think of this is that if there is a fixed number of long strings of length $\approx R_H$ per Hubble volume R_H^3 then the long string density scales with time

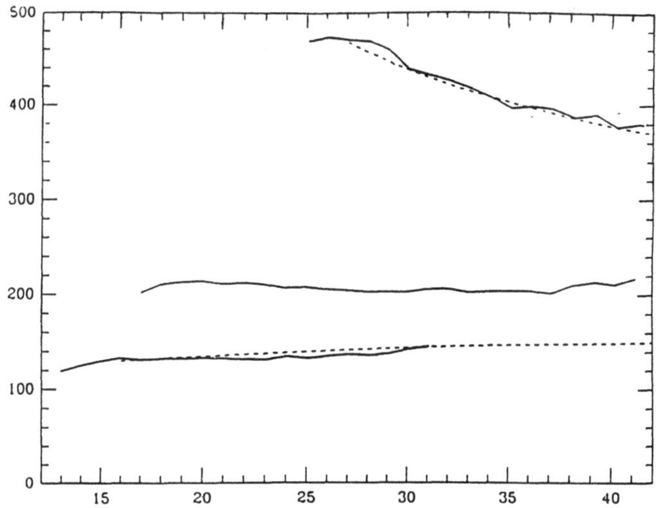

Fig. 2.12. The string density in the latest simulations by Albrecht and myself, for radiation dominated universes. The vertical axis shows the string density in units of the string tension divided by the Hubble radius squared ($\mu/(2t)^2$). The horizontal axis shows the comoving Hubble radius in units of the initial domain size ξ when the strings are formed. This is proportional to the scale factor. The scaling density appears to be at $200\mu/(2t)^2$. The upper and lower solid lines show simulation results starting at higher and lower than scaling density. The dashed line shows the prediction of the simple model explained in Section 2.12.

as R_H^{-2}, so the ratio ρ_s/ρ_b is constant. As can be seen this ratio does indeed appear to approach a constant and $\rho_s \approx 200/R_H^2$.

Why is this? The main idea of the 'scaling solution' [49, 50, 51] is that one should define a length scale ξ by

$$\rho_L = \frac{\mu}{\xi^2} \tag{54}$$

where ρ_L is the density in long strings. As long as reconnection keeps the network 'random' then ξ should also be the typical curvature scale on the string. Now as long as $\xi \ll R_H$, the strings should evolve as in flat space so that the energy in the long strings (ignoring chopping off loops) scales like matter. Also, the timescale for the long strings to chop off a fraction of their length into loops is simply ξ (recall the characteristic velocity of waves on the string is just the speed of light which we set equal to 1). Thus ignoring reconnections of loops we find

$$\dot{\rho}_L = -3\frac{\dot{a}}{a}\rho_L - c\frac{\rho_L}{\xi} \tag{55}$$

with c a constant (here dot refers to $\frac{d}{dt}$ not $\frac{d}{d\eta}$). The first term is simply a result of the expansion, the second the loss of energy into loops. Ignoring reconnections is reasonable in the light of the results that in flat space chopping off is heavily favoured by phase space over reconnection.

Equation (55) has a simple solution: for a radiation dominated background $a \propto t^{\frac{1}{2}}$ it is given by

$$\rho_{sc} = \frac{\mu}{c^2 R_H^2} = \frac{\mu}{c^2 (2t)^2} . \tag{56}$$

This is called the *scaling solution.*

In fact the scaling solution is a stable fixed point of (55) as may be seen by defining $y = (\rho_L/\rho_{sc})$ and calculating

$$\dot{y} = \frac{y}{R_H}(1 - \sqrt{y}) . \tag{57}$$

If y is greater than unity then it decreases, but if it is less than unity it increases. Thus $y = 1$ is a stable fixed point. Taking the initial scale factor a_i to be unity (57) may be solved for an initial value y_i to yield

$$y = \frac{y_i \sqrt{a}}{(1 - y_i(1 - \sqrt{a}))} . \tag{58}$$

Thus y approaches unity rather slowly.

The dashed lines on Figure 2.12 show the predicted approach to scaling from (58) with y_i chosen soon after the initial time (there is a short initial transient due to the fact that we start our strings off stationary in comoving coordinates). As can be seen the results fit (58) rather well. Figure 2.11 shows the string configuration in a carved out Hubble volume (actually $(R_H/2)^3$) in the scaling solution.

2.13 Loops

The distribution of loops produced in the scaling solution plays a very important role in the cosmic string theory of large scale structure formation. The importance of loops was first emphasised in [53] by Vilenkin, and the simple identification of one loop with one object today proposed in [54, 64, 66].

The energy lost from long strings in (55) goes into loops. Defining $\rho_l(e)$ to be the density contributed by loops of energy e to $e + de$ we have an equation

$$\dot{\rho}_l(e) = -3\frac{\dot{a}}{a}\rho_l + \frac{1}{R_H^4} f(e,t) \tag{59}$$

which defines the dimensionless *loop production function* f. In the scaling solution there is only one length and time scale in the problem, the Hubble radius $R_H = 2t$ in the radiation era. Thus f being dimensionless can only be a function of $e/(\mu R_H)$. Some of the loops produced will further selfintersect. I will ignore this complication here by regarding f as the loop production function for non-self-intersecting loops. Integrating (59) we find

$$\begin{aligned} \rho_l(e)de &= \left(\frac{\mu R_H}{e}\right)^{\frac{3}{2}} \frac{\lambda de}{R_H^3} \\ \lambda &= \frac{1}{2}\int_0^{x_C} dx x^{\frac{1}{2}} f(x) \end{aligned} \tag{60}$$

for loops with $e/(\mu R_H) \ll 1$ and we define $f(x)$ to be zero for $x > x_C \approx 1$ – such loops are simply part of the long string. Similarly the number of loops per unit volume with energy e to $e + de$ is

$$n_l(e)de = \left(\frac{\mu R_H}{e}\right)^{\frac{3}{2}} \frac{\lambda de}{e\, R_H^3} \,. \tag{61}$$

Our original simulations found $\lambda \equiv \nu(2\beta)^{\frac{3}{2}}$ (in the old notation; see [66]) approximately equal to 0.8. The results of our latest simulations will shortly be revealed [49]. Equation (60) shows that the loop density scales like matter – the small loops have been around longest so they dominate the total density in strings. This would be a disaster if the loops were totally stable. However they do slowly decay by emitting gravity waves, at a rate which is exactly calculable for simple loop trajectories [55, 57, 59]. The rate of energy loss is given by $\dot{e} = -\Gamma G\mu^2$ independent of the size of the loops, with Γ generically of order 50–100 for simple loops. The energy of a loop thus decreases slowly with time: $e(t) = e_i - \Gamma G\mu^2(t - t_i)$ where e_i and t_i are the initial values. For $t \gg t_i$ this yields the final loop distribution

$$n_l(e,t)de = \left(\frac{\mu}{R_H}\right)^{\frac{3}{2}} \frac{\lambda de}{(e + \Gamma G\mu t)^{\frac{5}{2}}} \,. \tag{62}$$

These loops act as gravitating 'seeds' which start accreting matter at around t_{eq} in the manner described in Section 1. The mass a loop accretes is proportional to its mass (with cold dark matter and as long as it was produced before t_{eq}, see below). Smaller loops are more numerous and give rise to smaller objects like galaxies. Larger loops are rarer and give rise to larger objects like clusters of galaxies. Thus (62) gives a simple prediction for number density versus mass for objects in today's universe (though this is in general altered by merging).

2.14 Strings as Density Perturbations

Cosmic strings of the simplest kind interact with matter mainly through gravity. As discussed in Section 1 density perturbations only start growing appreciably when the universe becomes matter dominated. To see how strings cause matter perturbations to grow in their vicinity we need to calculate their gravitational field.

For simplicity I will only discuss strings in a flat background in the weak field limit (the gravitational potential is of order $G\mu \ll 1$, so this is a very good approximation most of the time for cosmic strings. It may fail at 'cusps'.). The string stress tensor is given from the Nambu action as

$$\begin{aligned} T^{ab}(t,\boldsymbol{x}) &= -\frac{2}{\sqrt{-g}}\frac{\delta S}{\delta g_{ab}}\Big|_{g_{ab}=\eta_{ab}} \\ &= \mu \int d\sigma \begin{pmatrix} 1 & \dot{x}^i \\ \dot{x}^j & \dot{x}^i\dot{x}^j - x'^i x'^j \end{pmatrix} \delta(\boldsymbol{x} - \boldsymbol{x}(t,\sigma)) \end{aligned} \tag{63}$$

in the gauge (48) where η_{ab} is the Minkowski metric and $\boldsymbol{x}(t,\sigma)$ the string trajectory in space. For a straight static string along the z axis, $\boldsymbol{x}(t,\sigma) = (0,0,\sigma)$, we find $T_{00} = -T_{33} = \mu\delta^2(\boldsymbol{x})$ with other components zero. The sign of the T_{33} term corresponds to a *negative* pressure, i.e. a tension along the z axis.

To calculate the metric perturbation produced by the string we write the metric as $g_{ab} = \eta_{ab} + h_{ab}$ with $|\, h \,| \approx G\mu \ll 1$. In the harmonic gauge $g^{ab}\Gamma^c_{ab} = 0$ the linearised Einstein equations become [6]

$$\partial^2 h_{ab} = -16\pi G(T_{ab} - \frac{1}{2}\eta_{ab}T^c_c) \ . \tag{64}$$

The usual Newtonian potential is just $h_{00}/2$: for our straight string it is easily seen that the source for h_{00} vanishes exactly! The only regular solution is $h_{00} = 0$ everywhere, so there is no Newtonian potential outside the string. h_{33} also vanishes but there is a nonzero h_{11} and h_{22};

$$\begin{aligned} \nabla^2 h_{11} &= \nabla^2 h_{22} = 16\pi G\mu\delta(\boldsymbol{x}) \\ \Rightarrow h_{11} &= h_{22} = 8\pi G\mu \ln(r/W) \end{aligned} \tag{65}$$

choosing $h_{11} = h_{22} = 0$ at the string width W. Inside the string the true stress energy is not singular like the 'Nambu' approximation (63) and this removes the apparent singularity at $r = 0$ [63]. Clearly the weak field approximation breaks down at very large radii as well. Ignoring this for the moment we have in cylindrical coordinates

$$ds^2 = dt^2 - dz^2 - (1 - 8G\mu\ln(r/W))(dr^2 + r^2 d\phi^2) \, . \tag{66}$$

We can remove the prefactor to the dr^2 term by defining a new radial variable $dr' = dr(1 - 4G\mu\ln(r/W))$ or $r' = r(1 + 4G\mu(1 - \ln(r/W))$. Neglecting terms of order $(G\mu)^2$, consistent with our approximation, we then find

$$\begin{aligned} ds^2 &= dt^2 - dz^2 - dr'^2 - r'^2(1 - 8G\mu)d\phi^2 \\ &= dt^2 - dz^2 - dr'^2 - r'^2 d\phi'^2 \end{aligned} \tag{67}$$

upon defining $\phi' = (1 - 4G\mu)\phi$. This is not only regular everywhere, it is flat space! However the new angular variable ϕ' only runs from 0 to $2\pi - 8\pi G\mu$ – the geometry is that of 'flat space minus a wedge', or a cone [61, 60]. This is illustrated in Figure 2.13. The only difference is at radii of order W, the string width, where the apex of the cone is smoothly rounded out [62, 63].

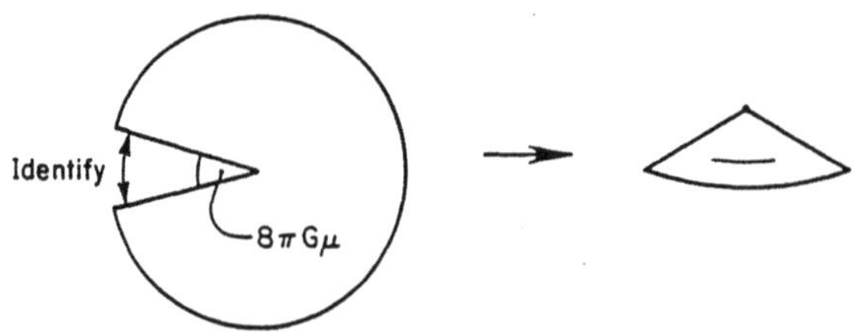

Fig. 2.13. The conical metric of a long straight string. If an angle $8\pi G\mu$ is 'cut out' of flat space (left) and the two edges identified, a cone is the result (right).

What about loops? These *do* produce a Newtonian potential – one finds

$$\partial^2 h_{00} = -16\pi G\mu \int d\sigma \dot{\boldsymbol{x}}^2 \delta^3(\boldsymbol{x} - \boldsymbol{x}(\sigma, t)) . \tag{68}$$

For an oscillating loop the right hand side oscillates periodically in the centre of mass frame of the loop. This leads to a peculiar pulsating gravitational field [54]. Its time average, however, is simple, because of the following identity. If the length of the loop is L and its period T we have

$$\begin{aligned}\int \frac{d\sigma}{L} \int \frac{dt}{T} \dot{\boldsymbol{x}}^2 &= \int\int -\boldsymbol{x} \cdot \ddot{\boldsymbol{x}} = \int\int -\boldsymbol{x} \cdot \boldsymbol{x}'' \\ &= \int\int \boldsymbol{x}'^2 = \int\int (1 - \dot{\boldsymbol{x}}^2) \\ &\Rightarrow \int\int \dot{\boldsymbol{x}}^2 = \frac{1}{2}\end{aligned} \tag{69}$$

where we used integration by parts and the equations (52). The average velocity squared for a closed loop is thus $\frac{1}{2}$. Now if we time average (68) we find the time averaged long range field of a loop is that of a shell of Newtonian dust with suface density proportional to $\dot{\boldsymbol{x}}^2$ and with total mass equal to μL, the mass of the loop. At distances from the loop much greater than its size we can treat it to a good approximation as a point mass.

Vachaspati [58] has tried to use the time variation of h_{ab} and the *beaming* effect mentioned earlier [55] to argue that an accreting loop will eject matter into an interesting shape. However the actual amount of mass which is in the region where h is so large that it moves the matter substantially during the existence of a cusp is tiny, so most of the matter around the loop simply sees the time averaged field, which shows no beaming effect [55].

2.15 Accretion by Loops

Accretion by moving loops has been looked at in some detail in [65] and [70]. In linear theory, the equation for the density perturbation $\delta\rho/\rho(\boldsymbol{x})$ is purely local in comoving coordinates, as noted in Section 1. This means that the surface 'traced out' by the loop as it moves through space will appear in the density perturbation it produces. This is only true while the perturbation is linear – as soon as collapse and virialisation occur the structure will be smoothed out and the object become more and more spherical. The shape traced out by a loop is in fact like a 'trumpet' [70]. This is seen as follows. The peculiar velocity of a loop $v_{pec} = a(t)\dot{\boldsymbol{x}}$ decays[3] as a^{-1}. This means that, in the matter dominated era, the comoving distance a loop moves is given by

[3] More precisely the momentum $\int d\sigma\epsilon\dot{\boldsymbol{x}}$ decays as a^{-1} as can be seen from (49), but if its velocity is substantially less than 1 and it is well inside the horizon then the energy is roughly constant.

$$\Delta x = \int_{t_i}^{t} v_i \left(\frac{a_i}{a}\right)^2 dt = 3v_i t_i \left(1 - (\frac{a_i}{a})^{\frac{1}{2}}\right) \tag{70}$$

where v_i is the initial velocity, at scale factor a_i and time t_i. Thus a loop slows down and stops in comoving coordinates in the matter era (in the radiation era Δx increases logarithmically with time). However the radius of the loop in comoving coordinates r_c shrinks (the *physical* radius is approximately constant, ignoring oscillations) so

$$r_c = r_i \frac{a_i}{a} \ . \tag{71}$$

So defining the distance from where the loop ends up as $z = 3v_i t_i - \Delta x$ we see that the shape traced out by the loop is

$$r_c \propto z^2 \tag{72}$$

the surface of revolution of a parabola – a 'trumpet' shape. As stated above, nonlinearity smears this out, making the accreted lump in the end rather spherical [65]. For galaxies this would be the case today. However larger clusters should still exhibit this shape – in fact the Perseus-Pisces filament [3] does look rather 'trumpet' like! Bingelli has made an attempt to measure the eccentricity for Abell clusters – this will serve as an interesting test in the future [10].

2.16 Loops as Seeds for Galaxies and Clusters

The string plus cold dark matter (SCDM) scenario is the easiest to understand. It is clear from Section 1 that the *total mass* accreted by a loop around at t_{eq} is today roughly $M_{loop}(1 + Z_{eq})$. In fact a precise calculation in the spherical collapse model yields [67]

$$m_{loop}(1 + Z_{eq}) = \frac{\xi}{5} M \left(\frac{\rho}{\rho_b}\right)^{\frac{1}{3}} \tag{73}$$

where M is the mass of the accreted object, ρ/ρ_b is the overdensity today and ξ is a 'loss of growth' factor, equal to 1 for loops laid down well before t_{eq}, equal to 4 for loops laid down at t_{eq} and greater for loops laid down still later.

It is now simple to calculate the required string tension. First we pick a class of objects, for example Abell clusters. Their number density is $1/(110 h_{50}^{-1}\,\mathrm{Mpc})^3$. Let me assume that the loops seeding them are all produced in the radiation era for simplicity (they are actually produced at or just after the matter-radiation transition according to our latest simulations). Now we take the loop distribution, given in equation (62), and evolve it through the matter-radiation transition to the present day. We then ask for what cutoff in energy will the integrated number density be equal to $1/(110 h_{50}^{-1}\,\mathrm{Mpc})^3$. Doing this [66, 67], one finds that the length of loops producing clusters is $L = e/\mu \approx 20\,\mathrm{kpc}$. We can then demand that loops of this size are massive enough to accrete clusters by today, from (73) and the observed mass and overdensities of clusters [67]. One finds

$$G\mu = 2 \cdot 10^{-6} h_{50}^{-1} \sigma_{700}^{\frac{8}{3}} d_{110}^{-2} \lambda^{-\frac{2}{3}} \xi_4 \ . \tag{74}$$

Here σ_{700} is the velocity dispersion in a cluster in units of $700\,\mathrm{km\,s^{-1}}$, d their mean separation in units of 110 h_{50}^{-1}Mpc, λ the constant in (62) and ξ the growth factor mentioned before. There are clearly large uncertainties in the normalisation of $G\mu$, probably of an order of magnitude either way. In particular there are several effects which might reduce $G\mu$. If clusters form by mergers of several loops, or directly from curved sections of the long strings, this would decrease the required value. This is one of the main motivations for the more detailed calculations now being performed in which we are directly evolving the matter distribution around a string network produced in the simulations [71].

Independently of $G\mu$ however, the *spatial distribution* of the loops and long strings should have testable consequences for the spatial distribution of galaxies and clusters, particularly clear cut on scales larger than those where gravitational clustering has been efficient (i.e. greater than 12 h_{50}^{-1} Mpc or so, see Section 1). Unlike theories like the ICDM model, there are no parameters to adjust here – the evolution of the strings is independent of the string tension. An early calculation [64] showed that remarkably enough the two-point correlation function of loops produced in a cosmic string simulation (in the radiation era) matched the observed cluster-cluster correlation function very closely. Subsequent simulations by Bennett and Bouchet, and those by Albrecht and myself using our improved code, have confirmed this correlation of loops when they are produced. The more detailed calculations using improved codes and including the radiation-matter transition and the effect of the long strings will show whether this apparently miraculous agreement with observation is confirmed.[4]

2.17 Cosmic String and Hot Dark Matter

It has recently been shown that cosmic string could solve the usual problems of the standard Hot Dark Matter (IHDM) scenario [25, 83] (see also the earlier work of [82]). I will call this the SHDM scenario. This is nice – the massive neutrino is still the best motivated dark matter candidate. Of course exactly why $m_\nu = 30 eV$ as is required is still a mystery (as are the other fermion masses!).

Without going into details, the reason for the improvement is that loops survive the free streaming era and begin accreting baryons slowly at around recombination. The baryon/loop lumps then start accreting neutrinos efficiently

[4] It has been argued, in particular by Bennett and Bouchet [68], that if one extrapolates the loop positions for several expansion times then the correlations are washed out by loop velocities. I believe this is a red herring – loops forming clusters are produced at or just after t_{eq} – accretion begins as soon as the loops are produced. Thus the *initial* positions of the loops is most relevant. Second, some (e.g. Scherrer and Press [69]) have argued that loop fragmentation produces fragments with fairly high average velocities $\approx 0.6c$, so one can use randomly distributed loops to calculate structure formation. However it is the larger loops around (and even those that will subsequently self-intersect) at or after t_{eq} that will form clusters, and these have much lower velocities [49]. String plus matter simulations will hopefully resolve the issue.

when the neutrino Jeans mass falls to galaxy scales. Flat rotation curves are obtained automatically (once heralded as a success of the standard ICDM scenario). The loss of growth on galaxy scales means that a higher value of $G\mu$ is needed – this brings the theory more into line with the observed large scale velocity fields [78]. Clusters and their correlations are similar to SCDM [25, 26, 83].

2.18 Cosmic String and Inflation?

How do cosmic strings fit in with inflation? Many cosmologists today take the idea of inflation very seriously (some even refer to it as a 'paradigm') because it 'explains' why the universe is so large, so homogeneous and so flat today.

However inflation is not a gospel. Many people including myself feel rather unhappy with the 'explanations' it offers. At a general level, it is certainly true that inflation will turn a wider class of initial conditions into a large homogeneous universe like our own. However it certainly will not turn *arbitrary* initial conditions into universes like ours and until it is possible to say precisely 'how much better' inflation makes things, one really cannot judge the idea.

It also seems very likely that it will be quite some time before we can say anything very definite about the initial conditions of the universe that is in any way testable or provable. The whole discussion may well simply remain in the realm of philosophy rather than physics.

In cosmology today there is a strong case for being conservative in our ambitions. It has always been easy in cosmology to get into the realms of untestable speculation. What is new today is that there is a rapidly increasing data base upon which theories can be tested. And there are now at least two rather precise theories to test. This is the most hopeful aspect of the subject. Now the aspect of inflation that *is* certainly a major advance from this point of view is that the fluctuation spectrum it 'predicts' (at least in the simplest version [73]) has definite consequences for the galaxy distribution, the microwave background fluctuations and so on. Cosmic strings are from this point of view also a very good theory, with clear testable predictions.

In fact one can make the case that cosmic strings are a much better justified extrapolation of low energy physics. The underlying successful idea in high energy physics has been that of gauge symmetry and spontaneous symmetry breaking. Cosmic strings are fairly generically (though certainly not always) produced in symmetry breaking phase transitions, and the precise details of the symmetry breaking (i.e. whether it occurs dynamically via a condensate or through a scalar 'Higgs field') are not crucial. On the other hand it is very difficult to make inflation work in a gauge symmetry breaking phase transition. Instead one has to use a gauge *singlet* inflaton field which 'allows' one to tune the coupling constant to very small values. This raises real questions about the initial distribution of the inflaton field – it was almost certainly not in thermal equilibrium. It also makes the subject very ad hoc.

I am often faced with the question 'well, how do cosmic strings explain the flatness and homogeneity of the universe?' to which I reply 'they don't'. It is perfectly legitimate in seeking to explain the large scale structure in the universe to use the fact that we *know* the universe was homogeneous and flat at early times, from the microwave background. This is exactly what we do in nucleosynthesis calculations for example.

Inflation does not go well with cosmic strings (as you might have guessed from my critical comments!). During inflation the string density would fall exponentially, so that after inflation there would be far fewer than one string per horizon volume. Indeed that is how inflation solves the monopole problem (to which there are other solutions [74, 75, 76]). For the reasons I have given, this does not bother me too much.

If you want to use inflation to solve the monopole problem, but have cosmic strings provide the fluctuations later on, you have to 'stand on your head'. You must first form the monopoles, inflate them away and then reheat to a temperature high enough to form strings but too low to form monopoles. This is tricky – in most inflationary models the universe heats up to temperatures well below the 10^{16} GeV scale typically needed to form strings. However in cosmic strings the mass per unit length $\mu \approx \eta^2$ is only weakly dependent on couplings whereas the phase transition temperature $T_c \approx \sqrt{\lambda/g^2}\eta$ can be made much lower by decreasing λ or imposing discrete symmetries which only allow high powers of Φ in the Higgs potential which effectively does the same thing [77].

Just as none of the models for inflation alone is compelling, none of the models for inflation plus strings is particularly inspiring. In fact they are quite ugly [81, 80].

2.19 Cosmic String Versus Inflation

Let me now turn to comparing the predictions of cosmic strings with those of 'inflation'. By inflation I mean the simplest prediction of quantum fluctuations from inflation – the scale-free Harrison-Zel'dovich spectrum [84, 85, 86].

The theories are really very different. In the case of inflation it makes sense to think of the density perturbation field as a linear superposition of plane waves with random phases

$$\frac{\delta\rho(\boldsymbol{x})}{\rho} = \sum_{\boldsymbol{k}} \delta_{\boldsymbol{k}} e^{i\boldsymbol{k}\cdot\boldsymbol{x}} \tag{75}$$

with the probability distribution for each $\delta_{\boldsymbol{k}}$ being a Gaussian: if $\delta_{\boldsymbol{k}} = r_{\boldsymbol{k}} e^{i\theta_k}$ then each mode is independent (except for the reality condition which forces $\delta_{-\boldsymbol{k}} = \delta_{\boldsymbol{k}}{}^*$) with probability measure

$$\prod_{\boldsymbol{k}} \frac{d\theta_{\boldsymbol{k}}}{2\pi} \frac{dr_{\boldsymbol{k}} r_{\boldsymbol{k}}}{P_k} \exp\{-[(r_{\boldsymbol{k}})^2/(2P_k)]\} \; . \tag{76}$$

This is a result of the inflaton field being essentially a free massless field during inflation. The power spectrum $P_k = <\mid \delta_{\boldsymbol{k}} \mid^2>$ completely defines the theory. The fact that the different $\boldsymbol{k}$ modes are decoupled means that any quantity like the excess mass in a ball of radius R

$$\delta M = \int_0^R d^3 r \delta\rho(\boldsymbol{r}) \tag{77}$$

is also Gaussian distributed, by the central limit theorem. This makes large fluctuations away from the 'typical' fluctuations very rare.

With strings the situation is quite different, and

$$\delta\rho(\boldsymbol{x}) = \sum_{loops,\ long\ strings} \delta\rho(\boldsymbol{x}) \tag{78}$$

is often the simplest way to look at the density perturbations. Many different $\boldsymbol{k}$ modes are correlated in a loop, for example. The probability distribution is nonGaussian as can be seen from (62) – in a volume V (ignoring correlations) the number of loops expected with mass M to $M + dM$ is

$$N(M)dM \propto M^{-\frac{5}{2}} dM \ . \tag{79}$$

When this number is much less than one, it is just the probability of finding a loop of mass M in the volume. Thus $P(\delta M) \propto (\delta M)^{-\frac{5}{2}}$ at large δM. Any one realisation of cosmic string looks like a collection of spikes and lines in the density distribution – small loops and long strings cause very high density perturbations locally. Fluctuations from inflation however look like a superposition of plane waves that is everywhere linear. This should make it easier in principle to identify objects with initial density perturbations in the cosmic string theory – the distinction between a lump and the surrounding matter is more clear-cut. In inflation there are mass fluctuations on every scale right down to the smallest scales. This makes the identification of 'galaxies' quite hazardous. Of course merging of lumps, which is important in the cosmic string plus CDM scenario [87] tends to destroy the identification of one loop with one object.

The 'typical' mass fluctuation on a given scale δ_M is actually quite similar in both theories – $\delta M/M \propto R^{-1.25}$ approximately for scales R corresponding to the mean separation of galaxies [72] for inflation, which leads to the successful prediction of the galaxy-galaxy correlation function. For strings, the mass m of the largest 'typical' loop in a volume R^3 is given from the loop distribution (62) by $n_>(m)R^3 \approx 1$ so $\delta M \propto R^2$ and $\delta M/M \propto R^{-1}$ instead. So the gross mass fluctuation spectrum on galaxy scales is not very different.

One of the virtues of the prediction from inflation is that it is much easier to calculate with. The linear phase of the evolution is relatively trivial. With strings, however, it is necessary to do a computer simulation of the strings even before including the growth of perturbations. This should not of course mistakenly be seen as a reason why one theory is 'better'.

The successes of the ICDM theory are mostly on small scales where clustering has gone nonlinear. With the correct *biassing* factor (initially put in by hand but now argued by some to emerge naturally [23]) the ICDM model matches the galaxy correlation function, pair velocity distribution, and gives flat rotation curves in good agreement with observation [21]. This is impressive but it must be remembered that the nonlinear dissipative baryonic physics has not yet been put in, and this is actually in determining whether the postulated bias actually happens.

In contrast, the ICDM model has fared rather poorly in matching the large scale observations. These are, as I argued in Section 1 the most unambiguous testing ground for theories of the origin of structure. The basic problem of the ICDM model is that it has very little power on large scales – the predicted cluster-cluster correlation function for example goes through zero at only 80 h_{50}^{-2} Mpc [22]. With suitable *biassing* it is claimed that it can produce 'bubbles' such as those seen in the 'slice of the universe' [21]. However with this biassing it fails by a long way to match the observed velocity field [17, 79]. While the theory is not yet clearly ruled out it certainly has enough problems to make the investigation of alternatives such as cosmic strings worthwhile. In particular the SHDM model offers particular promise.

Cosmic strings are, as I have said, harder to calculate with. However there are already indications that cosmic strings have a lot of power on large scales. Early calculations of the correlation function for loops showed remarkable agreement with that for Abell clusters [64]. The later simulations have confirmed these correlations are present and have the correct r^{-2} form. However they have also found a much higher scaling density and this makes the long strings relatively more important. The detailed calculation of large scale structure produced by long strings is underway. This will also be important in accurately normalising the theory.

2.20 Other Observational Tests for Cosmic Strings

Cosmic strings have several fairly direct observational consequences which give us some confidence that if they exist, and were massive enough to seed the observed large scale structure, they should be observable fairly soon.

The most clear-cut effect, as with other theories, is the predicted anisotropy of the microwave background. Observations of intrinsic anisotropies would be a major advance in cosmology – the statistics of the fluctuations would tell us a lot about the kind of processes that could have caused them.

With cosmic strings the effect is quite dramatic. Moving strings lead to linear discontinuity across the sky in the temperature of the microwave background [88]. This is a result of the conical metric discussed in Section 2.14. In the string's restframe there is a difference of $8\pi G\mu$ in the angle at which light from the same point on the last scattering surface of the microwave radiation is recieved by the observer. The observer thus appears to have a small component $8\pi G\mu v_s$ *towards* the last scattering surface behind the string, where v_s is the string's velocity perpendicular to the line of sight. This results in a Doppler shift of the radiation temperature observed of

$$\frac{\delta T}{T} = 8\pi G\mu v_s \tag{80}$$

of the order 10^{-5} if $G\mu = 10^{-6}$. This is not far below present sensitivities. The difficulty is that strings are actually quite rare on the last scattering surface, so one has to cover quite a large angle in order to be likely to see one at all. Recently Bouchet et al. have calculated the expected pattern from a string simulation [89] – they deduce the limit

$$G\mu < 5 \cdot 10^{-6} \tag{81}$$

from the experiment of Melchiorri et. al. [90].

Another unique signature of cosmic strings is their gravitational lensing effect. The conical metric leads, in the case of a straight string, to two identical images for objects behind the string with angular separation

$$\delta\theta = 8\pi G\mu \frac{d_o}{d_o + d_s} \tag{82}$$

where d_s and $d_o > d_s$ are the distances from us to the string and the object. This is of the order of a few arc seconds if $G\mu = 10^{-6}$. What is unique is that a string would produce a line of double images. In fact a candidate event involving four double galaxies was reported recently by Cowie and Hu [91]. Deeper observations in the same area of the sky to see whether galaxies behind the putative string connecting the objects are also lensed are currently underway.

A less direct but still powerful test of the cosmic string theory comes from the gravitational radiation produced by the strings during the radiation dominated era. The fact that the millisecond pulsar is so 'quiet' imposes strong limits on the level of gravitational background that can be around today. The expected contribution to Ω_{GW} today from cosmic strings with periods of the order of a year or so (the millisecond pulsar observations put the stringest constaint on waves with this period) is [92, 93, 94, 95]

$$\Omega_{GW} \approx 2 \cdot 10^{-7} \lambda \left(\frac{\Gamma}{50}\right)^{-\frac{1}{2}} \left(\frac{G\mu}{10^{-6}}\right)^{\frac{1}{2}} h_{50}^{-2} \Omega^{-1} \tag{83}$$

whereas the observations already constrain $\Omega_{GW} < 4 \times 10^{-7}$ [96]. Things are already getting tight! This limit is improving, and will rapidly improve with time should another quiet millisecond pulsar be discovered, and should be able to convincingly rule out the scenario with $G\mu > 10^{-8}$ in 10 years' time.

The other major effect of the gravity waves produced by strings in the radiation era is the disruption of standard nucleosynthesis. This gives the bound [97, 51, 95, 98]

$$G\mu < 10^{-5} \left(\frac{\Gamma}{50}\right) \lambda^{-2} . \tag{84}$$

The strength of these bounds can only be assessed when we have an accurate calculation of the normalisation of $G\mu$, as discussed in Section 1.16. One cautionary remark about the gravitational radiation bounds must be made. The calculations of the rate of loss of radiation were made using the Nambu action and equations. If there is more complicated dynamics, particularly in the late history of a loop, as I discussed in connection with corrections to the Nambu action, or if loops continue self-intersect late in their lifetime Γ could increase, and there could be other types of radiation emitted. Both of these effects would weaken the bounds (83) and (84). Finally there are many other large scale structure predictions waiting to be made and tested – higher correlation functions, void statistics, velocity field predictions and so on.

Acknowledgements

I would like to thank the organisers and participants of the Schladming School for making it such a stimulating one. I also thank A. Albrecht for a very enjoyable continuing collaboration. This work was supported in part by the DOE and by the NASA at Fermilab.

References

1. E. Kolb and M.S. Turner, The Early Universe, Addison Wesley, 1988.
2. T.W.B. Kibble, J. Phys. A9 (1976) 1387; Physics Reports 67 (1980) 183.
3. R. Giovanelli and M.P. Haynes, Astrophysical Journal 87 (1982) 1355.
4. R.F. Kirschner, A. Oemler, P.L. Shecter and S.A. Schectman, in 'Early Evolution of the Universe and its Present Structure', I.A.U. Symposium 104 (1983) 197.
5. V. De Lapparent, M. Geller and J. Huchra, Astrophysical Journal 302 (1986) L11.
6. S. Weinberg, Gravitation and Cosmology: Principles and Applications of the General Theory of Relativity, John Wiley, New York, 1972.
7. P.J.E. Peebles, The Large Scale Structure of the Universe, Princeton University Press, 1980.
8. J. Yang, M.S. Turner, G. Steigman, D.N. Schramm and K. Olive, Ap. J. 281 (1984) 493.
9. J. Primack, Dark Matter in The Universe, lectures presented at the International School of Physics, Varenna 1984, SLAC-PUB-3387.
10. B. Bingelli, Astron. Astrophys. 107 (1982) 338.
11. D. Koo, R. Kron, and A. Szalay, 13th Texas Symposium on Relativistic Astrophysics, Ed. P. Ulmer, World Scientific, 1986.
12. G.O. Abell, Astrophysical Journal Supp. 3 (1958) 211; for a nice review see J. Oort, Ann. Rev. Astron. Astrophys. 21 (1983) 373.
13. N.A. Bahcall and R.M. Soneira, Astrophysical Journal 270 (1983) 20.
14. M. Davis and J. Huchra, Astrophysical Journal 254 (1982) 437.
15. N.A. Bahcall and W.S. Burgett, Astrophysical Journal 300 (1986) L35.
16. S. Schectman, Astrophysical Journal Supp. 57 (1985) 77.
17. D. Lynden-Bell, S.M. Faber, D. Burstein, R.L. Davies, A. Dressler, R.J. Turlevich and G. Wegner, Astrophysical Journal, March 1988 in press.
18. A. Szalay and D.N. Schramm, Nature 314 (1985) 718.
19. J.R. Bond and G. Efstathiou, Astrophysical Journal 285 (1984) L45.
20. D. Burstein and V.C. Rubin, Astyrophysical Journal 297 (1985) 423.
21. M. Davis, G. Efstathiou, C. Frenk and S. White, Ap. J. 292 (1984) 371; S.D.M. White, in Nearly Normal Galaxies, From the Planck Time to the Present, Ed. S.M. Faber, Springer-Verlag, 1987.
22. J.M. Bardeen, J.R. Bond, N. Kaiser, A. Szalay, Ap. J. 304 (1986) 15.
23. C. Frenk, S. White, M. Davis and G. Efstathiou, Ap. J. 327 (1988) 507.
24. J.R. Bond and A. Szalay, Ap. J. 274 (1984) 443.
25. R. Brandenberger, N. Kaiser, D. Schramm and N. Turok, Phys. Rev. Lett. 59 (1987) 2371
26. R. Brandenberger, N. Kaiser and N. Turok, Phys. Rev. D36 (1987) 335.
27. S. Tremaine and J. Gunn, Phys. Rev. Lett. 42 (1979) 407.
28. D.A. Kirzhnits and A.D. Linde, Phys. Lett. 42B (1972) 745.

29. L. Dolan and R. Jackiw, Phys. Rev. D9 (1974) 3320;
S. Weinberg, Phys. Rev. D9 (1974) 3357.
30. A more detailed discussion is contained in K. Wilson and J. Kogut, Phys. Rep. 12 (1974) 75.
31. E. Copeland, D. Haws and R. Rivers, Fermilab preprint 1988.
32. T.W.B. Kibble, G. Lazarides and Q. Shafi, Phys. Rev. D26 (1982) 435;
D. Olive and N. Turok, Phys. Lett. 117B (1982) 193.
33. E. Witten, Phy. Lett. 149B (1985) 351.
34. J. Preskill, Phy. Rev. Lett. 43 (1979) 1365.
35. A. Vilenkin, Phys. Rep. 121 (1985) 263.
36. E.P.S. Shellard Nuc. Phys. B283 (1987) 624.
37. R. Matzner, University of Texas report, 1987.
38. E.B. Bogomolny, Sov. J. Nucl. Phys. 24 (1976) 449;
E.B. Bogomolny and M.S. Marinov, Sov. J. Nucl. Phys. 23 (1976) 355.
39. L. Jacobs and C. Rebbi, Phys. Rev. B19 (1979) 4486.
40. R. Davis, Phys. Rev. D32 (1985) 3172,
A. Vilenkin and T. Vachaspati, Phys. Rev. D35 (1987) 1138.
41. E. Witten, Nuc. Phys. B249 (1985) 557.
42. J. Ostriker, C. Thompson and E. Witten, Phys. Lett. 180B (1986) 231.
43. Y. Nambu, Proceedings of International Conference on Symmetries and Quark Models (Wayne State University, 1969); Lectures at the Copenhagen Summer Symposium, 1970.
44. K. Maeda and N. Turok, Phys. Lett. 202B (1988) 376.
45. K.B. Nielsen and P. Olesen, Nuclear Physics B61 (1973) 45.
46. J. Frieman, R. Scherrer, Phys. Rev. D33 (1986) 3556.
47. N. Turok and P. Bhattacharjee, Phys. Rev. D33 (1984) 1557.
48. N. Turok and T.W.B. Kibble, Phys. Lett. 116B (1982) 141.
49. A. Albrecht and N. Turok, Phy. Rev. Lett. 54 (1985) 1868; in preparation, 1988.
50. T.W.B. Kibble, Nuc. Phys. B252 (1985) 227.
51. D. Bennett, Phys. Rev. D33 (1986) 872.
52. D. Bennett and F. Bouchet, Phys. Rev. Lett. 60 (1988) 257.
53. A. Vilenkin, Phys. Rev. Lett. 46 (1981) 1169, 1496(E).
54. N. Turok, Phys. Lett. 123B (1983) 387.
55. N. Turok, Nucl. Phys. B242 (1984) 520.
56. T. Vachaspati and A. Vilenkin, Phys. Rev. D30 (1984) 2036;
D. Garfinkle and T. Vachaspati, Phys. Rev. D36 (1987) 2229.
57. T. Vachaspati and A. Vilenkin, Phys. Rev. D31 (1985) 3052.
58. T. Vachaspati, Phys. Rev. D35 (1987) 1767; Gravity Research Foundation Prize Essay, 1987.
59. C. Burden Phys. Lett. 164B (1985) 277.
60. A. Vilenkin, Phys. Rev. D23 (1981) 852.
61. An older reference is L. Marder, Proc. Roy. Soc. Lond. A252 (1959) 45.
62. J.R. Gott, Ap. J. 288 (1985) 422.
63. D. Garfinkle, Phys. Rev. D32 (1985) 1323.
64. N. Turok, Phys. Rev. Lett. 55 (1985) 1801.
65. E. Bertschinger Ap. J. 316 (1987) 489.
66. N. Turok and R. Brandenberger, Phys. Rev. D33 (1986) 2175.
67. N. Turok, in A Unified View of the Macro- and Micro-Cosmos, Ed. A. De Rujula, D.V. Nanopoulos and P.A. Shaver, World Scientific 1987.
68. F. Bouchet, LBL preprint, 1988.
69. R. Scherrer and W. Press, Smithsonian preprint, 1988.
70. N. Turok, in Nearly Normal Galaxies, From the Planck Time to the Present, Ed. S.M. Faber, Springer-Verlag, 1987.
71. A. Albrecht, A. Stebbins and N. Turok, in preparation, 1988.
72. P.J.E.Peebles, Ap. J. 263 (1982) L1.
73. L. Kofman and A.D. Linde, Nuc. Phys. B282 (1987) 555.
74. P. Langacker and S.-Y. Pi, Phys. Rev. Lett. 45 (1980) 1.
75. A. Everett, T. Vachaspati and A. Vilenkin, Phys. Rev. D31 (1985) 1925.
76. E. Copeland, D. Haws, T. Kibble, D. Mitchell and N. Turok, Nuc. Phys. B298 (1988) 445.
77. G. Lazarides C. Panagiotakopoulos and Q. Shafi, Phys. Lett. 183 (1987) 289.
78. R. Brandenberger, N. Kaiser, E.P.S. Shellard and N. Turok, Phys. Rev. D36 (1987) 335.

79. J.R. Bond, in Nearly Normal Galaxies, From the Planck Time to the Present, Ed. S.M. Faber, Springer-Verlag, 1987.
80. Q. Shafi and A. Vilenkin, Phys. Rev. D29 (1984) 1870.
81. E. Vishniac, K. Olive and D. Seckel, Nuc. Phys. B289 (1987) 717.
82. A. Vilenkin and Q. Shafi, Phys. Rev. Lett. 51 (1983) 1716.
83. E. Bertschinger and P. Watts Ap. J. 328 (1988) 23.
84. E. Harrison, Phys. Rev. D1 (1980) 2726.
85. Ya. B. Zel'dovich, M. N. R. A. S., 160 (1982) 1P.
86. S. Hawking, Phys. Lett. 115B (1982) 295;
A. Guth and S.-Y. Pi, Phys. Rev. Lett. 49 (1982) 1110;
J. Bardeen, P. Steinhardt and M. S. Turner, Phys. Rev. D28 (1983) 679.
87. R. Brandenberger and E.P.S. Shellard, Brown preprint, 1988.
88. N. Kaiser and A. Stebbins, Nature 310 (1984) 391;
A. Stebbins, Ap. J. 327 (1988) 584.
89. F. Bouchet, D. Bennett and A. Stebbins, Fermilab preprint, 1988.
90. F Melchiorri, B. Melchiorri, C. Ceccarelli and L. Pietranera, Ap. J. Lett. 250 (1981) L1.
91. L.L. Cowie and E.M. Hu, Ap. J. 318 (1987) L33.
92. A. Vilenkin, Phys. Lett. 107B (1981) 47.
93. E. Witten, Phys. Rev. D30 (1984) 272.
94. C. Hogan and M. Rees, Nature 311 (1984) 109.
95. R. Brandenberger, A. Albrecht and N. Turok, Nuc. Phys. B277 (1986) 605.
96. L.A. Rawley, J.H. Taylor, M.M. Davis and D.W. Allan, Science 238 (1987) 761.
97. R. Davis, Phys. Lett. 161B (1285) 285;
D. Bennett, Phys. Rev. D33 (1986) 872.
98. H. Hodges and M. Turner, Fermilab preprint, 1988.
99. D. Mitchell and N. Turok, Phys. Rev. Lett. 58 (1987) 1577; Nucl. Phys. B294 (1987) 1138.
100. P. Goddard, G. Goldstone, C. Rebbi and C. Thorn, Nucl. Phys. B56 (1973) 109.

This article was processed by the author using the TEX Macropackage from Springer-Verlag.

H. Latal, H. Mitter, Karl-Franzens-University, Graz, Austria (Eds.)

Concepts and Trends in Particle Physics

Proceedings of the XXV Int. Universitätswochen für Kernphysik Schladming, Austria, February 19-27, 1986

1987. 48 figures. IX, 325 pages. ISBN 3-540-17372-2

Contents: *M. Blau, W. Thirring, G. Landi:* Introduction to Kaluza-Klein-Theories. - *J. Wess:* Supersymmetry/Supergravity. - *W. Kummer:* Supersymmetric Yang-Mills Fields and Noncovariant Supergauges. - *G. Veneziano:* From Strings to Superstrings. - *G. C. Segrè:* Superstrings and Four-Dimensional Physics. - *R. D. Peccei:* Mass Issues in the Standard Model. - *H. Satz:* Critical Behaviour in Statistical QCD. - *H. Satz:* Critical Behaviour in Random Field Gauge Theory. - *L. M. Lederman:* Experiments Beyond the Standard Model.

G. Börner, Max-Planck-Institut für Physik und Astrophysik, Garching, FRG

The Early Universe

Facts and Fiction

1988. 180 figures, and 15 mostly colored plates. XIV, 439 pages. (Texts and Monographs in Physics). ISBN 3-540-16187-2

Contents: The Standard Big-Bang Model: The Cosmological Models. Facts - Observations of Cosmological Significance. Thermodynamics of the Early Universe in the Classical Hot-Big-Bang Picture. Can the Standard Model be Verified Experimentally? - Particle Physics and Cosmology: Gauge Theories and the Standard Model. Grand Unification Schemes. Relic Particles from the Early Universe. Baryon Synthesis. The Inflationary Universe. - Dark Matter and Galaxy Formation: Typical Scales - From Observations and Theory. The Evolution of Small Perturbations. Computer Simulations and the Large-Scale Structure. - Appendix. - References. - Subject Index.

N. Straumann, University of Zurich, Switzerland

General Relativity and Relativistic Astrophysics

1984. 81 figures. XIII, 459 pages. (Texts and Monographs in Physics). ISBN 3-540-13010-1

Contents: Differential Geometry. - General Theory of Relativity: Introduction. The Principle of Equivalence. Einstein's Field Equations. The Schwarzschild Solution and Classical Tests of General Relativity. Weak Gravitational Fields. The Post-Newtonian Approximation. - Relativistic Astrophysics: Neutron Stars. Rotating Black Holes. Binary X-Ray Sources. Accretion onto Black Holes and Neutron Stars. - References. - Subject Index.

Springer-Verlag
Berlin Heidelberg New York
London Paris Tokyo Hong Kong

Springer

MIX
Papier aus verantwortungsvollen Quellen
Paper from responsible sources
FSC® C105338

If you have any concerns about our products,
you can contact us on
ProductSafety@springernature.com

In case Publisher is established outside the EU,
the EU authorized representative is:
Springer Nature Customer Service Center GmbH
Europaplatz 3, 69115 Heidelberg, Germany

Printed by Libri Plureos GmbH
in Hamburg, Germany